ASTRONOMY

ASTRONOMY

Franklyn M. Branley

Astronomer Emeritus,
The American Museum–
Hayden Planetarium

Mark R. Chartrand III

Chairman,
The American Museum–
Hayden Planetarium

Illustrated by

Helmut K. Wimmer

Art Supervisor,
The American Museum–
Hayden Planetarium

Thomas Y. Crowell Company

New York
Established 1834

Acknowledgments. The illustrations by Helmut K. Wimmer listed according to their figure numbers in **Astronomy** appeared formerly in the following books written by Franklyn M. Branley and published by the Thomas Y. Crowell Company. These illustrations are used by the courtesy of the author, illustrator, and publisher.

Comets, Meteoroids, and Asteroids: Mavericks of the Solar System. Illustrations copyright © 1974 by Helmut K. Wimmer. Figs. 11-2, 18-1, 18-2, 18-3, 18-4, 18-5, 18-7, 18-10, 18-11, 18-12, 18-13, 18-14, 18-15, 19-3.

The Earth: Planet Number Three. Illustrations copyright © 1966 by Helmut K. Wimmer. Figs. 12-5, 12-7, 12-8, 12-11, 12-12, 12-14, 15-3, 15-5.

Mars: Planet Number Four. Illustrations copyright © 1975, 1966, 1962 by Helmut K. Wimmer. Figs. 13-3, 13-4, 13-5, 13-6, 13-7, 13-8.

The Milky Way: Galaxy Number One. Illustrations copyright © 1969 by Helmut K. Wimmer. Fig. 14-6.

The Moon: Earth's Natural Satellite. Illustrations copyright © 1972, 1960 by Helmut K. Wimmer. Figs. 13-3, 13-4, 13-5, 13-6, 13-7, 13-8.

The Nine Planets. Illustrations copyright © 1971, 1958 by Helmut K. Wimmer. Figs. 17-20, 17-23, 17-26.

The Sun: Star Number One. Illustrations copyright © 1964 by Helmut K. Wimmer. Figs. 11-1, 11-2, 11-3, 11-8, 11-10, 16-6.

Library of Congress Cataloging in Publication Data

Branley, Franklyn Mansfield, 1915-
    Astronomy.

    Includes index.
    1. Astronomy. I. Chartrand, Mark R., joint author.
II. Wimmer, Helmut K., ill.
QB43.2.B72 1975   520   74-28364
ISBN 0-690-00760-4

Typography by W. P. Ellis and Robert Carola
Cover design by Robert Carola

Manufactured in the United States of America

PREFACE

Ancient people were sky observers—they had to be. They had had no clocks and so regarded, instead, the face of the sky. The stars were their almanac, the calendar of the year that indicated anniversaries and special events, the times to hunt and fish, to sow and reap. Today sky-observing is not a common concern. Still, the lives of all of us are strongly affected by the cycle of day and night, and the succession of seasons, both of which are astronomical events. And, as this book will uncover, we are affected by many other astronomical phenomena.

Some 2500 years ago, all educated persons, whether observers or not, were expected to have a knowledge of astronomy. Since then, astronomy has grown almost immeasurably, and today continues to assume an increasingly important place in the lives of laymen. In preparing this book, we have attempted to provide students with a feeling for the patterns, surprises, and, indeed, excitement of present-day astronomical investigation.

Rather than treat the subject matter in the order of historical discovery—Earth to Moon to planets to stars, to galaxy and Universe—we have chosen to begin with the universe, what it is, how we know what we know, and to consider the matter in space that leads to the stars. Planets and satellites are better understood after the student has a feeling for their place among the more massive bodies. Throughout, the people who have contributed to astronomy are emphasized, providing a considerable amount of biography that students may wish to delve into more deeply.

Newton acknowledged the contributions of his predecessors when he wrote in a letter to Robert Hooke: "If I seem to see farther than other men, it is because I am standing on the shoulders of giants." Similarly, understanding of today's astronomy cannot be complete without a knowledge of traditional astronomy—the reader will find it here. But he will also find that astronomy is very much alive and viable in these closing decades of the twentieth century. Pulsars, quasars, white dwarfs, black holes, and neutron stars are newspaper topics. So are space telescopes, the cratered surfaces of Mercury and Mars, the red spot of Jupiter, the highlands of Venus, interstellar molecules, moon rocks and what they tell us about our satellite.

The text, as clear and direct as we can make it and, we hope, not without a degree of verve and enthusiasm, is enhanced by superb astronomical illustrations, the bulk of which were painted especially for this book. Helmut Wimmer brings to the subject techniques that permit perspectives never seen before in astronomy illustration and which serve to clarify ideas that, in spite of the fullest written explanation, may otherwise remain obscure.

The book includes more than 150 paintings (and more than 100 photographs). Art students and those especially interested in definitive illustration will be fascinated by the techniques and execution employed, and all students will better appreciate the beauty in the subject. The human eye is certainly man's most powerful device for obtaining information about his world. Here are sights that contain vast amounts of information and should prove to be delights to both beginning and advanced students.

It is hoped the student will leave this book with the feeling that science is man's continuing quest to understand himself and his surroundings, not because the authors say so, but because the thought permeates the text.

In preparing this book, many people were involved, all of whom made important contributions to its creation. We particularly wish to thank those teachers who generously read the manuscript to search out flaws in conceptualization and fact, and who gave valuable suggestions for making the book fit their classroom needs more precisely. We also owe a special debt to Dr. William P. Bidelman, Dr. Robert E. Murphy, Mr. Allen Seltzer, and Dr. Gerrit Verschuur for their most helpful contributions to the book.

F.M.B.

M.R.C.

CONTENTS

Contents

ASTRONOMY

THE STUDY OF THE UNIVERSE

CHAPTER 1

Astronomy, the oldest of the sciences, is the study of the universe. It is concerned with the origin of the universe and its structure, how the universe evolved, how it is organized, and how its various parts interact. In his search for answers to questions posed by such topics the astronomer uses techniques and principles of other disciplines—physics, chemistry, mathematics. The history of astronomy is the history of man's discovery of his place in the universe.

Originally, watchers of the skies were astrologers, priests who tried to foretell events on Earth by studying motions of objects in the sky. They had reason to believe there was a relationship, for the rhythms of the sky dominated

Leonardo da Vinci's man stands in the middle of the universe, but not in the geometric center as the ancients thought, for there is no center. Instead, man is about halfway between the size of the atom and the size of the universe, but he reaches for the stars.

their lives in many ways: the Sun governed crops and weather, the Moon was related to the tides; the Moon provided light for hunting; the appearance of Sirius in the morning sky meant to the Egyptians there would soon be life-giving floods. Astrologers taught that motions of the Sun, Moon, and planets were of more universal influence and regulated every facet of a person's life, determining the length of his life, his success or failure, whether he would be happily married, the number of children he would sire. Today astrology remains popular. Many still believe they are influenced by the positions and motions of the planets. However, astronomy has developed far beyond its early stages. It has become a science, a rational, objective study of the planets, stars, and galaxies, of the entire universe in which we live.

THE DEFINITION OF THE UNIVERSE

It would seem that the first step in the study of the universe is to define it. Scientists attempting to study the fundamental structure of the universe must ask if it is everything we can see, if the universe is all those things which obey physical laws as we know them whether we can see the objects or not, or if the universe is something else, something that has escaped definition. Are there other universes? Here we tread the boundary between astronomy and what one might call "astro-poetry"—we enter the realm of metaphysics, of abstractions and philosophical ideas. Our working definition will be that the universe is *everything that we can observe* in one way or another.

Why the universe exists is quite another question. Despite centuries of study, the answer to that question still eludes us.

THE SCALE OF THE UNIVERSE

It is questionable if any layman or professional astronomer can grasp the overwhelming size of the universe in either space or time, and comprehend the vast emptiness between lumps of matter. Figures 1–1 and 1–2 on pages 6 to 9 may help, since they provide a visual scale of the universe in space. Each view is 100 times the one preceding. The universe covers a spread of sizes in the order of 10^{40}, from the smallest atomic particles to the gigantic clusters of galaxies. Man is about in the middle.

Not only is the universe immense in the scale of distance, but also in the scale of time. It covers some forty powers of ten; the "lifetime" of a helium-5 nucleus is about 10^{-21} seconds (0.000 000 000 000 000 000 001), while the "flip-over" of the electron (to be discussed on page 98) in a hydrogen atom happens only once in about 10^{18} seconds, or 100 billion years. (When this happens, energy is released. In the case of hydrogen the energy appears as the 21-centimeter line, which we will discuss later.) Stars which seem to be unchanging really do change, our Sun included. However, because changes in structure and appearance may appear only after millions or billions of years, the stars seem to be everlasting.

Because of the huge spread of space and time we must use the power-of-ten notation—10^3 for a thousand, 10^6 for a million, and so on. Without it we would find the writing and counting of zeros incredibly awkward and cumbersome. Also, we will use the metric system of measurement. It is the system of science, is much easier to manipulate, and will soon be the standard for the entire world (see the appendices).

THE LARGE-SCALE STRUCTURE OF THE UNIVERSE

The Expanding Universe

A *galaxy* is basically a collection of stars, each bound to the others by gravity. Galaxies are the largest forms, or "objects," in the universe.

Even before the composition of galaxies had been determined, the American astronomer V. M. Slipher (1875–1969) discovered a most curious thing. Using the principle of the Doppler shift (to be discussed in Chapter 2), Slipher found that, except for a few nearby galaxies, every galaxy was moving away from us. Even stranger, the farther away the galaxy was, the faster it was moving. The entire universe was expanding, and we *seemed* to be at the center of the expansion. Figures 1–3 and 1–4 show what we observe.

In the past, whenever man has thought he was in any special position in the universe, he has had to make corrections later on. Early astronomers thought the visible Earth marked the boundaries of the universe. More informed investigators knew there was more to it than a single planet, but they believed, as Ptolemy (fl. 127–157) recorded, that Earth was the center and all things moved about Earth. Nicholaus Copernicus (1473–1543) and others after him declared this was not so; the Sun was the center of the universe. Later, when the Milky Way Galaxy (our "local" galaxy) was studied, William Herschel (1738–1822) said the Sun was the center of the Galaxy—stars were distributed around the Sun. It was not until the 1920s that such

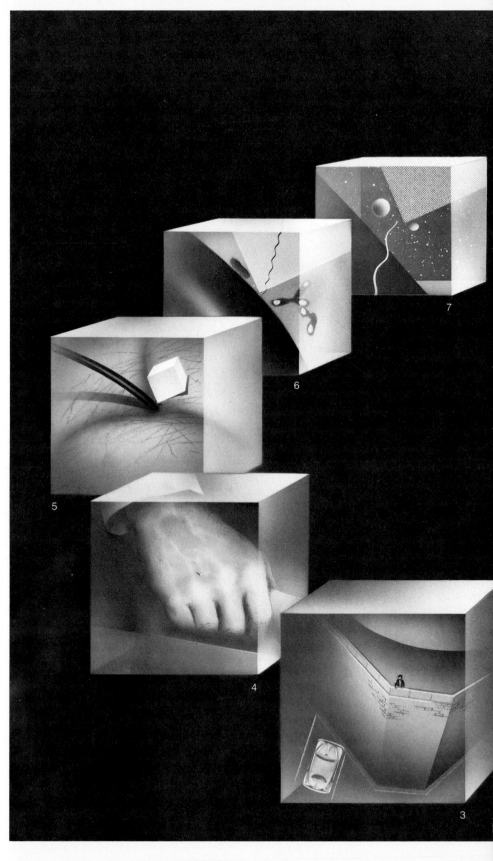

FIGURE 1–1

From Earth to atom, adapted from an idea by Kees Boeke. Each view is about 100 times smaller than the one preceding. Cube 1 shows a portion of the eastern seaboard of the United States. Cube 2 contains a part of New York City near Central Park. Cube 3 is the corner of a building with a man standing on the roof. Cube 4 contains just the man's hand, with a small crystal of salt on the surface of the skin(5). Cube 6 magnifies the corner of the crystal and shows some large bacteria on the skin. Shown schematically to scale is a ray of visible light. Cube 7 enlarges the volume and reveals a couple of viruses, the smallest living organisms. Molecules of oxygen and nitrogen in the air can now be seen, as well as a suggestion of the orderly arrangement of sodium and chlorine atoms in the salt crystal. Cube 8 shows the crystal lattice structure. Cube 9 concentrates on a single sodium atom. The long and short waves are ultraviolet and X rays respectively. Cube 10 shows the nucleus of the atom, about 20 orders of magnitude smaller than Earth.

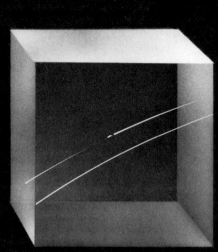

FIGURE 1–2
From Earth to the universe in ten steps. Each view is about 100 times larger than the previous one. Cube 1 contains Earth and the Moon, and Cube 2 includes the orbit of Venus. Cube 3 contains our entire solar system and shows the orbits of a few long-period comets which orbit the Sun much as electrons orbit an atomic nucleus. Cube 4 shows only a few of the outermost comet orbits, for the solar system has shrunk to a dot. Cube 5 encompasses the Sun and the nearest star system, Alpha Centauri. Cube 6 is our "solar neighborhood" in the Galaxy, while Cube 7 contains the Galaxy and its retinue of globular clusters. Cube 8 takes in our Local Group, a cluster of a couple dozen galaxies near the Milky Way. Cube 9 shows clusters of clusters of galaxies. Cube 10 encompasses clusters of galaxies, a scale we cannot yet see with our instruments. Another step upward in size might show still higher order groups.

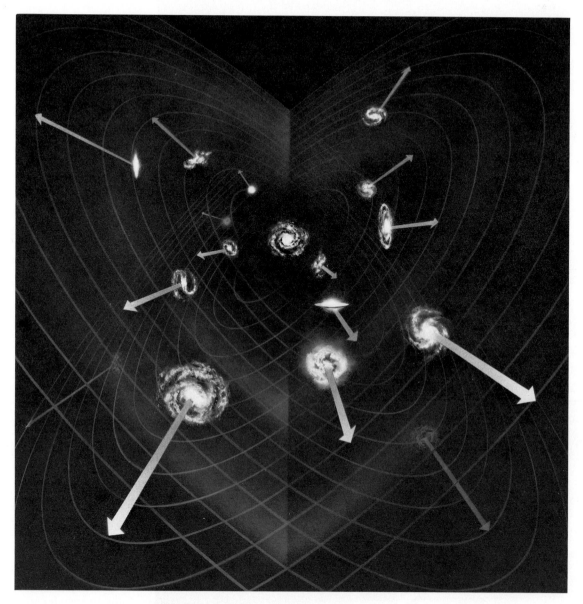

FIGURE 1–3 *The expansion of the universe. In all directions the galaxies are receding from us. The speed of recession, symbolized by the lengths of the arrows, is proportional to the distances of the galaxies, shown by the circles. Our galaxy is not in any preferred location. Such a picture centered on any other galaxy would appear similar.*

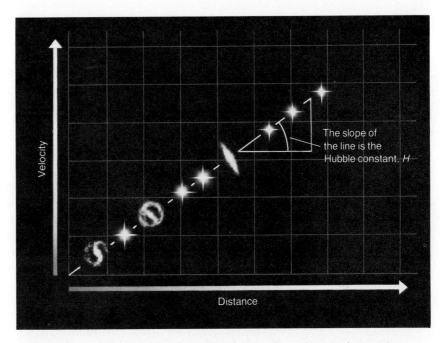

FIGURE 1–4   *When a graph is made plotting the velocity of recession of a galaxy and its distance, a straight line is obtained, indicating that velocity and distance are proportional. The slope of the line is the Hubble constant, H (pp. 246, 257), and is the constant of proportionality relating velocity and distance. This figure is only schematic, since real data would not fall exactly along a single line but would be scattered about the line slightly owing to inaccuracies in observation and uncertainties in distance measurements.*

ideas were shown to be incorrect. We are therefore suspicious whenever an observation seems to indicate we are in the center of anything.

Slipher's observations can be described by using the example of a nutcake baking in an oven. As the dough rises, every nut gets farther from every other nut. Even if we are observing from a nut off to one side, we will see an expanding universe of nuts. Furthermore, each nut will speed away from us at a rate proportional to its distance from us—the farther away it is, the faster it moves away.

The universe is similar to that nutcake, except it doesn't have a center. A center isn't needed to explain the expansion. The universe is expanding—the reason why raises perplexing questions. To attempt an answer we must seek explanations of how the universe might have originated.

The Origin of the Universe

Presently there are two main schools of thought about the evolution of the universe. One holds that at some time, about 10 to 20 billion years ago, all present matter and energy were compressed into a small ball only a few kilometers across. It was a super-atom, one that contained in the form of pure energy the total components of the entire universe. How this "cosmic egg" got there, and what came before it, are not known, and probably are not even knowable. For some reason, at an instant which astronomers call $T=0$, the ball exploded, sending the energy into space. As expansion continued, and the "cosmic fireball" cooled, most of the energy became matter. This is entirely feasible, for, according to Einstein, energy and matter are related—when one is destroyed, the other is created. The matter produced first was protons, neutrons, and electrons. During the first half-hour after the explosion, these original particles had combined to produce hydrogen, deuterium ("heavy" hydrogen—discussed in Chapter 2), and helium. As expansion continued, the matter formed into galaxies and stars formed within the galaxies. Now, sometime between 10 and 20 billion years later we see a universe of star-filled galaxies flung across space. In one average spiral galaxy is Earth, a very special planet going around a relatively cool, average star.

This explosion theory is known as the *Big Bang model* of the universe.

The Steady-State model. The leading rival model of the universe maintains that the universe did not have a beginning. Its proponents say that it makes no difference at what time one makes an observation, or from where, the average properties of the universe will be the same. One of these properties is the *density* of the universe, the amount of matter per cubic centimeter. Since observations indicate that the universe is expanding, matter must be created to compensate for the spreading out of material; otherwise the density would go down throughout the universe. It is not known, or even theorized, where the new matter comes from. If there is a process that keeps density uniform, the rate of creation of new matter must be very slow. It would seem that the Steady-State model calls for matter to be created out of nothing, a seemingly impossible event. However, the amount of material which must be created is very small, less than one hydrogen atom in a cubic meter every 100 000 years. (Consider that in a cubic meter of the near-vacuum of interstellar space there are about 100 000 hydrogen atoms.)

Cosmology. Cosmology is the study of the universe as a whole, its origin (if any) and its future. The main question at present is which

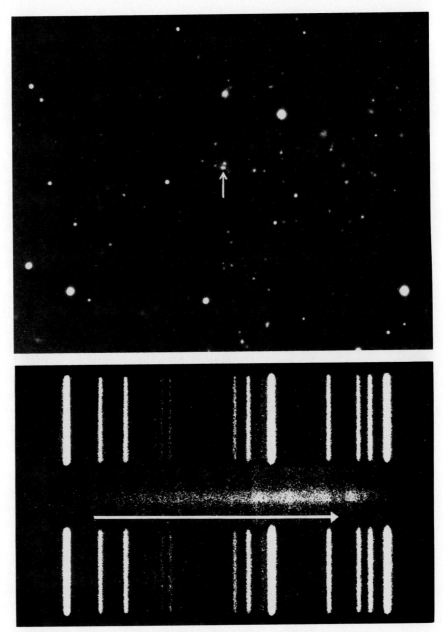

FIGURE 1–5   *A cluster of galaxies in the constellation Hydra. The spectrum of one of the galaxies in the cluster is shown, and the length of the arrow indicates the amount of red shift. This corresponds to a velocity of recession of 61 000 kilometers per second (38 000 miles per second). (Hale observatories)*

model of the universe is correct. Present evidence, which is by no means conclusive, tends to support the Big Bang model.

There are several reasons. First, the observed expansion is easier to explain as the result of an explosion. Second, recent observations have provided additional evidence of such an explosion long ago. Scientists at the Bell Telephone Laboratories discovered weak radio radiation coming from all directions in space. The radiation is believed to have been created at the time of the explosion and has been traveling in space for billions of years—a remnant of creation. It is referred to as the "3-degree background radiation." (For an explanation of why it is called this, see page 37 in Chapter 2.) More observations are being made. Perhaps they will confirm this observation and will provide more definite answers than are available presently.

Other evidence that helps one decide between models of the universe is found at the extreme reaches of observation—some 10 billion light-years. [A *light-year* is the distance that light, moving at almost 300 000 kilometers per second, will travel in a year (about 31 million seconds). One light-year is about 10 trillion (10^{13}) kilometers, or 6 trillion miles.] We can plot a graph of the speed of a galaxy away from us versus its distance. Until we reach very distant galaxies, the line is straight, indicating a direct relationship between distance and speed. At the tip of the line, we would expect it to curve upward or downward, as shown in Figure 1–6, depending upon the kind of universe we live in. When we look far out into space, we are looking far back into time to objects that may no longer exist. We see them as they were billions of years ago. If the universe is evolving as in the Big Bang, the properties of the universe would have been different long ago from what they are now. Recent observations show that the density of galaxies is higher at great distances than near our galaxy, supporting the Big Bang model.

Another argument against the Steady-State model comes from *quasars* (which is short for quasi-stellar radio sources). If, as many astronomers believe, quasars are out among the galaxies, or are actually unusual galaxies themselves, then they violate the principal assumption of the steady-state theory which says that the universe is the same now as it always has been. The quasars are seen at great distances, which means that we see them as they were long ago. Since we don't see any nearby ones, they must have been more common in the past than now, which violates the assumption.

Observations indicate, however slightly, that the line of the graph curves in a manner expected if the universe is evolving. However, observations at distances of billions of light-years are very difficult, and there is much uncertainty. Many more data are needed to be sure.

The future of the universe. What will be the fate of the universe, especially if it is expanding as observations indicate? Will the expansion continue, eventually dispersing all matter? Some cosmologists believe that it will, others don't think so.

The crucial piece of data needed is the average density of the universe now thought to be about 10^{-30} grams per cubic centimeter. It can be computed that if there is sufficient mass, expansion of the universe will eventually stop owing to gravity, reverse its motion,

FIGURE 1–6   *The Hubble law graph, similar to Figure 1–4, but extended to include data from very distant galaxies and quasars. At the upper end we would expect the line to curve. The shape and amount of curve yield information on what kind of universe we live in. So far data from these great distances are incomplete and inconclusive.*

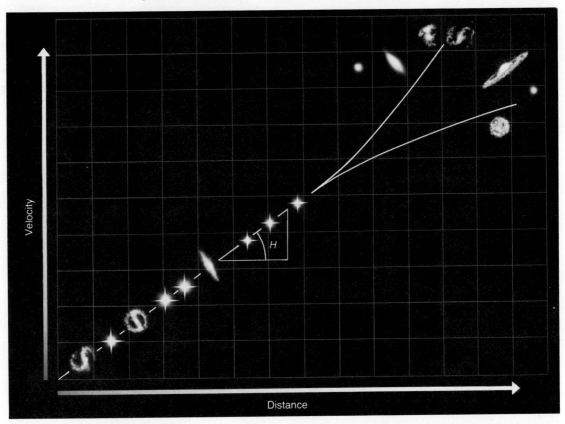

FIGURE 1–7
At the present time the universe is at the crossroads of the past and the future, illustrated by the cube at the intersection of the rows of cubes. Our problem is to determine what the past was like and what the future will be like. Three alternatives are shown here. The row from top to bottom represents the Steady-State universe: the average density of the universe is the same for all time. The row from upper left to lower center is the Big Bang universe: a single explosion producing the expanding universe which will continue to disperse. The row of cubes from upper right to lower left is the Oscillating universe: a series of big bangs—expansion, contraction, and another explosion.

and become a contracting universe. Observers in the distant future would then see all the galaxies coming toward them. Perhaps they would end up in another "super-atom" and there would be another "Big Bang." The idea of an alternately expanding and contracting universe is known as the *Oscillating Universe model.* Some call it the Bang-Bang-Bang model.

More questions. No model of the universe satisfies all requirements; all models have holes in them. One problem of the Steady-State model is the creation of new matter. The Big Bang model and the Oscillating Universe model do not explain the origin of the super-atom; both just push the question of ultimate origin one step farther back. Figure 1–7 shows the three alternatives.

Nevertheless, at present we can say that we probably live in a Big Bang universe having an age of between 10 and 20 billion years. Whether this universe will prove to be oscillating we cannot say, for the future remains obscure.

Questions that have been raised in this chapter are among the most interesting and fundamental we can ask. Much of the excitement of astronomy comes from the continuing quest for answers. Once obtained, the answers will give everyone, astronomers and laymen alike, wider perspectives on the nature of the universe in which we live.

Presently only a few of the questions can be answered. But astronomers will continue their search. Observations made next month, next year, or maybe in the next century may lead them to radically different conclusions from those we have today. This is the way any science progresses, by throwing away those ideas that do not agree with observations that are made after the theory has been formulated. We cannot predict what those observations will be, or even what they might be, but it is more than likely that astronomers a hundred years from now will regard our ideas of the universe much as we regard the ideas that Copernicus, Galileo, and so many others had about the solar system. Their explanations of observations were perceptive, but not always correct. Similarly, our explanations of observations of the universe may need considerable modification.

REVIEW

1   Astronomy is the oldest of the sciences. It grew from a beginning in superstition to a rational inquiry into the structure of the universe and the laws which govern it.

2 The universe is expanding. This is seen from the observation that, except for a few nearby galaxies, all galaxies in all directions are moving away from us with velocities proportional to their distances.

3 The two principal theories of cosmology are (a) the Big Bang, or Evolutionary, model, which states that the universe had a beginning in an explosion 10 to 20 billion years ago; and (b) the Steady-State model, which states that the universe was and always will be the same.

4 At present, observational evidence supports the Big Bang model. This evidence consists of (a) the existence of quasars, (b) the radio radiation coming uniformly from all parts of space, and (c) the change in the speed of recession of distant galaxies.

5 The physical quantity which determines whether or not the expansion will stop is the average density of the universe, now thought to be about 10^{-30} g/cm^3. If the density is much less than this, the universe will continue to expand. If it is much larger, the expansion will eventually stop.

QUESTIONS

1 Why does the graph of speed of recession of the galaxies not give us an accurate picture of the universe *now*? What sort of correction could we make, and what assumptions would we have to make?

2 Why would we not be able to tell anything about what came before the Big Bang?

3 Why should we expect the 3-degree background radiation to be coming equally from all directions?

4 How can we attempt to find out if quasars really are very far away from us, and if they are associated with distant galaxies?

5 Explain in detail why the average density of the universe determines its fate.

READINGS

Bergmann, P. *The Riddle of Gravitation.* New York: Charles Scribner's Sons, 1968.

Charon, J. *Cosmology: Theories of the Universe.* New York: McGraw-Hill Book Company, 1970.

Jaki, S. *The Paradox of Olber's Paradox.* New York: Herder and Herder, Inc., 1969.

Kaufmann, W. J. *Relativity and Cosmology.* New York: Harper & Row, Publishers, 1973.

Ley, W. *Watchers of the Skies.* New York: The Viking Press, Inc., 1966.

National Academy of Sciences. *Astronomy and Astrophysics for the 1970s.* Washington, D.C.: National Academy of Sciences, 1973.

Pannekoek, A. *A History of Astronomy.* New York: Wiley-Interscience Publishers, 1961.

Peebles, P. J. E., and D. T. Wilkinson. "The Primeval Fireball." *Scientific American,* June 1967.

Sciama, D. W. *Modern Cosmology.* New York: Cambridge University Press, 1971.

Struve, O., and V. Zebergs. *Astronomy of the Twentieth Century.* New York: The Macmillan Company, 1962.

WAVES
FROM SPACE

CHAPTER 2   Astronomy is, for the most part, not an experimental science. Astronomers cannot arrange observations and conditions to suit some particular need for data, but must gather what they can get from the light received from celestial objects. Astronomical events which may give important information may happen at inconvenient times and places.

    Except for very recent techniques, such as bouncing radar waves off the Moon and planets, or the advent of manned space flight, astronomers attempting to study the universe had nothing to work with except the light which they receive. The light is often weak, and disturbed by its passage through the atmosphere and through space. In fact, there are only four basic things we can

The astronomer finds out about the universe by analyzing the light coming to his instruments. The methods of spectroscopy, breaking the light into its component frequencies or colors, allow him to determine important physical properties of the objects in the sky.

tell about this light coming to us: *direction, intensity, spectrum,* and *polarization.* We can also measure changes of these properties with time. It is thus somewhat amazing that we have been able to find out so much about our universe, analyzing each beam of light for the information it contains. Therefore, it is very important to understand the nature of light.

Direction, or *position,* of a light source is readily apparent. Many objects are point sources: owing to their great distances they show no size whatever. Planets, nebulae, and galaxies are extended objects. The position of an object in the sky is measured either in reference to other objects (similar to the way one would say that a town is near some city) or with respect to some coordinate system (similar to the way one would give the latitude and longitude of a point on Earth). In the sky, the coordinates are called *right ascension* and *declination;* they are described in Chapter 14.

The *intensity* of a source of light is its brightness, and even a casual glance into the sky shows a great range of brightnesses of astronomical objects visible to the unaided eye, ranging from the bright Sun to the faint stars.

With telescopes astronomers can see and photograph objects whose brightness is 16 million times less than the faintest visible without aid. The Sun is 160 billion times brighter than the faintest stars seen with the unaided eye. This great range makes it necessary to use a rather unusual scale, the magnitude scale, for measuring brightness. It is discussed on page 59.

The *spectrum* of light is the distribution of the light into colors, including those not visible to the eye. Everyone has seen the band of colors produced by holding a prism in a beam of light, or a rainbow. This is the visible spectrum. The spectrum, both visible and invisible, reveals considerable information about the physical conditions (temperature, motion, composition) of the source of light.

The *polarization* of a light beam is a property having to do with the orientation of the waves of light. To understand this more fully, we must look into the nature of light and how it is produced.

THE ELECTROMAGNETIC SPECTRUM

Light is energy. It is a form of *radiant energy,* which means that it can travel through empty space and needs no *medium* to carry it in the way that, say, sound waves are carried by air or rock. Light can be thought of either as a stream of particles or as a wave. Which model we use depends on which specific properties of light we wish to study. (A *model* is a scientist's way of describing a phenomenon. Our model of light as a wave or a particle does not say that light *is* either a wave or a particle, only that its properties may be conveniently described by thinking of it as such.) Some people call light a "wavicle."

If light is radiated from a point source in all directions, the intensity, or the energy, falling on each square centimeter will decrease as distance from the source increases. At any distance, the light will cover the surface of a sphere. Since the surface area of a sphere increases with the square of the radius of the sphere, the intensity of light will decrease as the inverse square. This is known as the *inverse square law*. For example, if the distance is doubled, the intensity will be $(\frac{1}{2})^2 = \frac{1}{4}$ the original amount. If the distance is tripled, the intensity will be reduced to $\frac{1}{9}$. This effect is shown in Figure 2–1.

We know that waves, such as light waves, have three basic properties: *frequency* (number of waves, or crests, per second), *wavelength* (the distance between wave crests), and *velocity* (how

FIGURE 2–1
The inverse square law of light intensity. Light from a point source will spread out in all directions. If a certain amount covers a unit area at a unit distance, then the same amount of energy will cover 4 times the area at twice the distance, and 9 times the area at 3 times the distance. Thus, the intensity varies as 1 divided by the square of the distance from the source.

fast energy is transmitted). These properties are related in a simple way (Figure 2–2). We will let V be the velocity of the wave, W the wavelength, and f the frequency. Then the relation between them is

$$V = W \times f$$

Thus, if we know two of these numbers, we can figure out the third.

Any time we write an equation, the units of measurement must be consistent. In this case, the speed of light is in kilometers per second (km/sec), so wavelength must be in kilometers, and frequency in cycles per second, also called *Hertz* (Hz). Since the speed of light in empty space is constant, equal to about 300 000 km/sec, the wavelength and frequency of light (or any electromagnetic radiation) are quite simply connected. To take an example, a radio station in the broadcast band with a frequency of 1000 kilohertz (kHz, or 1000 Hz) will have a wavelength of

$$W = \frac{V}{f} = \frac{300\ 000}{1\ 000\ 000} = 0.3 \text{ km (not quite 1000 feet).}$$

In the early days of radio experimentation, the frequencies and wavelengths used were in this range. Later, when higher frequencies became possible, the next region of the electromagnetic spectrum was called "short wave" or "high frequency." Later still, VHF (very high frequency) and UHF (ultrahigh frequency) radio became pos-

FIGURE 2–2
Light may be thought of as a wave of electromagnetic energy. The distance between crests is the wavelength, W. The number of vibrations, or crests, per second, is the frequency of the wave, f. The wave is traveling in the direction shown by the arrow at a velocity V. The three properties of the wave are related by the equation V = W × f.

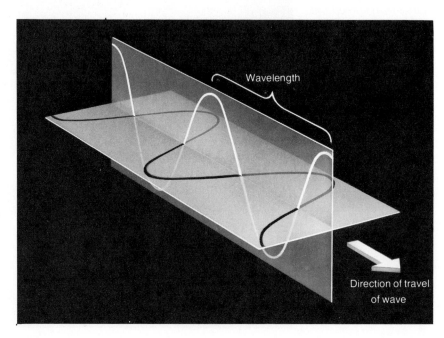

Wavelength

Direction of travel of wave

sible. Now we are experimenting with EHF and SHF ("extremely" and "super," respectively) radio. In a similar way, lower frequency ranges became possible, and today the range of the electromagnetic spectrum in use extends over many decades of frequency. (A *decade* is one power of 10.)

You will have noticed that we have been talking of radio, not light. The two are different forms of the same thing, different frequencies of vibration of *electromagnetic waves*. This identity was first shown around 1889 by Heinrich Hertz (1857–1894) whose name we honor with the unit for frequency. Hertz proved what had been only a suggestion made by James Clerk Maxwell (1831–1879) in 1862. Maxwell had noted that radio waves and light waves travel at the same speed. This seemed to be too much of a coincidence, and he suggested that they were really the same thing. Even before Maxwell, however, the German-English astronomer William Herschel (1738–1822) had shown that there are waves that lie outside the visible region of the spectrum.

Herschel's experiment can be repeated by anyone and with very little equipment. He placed a prism in a beam of sunlight, and placed thermometers at different colors in the spectrum as well as *beyond* the band of visible colors on either end. He noticed that the thermometer in the red region went up more than the one in the violet region. The thermometer just outside the red end, where no illumination was visible, read highest. He concluded that there must be some sort of radiation in the infrared (below red) wavelengths.

Later, other scientists showed that there are invisible waves of higher frequency than violet (ultraviolet), and even beyond to X rays and gamma rays. Today we study and use the electromagnetic spectrum (Figure 2–3) over a range of frequencies or wavelengths of more than 10^{18}, ranging from the radio waves to the gamma rays.

Particles of Light

For many purposes it is convenient to talk of light as a wave. When we wish to discuss the energy carried by light, it is a bit easier to think in terms of particles of light, each having a certain amount of energy.

A "particle" of light is called a *photon*. Each photon has a particular energy corresponding to the frequency (or wavelength) of a light wave. The higher the frequency (in the visible region, the bluer the light), the more energy the photon will have. The energy E and the frequency f are related by

$$E = h \times f$$

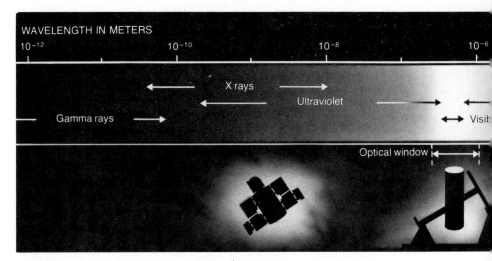

FIGURE 2–3   The electromagnetic spectrum. Visible light is only a
small part of the spectrum of electromagnetic energy, which extends to
waves of all frequencies. Only visible light and some radio waves pass
through Earth's atmosphere to the surface. The other wavelength regions
of the spectrum must be observed from space. Each wavelength
corresponds to a certain frequency and energy of the photons, and each
region of the spectrum yields different information about the source of
the radiation.

where h is called *Planck's constant* (after the German physicist Max
Planck), and is a very small number, equal to 6.62×10^{-27} when f is in
Hertz and E in ergs. Even at very high frequencies, each photon has
very little energy. The energy of a photon of yellow light is about
10^{-13} ergs (an *erg* is about the energy of a mosquito flying full speed
ahead). The intensity of a beam of light is the number of ergs per
second.

The Spectrum

In the most general terms a *spectrum* is the distribution (think of a
graph) of the intensity at each frequency, from highest to lowest.
Or it might be just some portion, such as the visible spectrum. If
some of the waves lie in the visible region which extends from a
wavelength of 4000 *angstroms*, violet, to about 7000 angstroms
($1\text{A} = 10^{-10}$ meters), red, then we will see the object with some particular color, depending upon the dominant wavelength of light. The
peak in the Sun's spectrum is in the yellow region, so the Sun
appears yellow (Figure 2–4). Some cooler stars have a spectrum
which peaks in the infrared, which we cannot see, but the stars

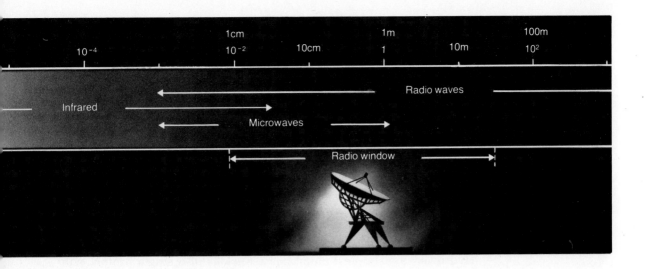

FIGURE 2–4 *The distribution of energy (spectrum) emitted by two objects (stars) of different temperatures. Such a graph is called a Planck curve. The lower line is the spectrum of the Sun, an object with a surface temperature of about 6000°K. The maximum intensity is at about 5000 A, so the Sun appears yellow-white in color. The upper line is for a star with a surface temperature of 24 000°K. Note that the line for this hotter object is everywhere above the line for the cooler one, indicating that it is brighter at every wavelength. Note also that the peak intensity for the hotter object is at a shorter wavelength than for the cooler object. We would see this star as blue-white in color.*

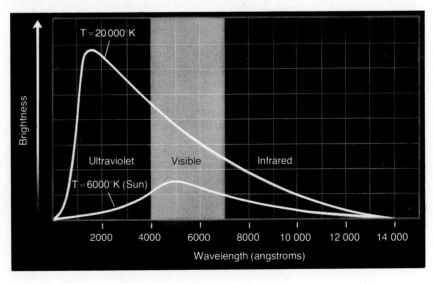

appear as red in color. In order to understand how the spectrum gives us information, we must look into how light is produced.

LIGHT AND ATOMS

A useful model of an *atom* is to think of a *nucleus* containing positively charged *protons* and chargeless *neutrons*, with one or more negatively charged *electrons* in orbit about the nucleus (Figure 2–5). Note that the word "model" appears again; we are saying that this description of an atom works (for some purposes), not that an atom is actually like this.

Each *chemical element* has an atom of a distinct structure. Indeed, it is the structure of the atom which makes it that element. Hydrogen, the simplest atom, has a nucleus of a single proton, with a single electron in orbit around it. There are also "heavy hydrogen" atoms (deuterium and tritium), with one or two extra neutrons added respectively to the nucleus. Since it is the number of protons and electrons which determines which element it is, the chemical properties of heavy hydrogen are about the same as those of normal hydrogen. The different atoms of the same element but with different numbers of neutrons in the nucleus are called *isotopes*. The *atomic*

FIGURE 2–5
A model of a carbon atom. The nucleus contains 6 protons, so the atomic number of carbon is 6. The nucleus also contains 6 neutrons, giving a total of 12 nucleons, so the atomic weight of carbon is 12. Surrounding the nucleus are 6 electrons in orbits of certain sizes.

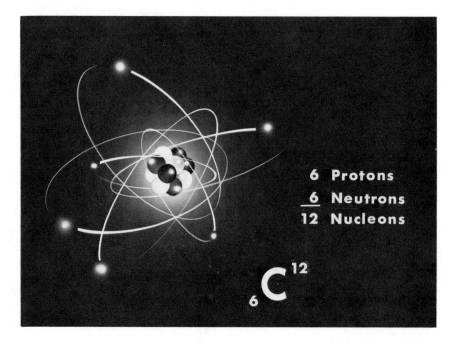

6 **Protons**
6 **Neutrons**
12 **Nucleons**

$_6C^{12}$

PERIODIC CHART OF THE ELEMENTS

Group IA

Atomic number
11
Sodium ← Name
Na ← Symbol
23 ← Approximate atomic weight (to nearest whole number)

Period

																	VIII
1 Hydrogen **H** 1	IIA											IIIA	IVA	VA	VIA	VIIA	2 Helium **He** 4
3 Lithium **Li** 7	4 Beryllium **Be** 9											5 Boron **B** 11	6 Carbon **C** 12	7 Nitrogen **N** 14	8 Oxygen **O** 16	9 Fluorine **F** 19	10 Neon **Ne** 20
11 Sodium **Na** 23	12 Magnesium **Mg** 24	IIIB	IVB	VB	VIB	VIIB	VIII	IB	IIB			13 Aluminum **Al** 27	14 Silicon **Si** 28	15 Phosphorus **P** 31	16 Sulfur **S** 32	17 Chlorine **Cl** 35	18 Argon **Ar** 40
19 Potassium **K** 39	20 Calcium **Ca** 40	21 Scandium **Sc** 45	22 Titanium **Ti** 48	23 Vanadium **V** 51	24 Chromium **Cr** 52	25 Manganese **Mn** 55	26 Iron **Fe** 56	27 Cobalt **Co** 59	28 Nickel **Ni** 59	29 Copper **Cu** 64	30 Zinc **Zn** 65	31 Gallium **Ga** 70	32 Germanium **Ge** 73	33 Arsenic **As** 75	34 Selenium **Se** 79	35 Bromine **Br** 80	36 Krypton **Kr** 84
37 Rubidium **Rb** 85	38 Strontium **Sr** 88	39 Yttrium **Y** 89	40 Zirconium **Zr** 91	41 Niobium **Nb** 93	42 Molybdenum **Mo** 96	43 Technetium **Tc** 99	44 Ruthenium **Ru** 101	45 Rhodium **Rh** 103	46 Palladium **Pd** 106	47 Silver **Ag** 108	48 Cadmium **Cd** 112	49 Indium **In** 115	50 Tin **Sn** 119	51 Antimony **Sb** 122	52 Tellurium **Te** 128	53 Iodine **I** 127	54 Xenon **Xe** 131
55 Cesium **Cs** 133	56 Barium **Ba** 137	57 *Lanthanum **La** 139	72 Hafnium **Hf** 178	73 Tantalum **Ta** 181	74 Wolfram (Tungsten) **W** 184	75 Rhenium **Re** 186	76 Osmium **Os** 190	77 Iridium **Ir** 192	78 Platinum **Pt** 195	79 Gold **Au** 197	80 Mercury **Hg** 201	81 Thallium **Tl** 204	82 Lead **Pb** 207	83 Bismuth **Bi** 209	84 Polonium **Po** 210	85 Astatine **At** 210	86 Radon **Rn** 222
87 Francium **Fr** 223	88 Radium **Ra** 226	89 **Actinium **Ac** 227	104 Kurcha-tovium **Ku**	105 Hahnium **Ha**													

Lanthanide Series*

6

58 Cerium **Ce** 140	59 Prase-odymium **Pr** 141	60 Neodymium **Nd** 144	61 Promethium **Pm** 147	62 Samarium **Sm** 150	63 Europium **Eu** 152	64 Gadolinium **Gd** 157	65 Terbium **Tb** 159	66 Dysprosium **Dy** 163	67 Holmium **Ho** 165	68 Erbium **Er** 167	69 Thulium **Tm** 169	70 Ytterbium **Yb** 173	71 Lutetium **Lu** 175

Actinide Series**

7

90 Thorium **Th** 232	91 Protactinium **Pa** 231	92 Uranium **U** 238	93 Neptunium **Np** 237	94 Plutonium **Pu** 242	95 Americium **Am** 243	96 Curium **Cm** 247	97 Berkelium **Bk** 247	98 Californium **Cf** 251	99 Einsteinium **Es** 254	100 Fermium **Fm** 253	101 Mendelevium **Md** 256	102 Nobelium **No** 254	103 Lawrencium **Lr** 257

FIGURE 2–6 When the elements are arranged in order of increasing atomic number, similar chemical properties recur periodically. This arrangement is known as the periodic chart. For example, boron (B), aluminum (Al), gallium (Ga), indium (In), and thallium (Tl), all in one group, behave similarly in the way they react with other elements to form compounds. The elements in group VIII, the "noble gasses," do not ordinarily react with other elements. Elements 57 through 71 (the lanthanide series) have similar chemical properties, as do elements 89 through 103 (the actinide series). Elements above uranium, atomic number 92, are artificially produced. The periods order the elements according to the configuration of their electrons. The atomic weights, given to the nearest whole number for rapid comparison, represent the average of the masses of all of each element's isotopes as they naturally occur. For a number of heavier elements, the atomic weight of the most stable isotope is given. Detailed information about periods, atomic weights, and other aspects of the periodic chart can be found in introductory chemistry textbooks. (NOTE: Element 106 has been detected, but its name and characteristics have not been established.)

number of an element is the number of protons in the nucleus. Since an atom is usually electrically neutral (equal number of positive and negative charges), there will be a number of electrons distributed in various orbits about the nucleus, and the number of electrons will be equal to the number of protons in the nucleus. The total number of protons and neutrons in the nucleus is called the *atomic weight.* Thus isotopes will have different atomic weights, even though they have the same atomic number.

There are 92 natural elements, and another dozen or so transuranium elements not found in nature due to their short lifetimes but which can be produced in the laboratory (Figure 2–6). These last dozen elements are *radioactive;* they "decay" into other elements in a short time. Some of the natural elements are also radioactive, notably uranium, number 92, the heaviest natural element. The transuranium elements are produced in nuclear accelerators, and it is probable that as research continues new elements will be created in the future, perhaps even ones more stable than some of the transuranium elements now known.

An electron in orbit around the nucleus has a certain amount of energy. The interactions of the electrons of the various elements are responsible for the chemical properties of the elements. To explain the behavior of atoms, the electrons cannot be in just any orbit, but they must occupy only certain orbits which correspond to certain energies. We say that the orbits, and hence energies, are discrete, or *quantized.* This is the basis of the branch of physics known as *quantum theory.*

When an atom is left alone, each electron tends to be in the lowest permitted state of energy, or lowest orbit. If the atom is *excited,* that is, if it has absorbed energy from a photon (one might think of a photon hitting it) or from a collision with another atom, the electron moves into a higher orbit. When the atom is left alone again, the extra energy is released in the form of a radiated photon. The photon removes just the right amount of energy for the electron to jump back down to a lower orbit. The photon may be emitted in any direction. Since the orbits are fixed, the energy differences between orbits are also fixed, and so are the frequencies of light emitted. Thus at one time, a single atom will emit or absorb light of only one color. The set of all possible energy differences between orbits will be a set of individual colored lines of emitted light. This is the spectrum of that element (Figure 2–7). Since the orbits of the electrons of each element are different, each element will have its own characteristic set of lines, a characteristic spectrum. It is usual to compare this spectrum with a set of fingerprints, uniquely identifying the culprit responsible for the radiation (Figures 2–8 and 2–9).

If the electron has been given too much energy by a collision of an incoming photon, an electron may be ejected from the atom. If

FIGURE 2–7   How a spectral line is produced. If an electron is in an orbit having more energy than the minimum, the atom is said to be excited. The orbits have definite energy levels, and are said to be quantized. When an electron "falls" from an orbit of higher energy to one of lower energy, a photon is emitted which has an energy exactly equal to the difference in energy between the orbits. In the upper left of the figure are three hydrogen atoms of different excitations. In the upper right is an energy-level diagram for hydrogen. A change in orbit of an electron corresponds to a jump from one energy level to another. Along the bottom is a part of the spectrum of hydrogen. Lines A, B, and C correspond to orbit and energy-level changes shown above. The region where there are no lines is called the continuum, produced by electrons falling from outside the atom.

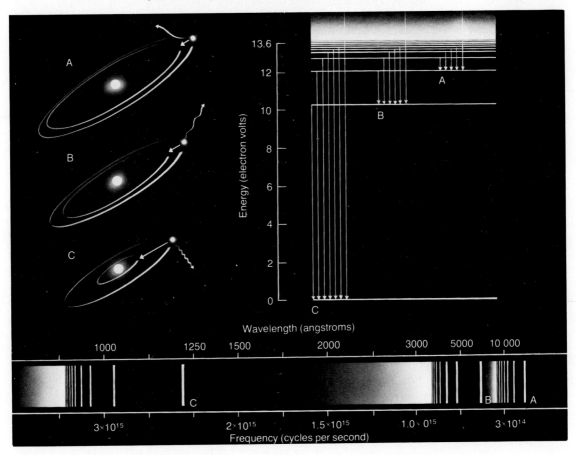

Wavelength
Iron emission spectrum
Sun absorption spectrum

FIGURE 2–8   *A portion of the spectra of the Sun and of iron. The iron spectrum consists of emission lines (see p. 37), which are placed above the absorption spectrum of the Sun. Notice that some of the dark lines in the solar spectrum are at the same wavelength as some of the iron emission lines, indicating that iron is present in the solar atmosphere. The wavelengths are expressed in angstroms. (Hale Observatories)*

FIGURE 2–9   *An example of the spectra of several stars of different surface temperatures. (Yerkes Observatory)*

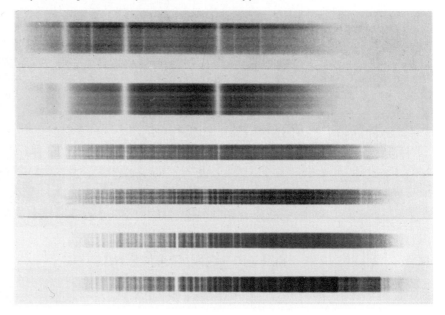

this happens, the atom is no longer electrically neutral, and is said to be an *ion*. It has been *ionized*. In this state the orbits of the remaining electrons will be different from those of the neutral atom, and so an ionized atom will show a different spectrum.

Each element is excited and ionized by different amounts of energy. Since this energy is usually in the form of heat energy (energy of motion), different elements will radiate at different temperatures. Thus, hydrogen is prominent if the temperature is about 10 000° Kelvin (K), but is weak at 4000°K, where radiation from ionized calcium is strong (Figure 2–10). See Appendix B for discussion of temperature conversions.

When atoms combine, a *molecule* is formed. Molecules also radiate and absorb energy; however, they can store energy in more ways than an individual atom. Not only are there electrons in orbits around the nuclei of the atoms, but the entire molecule may vibrate or rotate. It can change its energy of vibration or rotation by emitting or absorbing a photon. Most of these changes in energy are smaller than those which occur when an electron jumps from one orbit to another, so the spectra of molecules are often in the infrared or radio portions of the spectrum, since these waves have less energy than visible light waves.

FIGURE 2–10   The distribution of energy emitted by two stars of different surface temperatures. This is known as a "black body" spectrum (page 37), or a Planck curve (also see Figure 2–4).

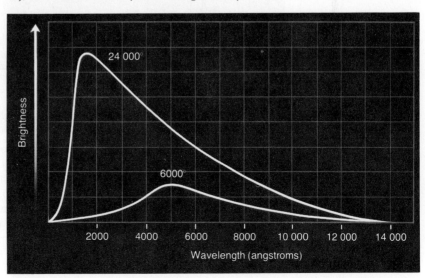

FIGURE 2–11   The three types of spectra. The top
illustration shows that a hot, dense object, such as the filament of a light
bulb, produces a continuous spectrum. Such a spectrum might be graphed
as in Figure 2–10. In the center illustration, a hot gas produces a bright-line,
or emission, spectrum. The two lines seen are characteristic of sodium
from the salt crystal being held in the flame. At the bottom, a continuous
spectrum is passed through a gas containing sodium vapor. In this case the
sodium removes light from the wavelengths where sodium would have
emitted light if the gas were hot enough. This produces a dark-line, or
absorption, spectrum.

Over a hundred years ago, Gustav Kirchoff, the German physicist, found that there are three types of spectra that an object can emit (Figure 2–11). The type depends on the physical conditions of the source.

Bright-line Spectrum

If a thin gas is heated it will show a spectrum consisting of bright lines. We know from the discussion above that this is due to the excitation of the atoms by heat energy, and that the lines seen will be characteristic of the chemical composition of the gas. This is the sort of spectrum emitted by some of the gas clouds in space. It is sometimes called an *emission spectrum* (Figure 2–7).

Continuous Spectrum

If the density of the gas is made higher, or if a solid or liquid is heated, the spectrum will be a continuous band of colors from red to blue and will extend beyond the visible colors on both ends. This is known as a continuous spectrum, or *continuum*. We can think of many, many atoms so close together that their energy levels are disturbed by one another; we have an infinite number of individual lines, so many that they blend into a continuous band. A continuous spectrum is characteristic of a dense body, but the body does not have to be very dense. An opaque compressed gas will show a continuous spectrum.

If the object absorbs and radiates energy perfectly (a so-called *black body*), the spectrum is said to be a black-body spectrum, shown in Figure 2–10. Both the peak of the spectrum, the wavelength where most of the light is emitted, and the exact shape of the spectrum are determined by just one thing, the temperature of the object. If we compare the radiation from two objects at different temperatures, we notice two important facts. First, the one with a higher temperature emits more energy, as we might expect. Second, the wavelength at which the peak occurs is shorter for the higher temperature object. Thus, investigation of the spectrum of a hot dense object tells us its temperature, but not its composition, for all dense objects having the same temperature radiate the same spectrum.

For example, the cosmic radio radiation mentioned in Chapter 1 as evidence for the Big Bang is characteristic of a black body at 3° Kelvin. This is the temperature to which the primordial fireball has cooled in the billions of years since the explosion.

Dark-line Spectrum

If we place a cool gas in front of a hot dense object which is giving off a continuous spectrum, we will observe a continuous spectrum, broken by dark lines where some wavelengths have been removed by the gas. These dark lines are seen at the same frequencies at which the cool gas would show bright lines if it were hot enough. What is happening is that each particular atom of the gas can accept photons of those particular frequencies and the atoms become excited. They soon radiate away this extra energy. It is radiated in *all* directions, and so most is removed from the specific direction toward the observer. Thus, a dark line appears at that color or wavelength in the spectrum. The line is not completely dark, however, since not all of the light is removed from the beam. Quite a bit can be left, but it looks dark by contrast to other parts of the spectrum. (In fact, the light which does get through can still give important information about the source.)

This is the type of spectrum shown by the Sun and the other stars. The hot photosphere (the "surface" of the Sun, see Chapter 11) is a dense gas and gives off a continuum. Overlying it is a thin cooler gas, which absorbs many frequencies. These dark lines are called *Fraunhofer lines*, after the first scientist to see them. There are over 20 000 dark lines in the Sun's spectrum.

Other Regions of the Spectrum

In all of the preceding we have talked of light, but the discussion applies equally to the "invisible" frequencies of electromagnetic radiation. The differences lie in the methods of production of the energy. Instead of different energy levels of the electrons, for gamma rays we would talk of energy levels of the *nucleus,* and for radio waves we would talk of vibrations of *molecules* or the direction of spin of electrons. Thus each region of the electromagnetic spectrum reveals the details of different physical processes in the universe, and each gives information which it may not be possible to gather in any other wavelength region. A better view of the spectrum of a celestial object, or a broader view of more of the spectrum, reveals new information. The development of radio astronomy, infrared astronomy, X-ray astronomy, and gamma-ray astronomy has expanded our knowledge of the universe. The ability to observe outside Earth's atmosphere has also provided us with data free from the interfering effects of our turbulent air.

The regions of the electromagnetic spectrum actually have no sharp boundaries. Each region gradually blends into those on either side, but we often adopt certain wavelengths as the dividing points between portions of the spectrum. Thus, we often use 7000 A as the division between visible and infrared, and a wavelength of 1 mm (10^7 A) as the dividing line between infrared and radio. A wavelength of 4000 A marks the short wavelength edge of the visible spectrum, below which is ultraviolet.

THE DOPPLER EFFECT

If a source of light is moving toward or away from an observer, he will see the spectrum of the light changed in wavelength (or frequency). This is known as the *Doppler effect* or *Doppler shift*, after the Austrian physicist Christian Doppler (1803–1853), who discovered it.

The true motion of an object in space with respect to an observer is called its *space motion*. It can be thought of as made up of two parts: (1) a motion toward or away from the observer, called *radial velocity*, and (2) a motion across the line of sight, called *transverse velocity*. If the radial velocity is away from us, it is called positive; radial velocity is negative if the motion is toward us. We cannot directly measure the transverse velocity, but we can measure the changing position of the object in the sky. This change is a small angle each year and is called *proper motion*.

The Doppler effect allows us to measure directly the radial velocity of an object. Due to the Doppler effect, a spectral line at wavelength W in the spectrum will be shifted by an amount ΔW. The relation between ΔW and the radial velocity V is

$$\frac{\Delta W}{W} = \frac{V}{c}$$

where c is the speed of light. For example, one of the strongest spectral lines for some types of stars is due to hydrogen at a wavelength of $W = 6562.85$ A. Suppose a star has a radial velocity of $+75$ km/sec. Then the amount of Doppler shift is

$$\Delta W = W \times \frac{V}{c} = 6562.85 \times \frac{75}{3 \times 10^5}$$
$$= 1.64 \text{ A}$$

In this case the line will appear at a wavelength of $6562.85 + 1.64 = 6564.49$ A. Notice that since the star is moving away, the shift is toward the red (longer wavelength) end of the spectrum. In actual practice, astronomers would measure the wavelength of this and other lines in the spectrum, and conclude that the star has a radial velocity of $+75$ km/sec. The ability to measure radial velocities directly is a very powerful tool for the astronomer.

Position Measurement

We have discussed the intensity of light and the spectrum. A third property of light is its direction, or source. It is the area studied by the branch of astronomy called *astrometry.* Special telescopes and instruments are used to get position observations as accurate as possible. A few thousand stars may be selected as standards of position, in the way that benchmarks are used for surveying the surface of Earth. These stars are observed carefully for many years. They are used to define a *coordinate system* in the sky. The positions of other objects in the sky are compared to the standards. It is a very exacting and tedious field, and not a very exciting one to the layman. Yet, it is extremely important in supplying data for other branches of astronomy. The positions, and, even more important, the changes in position of astronomical objects, give important information about the objects.

For example, the careful measurement of the very small motions through the sky of the stars in a nearby star cluster enables us to find the distance to the cluster, and this in turn allows us to determine the actual brightness of the stars in the cluster. Knowledge of star brightness then enables us to get the distance to very remote stellar systems. (See Chapters 3 and 8 for more details of stellar brightness and star clusters.)

Polarization

Polarization is the fourth property of light we spoke of earlier. If we think of light as a wave, like a wave on a rope when it is shaken, then the wave vibrates in a plane. This is called the *plane of polarization* (Figure 2-12). Each time an atom produces a wave of light, the wave has some particular polarization. When many atoms radiate at random, the planes may be at any angle, and so there is no net polarization. We say that the light is *unpolarized.*

There are a number of physical processes which can produce polarized light. If the atom is in a magnetic or electrical field, it may produce a net polarization. Or, unpolarized light passing through some kinds of substances may have light vibration in some planes absorbed, leaving only light of one plane of polarization. This latter is the effect of the lenses in polarized sunglasses. There are also special types of radiation not produced by the sort of process we talked of earlier which are intrinsically polarized. Observations of polarization in the light from astronomical sources give information about the physical conditions of the source or the region between the source and the observer.

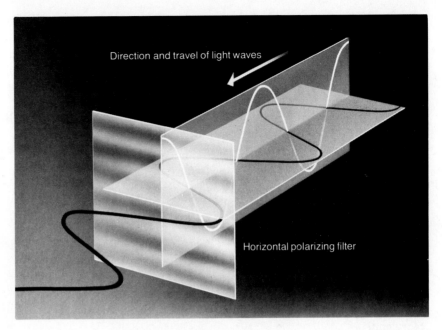

Direction and travel of light waves

Horizontal polarizing filter

FIGURE 2–12
Most sources produce light which is unpolarized; that is, there are equal numbers of waves vibrating in all possible planes. If one plane of vibration is predominant, the light is said to be polarized. We can produce polarized light by placing a filter in the light path which absorbs light from one plane of polarization, leaving other planes unaffected.

NONTHERMAL RADIATION

So far we have been talking about radiation from hot objects or gases which is characterized by the temperature of the source. There are a few types of radiation that are not caused by thermal effects. One important type is known as *synchrotron radiation*, because it was first observed in nuclear accelerators. The spectrum of synchrotron radiation is very different from that of thermal radiation. Also this kind of radiation is strongly polarized. These facts give us the means of identifying this type of radiation.

Synchrotron radiation is produced when charged particles (usually electrons) are accelerated to nearly the speed of light. The theory of relativity predicts that these accelerated particles will lose energy by radiation, and this is just what is observed. The radiation is directed outward into a cone in the direction of travel of the particle. The faster the electron is moving, the narrower the cone. Sources of synchrotron radiation are very energetic objects, such as the Crab nebula (Figure 4-27).

GRAVITY AND GRAVITY WAVES

Sir Isaac Newton (1642-1727) formulated the *law of gravity* in the middle of the seventeenth century. According to this law, every object has a gravitational field surrounding it. The gravitational force between two objects is proportional to their masses, and inversely proportional to the square of the distance between them. The constant of proportionality is called G, the *gravitational constant*. Thus, if m_1 and m_2 are the masses of two objects, and d is the

FIGURE 2–13
Physicist Joseph Weber is shown with his invention, the gravitational radiation detector, at the University of Maryland, College Park. (University of Maryland)

distance between them, the force of attraction each for the other can be written as

$$F = G \times \frac{m_1 \times m_2}{d^2}$$

The gravitational field of an object extends to infinity in all directions, so every piece of matter in the universe is attracted to every other piece, no matter how distant. There is no such thing as being "outside the gravitational field of Earth." The gravity field of our planet goes on forever. Of course, at great distances it is very weak and may well be negligible, but it is still there.

In this century, we have come to know that matter may be converted into energy, or vice versa. This is one result of the *theory of relativity* first worked out by the German-American scientist Albert Einstein (1879-1955), and later expanded by many others. In this theory (which has so far been borne out by experiments), light also has mass, the mass into which it could be converted. The famous relationship between mass and energy is

$$E = m \times c^2$$

in which c is the speed of light.

The force of gravity is the most important force we know of for large objects. It holds together stars, star clusters, galaxies, and the universe itself. We will study its effects in more detail later.

One prediction of the theory of relativity is that when large amounts of mass are moved about or converted to or from energy, *gravitational waves* should be produced. Some scientists have attempted to detect these waves using detectors made of large cylinders of metal (Figure 2-13). If a gravity wave passes through such a cylinder, the cylinder should vibrate a very tiny amount, and sensitive instruments should be able to detect the vibration. Some researchers claim to have detected gravity waves, while others have tried and failed. The detection is complicated, since vibrations could be caused by other means, such as earthquakes. As of now, it is not certain whether gravity waves really have been discovered. If these waves do exist, and we can learn to observe and interpret them, they may give us additional information about the universe.

REVIEW

1   There are four things we can determine about the light from a point source: (1) direction from which it comes, (2) intensity, (3) spectrum, and (4) polarization.

2   The intensity of light from a point source decreases as the square of the distance from the source (inverse square law).

3   A wave may be described by its velocity, V; wavelength, W; and frequency, f. The three are related by $V = W \times f$.

4   The electromagnetic spectrum includes light, as well as radio, ultraviolet, X-ray, and gamma-ray radiation.

5   An atom is composed of a nucleus made of positively charged protons and neutral neutrons, surrounded by an orbiting cloud of negatively charged electrons. The number of protons is the atomic number of the element; the number of protons plus neutrons is the atomic weight.

6   Atoms with the same number of protons (same atomic number) but different atomic weights are called isotopes.

7   A neutral atom has a number of electrons equal to the number of protons. If the number of electrons is different from the number of protons, the atom is said to be ionized.

8   An atom emits or absorbs photons when an electron jumps from one orbit to another. Such jumps are quantized; that is, there are only certain orbits the electron may occupy.

9   If a thin gas is heated, it will emit a spectrum composed of bright lines which are characteristic of the composition of the gas.

10 A solid, liquid, or dense gas will emit radiation of all wavelengths. This is known as a continuous spectrum, or continuum.

11 If a continuous spectrum passes through a thin gas, the gas will absorb certain wavelengths, producing a dark-line spectrum. The wavelengths absorbed will be the same as those which are emitted when the gas is heated.

12 If a source of light (radio, etc.) is moving toward or away from the observer, he will see the spectral lines shifted toward the blue or red end of the spectrum, respectively. This is the Doppler effect.

13 Gravity is an inverse-square-law force.

14 Mass and energy are convertible one into the other. The relation between them is $E = m \times c^2$ (or $E = mc^2$).

QUESTIONS

1   What additional information might we be able to obtain from an object that is an extended object, not a point source?

2 To measure the position of an object, we must have a system of coordinates as a reference. How can we go about establishing such a coordinate system in the sky? Is there more than one coordinate system we can use?

3 Without going there, how can we determine the total amount of energy coming out of a square centimeter of the "surface" of the Sun?

4 Explain the difference between excitation and ionization of an atom. Why is there no such thing as doubly ionized hydrogen?

5 Figure 2–10 shows in the form of a graph the continuous spectrum from a hot, dense object. Draw in the form of a graph (a) a bright-line spectrum and (b) a dark-line spectrum.

6 If you assume that, on the average, the space motions of all stars are the same, how could you use measured proper motions of stars as an indicator of their distances?

7 Suppose a small cloud of electrons is in orbit about an object at nearly the speed of light, and that we are near the plane of the orbit, but very far away from the object. The cloud will emit synchrotron radiation. Use what you know about this type of radiation to predict what we would observe. Does this suggest any known type of astronomical object?

READINGS

Friedman, H. "X-ray Astronomy." *Scientific American*, June 1964.

Goldberg, L. "Ultraviolet Astronomy." *Scientific American*, June 1969.

Hewish, A., ed. *Seeing Beyond the Visible*. New York: American Elsevier Publishing Co., Inc., 1970.

Neugebauer, G., and R. L. Leighton. "The Infrared Sky." *Scientific American*, August 1968.

Page, T., and L. W. Page, eds. *Starlight*. New York: The Macmillan Company, 1967.

Rublowsky, J. *Light, Our Bridge to the Stars*. New York: Basic Books, Inc., Publishers, 1964.

Scientific American. *Lasers and Light*. San Francisco: W. H. Freeman and Company, Publishers, 1969.

OBSERVING
THE UNIVERSE

CHAPTER 3

Just as different physical processes produce radiation of different frequencies, so different receivers must be used by astronomers to study these different frequency or wavelength regions. We are most familiar with telescopes which collect light, but there are others which collect radio waves, infrared waves, ultraviolet waves, X-rays, and gamma rays. Each wavelength region requires a particular technique and particular equipment, which enable us to gather information unobtainable in any other way,

A view of the Kitt Peak National Observatory, a national facility open to all astronomers, near Tucson, Arizona. In the foreground is the 4-meter Mayall Telescope. In the background are the domes of other optical telescopes. The McMath Solar Telescope, not shown in this photograph, is also a part of this observatory.

IMPLEMENTS FOR OBSERVING

Optical Telescopes

Telescopes for collecting visible and near-visible wavelength are usually classified as either refractors or reflectors, depending upon whether the main light-gathering element or objective is a lens (Figure 3-1) or mirror (Figure 3-2). The objective of a telescope collects the light from a distant object and brings it to a point where it may be studied. When used in this manner, the objective does not magnify; it merely *gathers light*. The telescope may be thought of as a "light-bucket." What is done with the light after it is collected is up to the astronomer.

Since the larger the telescope, the bigger the beam of light it can intercept, larger telescopes can detect fainter objects than can smaller ones. Another advantage of size is in the telescope's ability to distinguish between objects in nearly the same direction, or to see detail. This function is called *resolution,* or *resolving power.* The larger the telescope, the smaller may be the angular distance between two objects that nevertheless still appear as separate.

Most refractors are similar in design and appearance. Reflector-type telescopes show a greater range in sizes and configurations, in order to take the best advantage of the light, or to do certain things with the collected light. The world's largest telescopes are reflectors for several reasons. A mirror is easier to construct than a lens. It may be supported around the edges and back, whereas a lens can be held only by its edges. And, important for color measurements, a mirror reflects all colors in the same way, whereas a lens acts like a series of small prisms, focusing red and blue light at slightly different points.

The distance from the lens or mirror to the focus is called the *focal length.* The telescope objective brings the collected light to a *focus,* or *focal plane.* If the object being observed is to be viewed, then an *eyepiece* is placed at the focus. It is the eyepiece which magnifies the image produced by the objective. Since the image size produced by the objective will be larger the longer the focal length of the objective, and will be magnified more the shorter the focal length of the eyepiece, the magnification M will be

$$M = \frac{\text{focal length of objective}}{\text{focal length of eyepiece}}$$

FIGURE 3–1   The 40-inch refractor telescope at the Yerkes Observatory. This is the largest refractor ever built, and is likely to retain that record since it is easier to build large reflectors. The inset shows the light path through a typical refractor. (Yerkes Observatory)

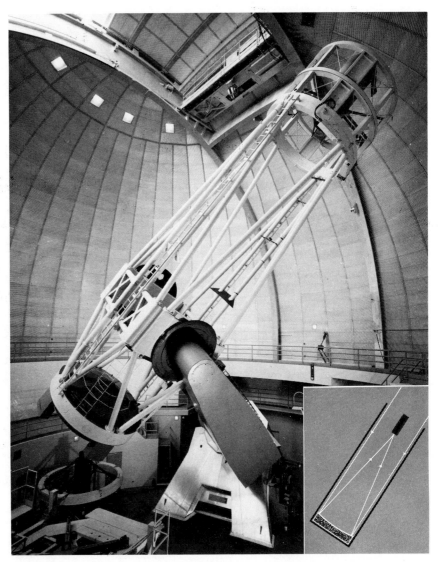

FIGURE 3–2   The 120-inch reflector telescope at the Lick Observatory. Note that there is no solid tube, but only an open gridwork of supporting structures holding the mirrors at the top of the telescope. The inset shows the light path through a typical reflector such as this. (Lick Observatory)

If the object is to be photographed, a photographic plate is placed at the focal plane. Photography is capable of detecting objects which cannot be seen by the eye even through the telescope, since the astronomer may take a time exposure of many hours, storing up the light from faint sources. The eye stores light for only about $\frac{1}{10}$ of a second. If some particular color band of light is desired—say, just the orange light coming from the object—then a combination of a filter which lets through only orange light and a photographic plate sensitive to orange light may be used together. Photographic plates are available for many different regions of the spectrum, including portions of the infrared and ultraviolet.

If the intensity of the source of light is to be measured, a *photometer* is used. Filters may be used with it to study the brightness of the object in different color ranges. A photometer is composed of a light-sensitive device, usually some sort of vacuum tube, and amplifiers and recorders. A camera light meter is a simple form of photometer. These techniques yield very accurate measurements of brightness.

If the detailed spectrum of the source is desired, then a *spectroscope* or *spectrograph* may be attached to the focus to break up the light into its component colors. The part which *disperses* the light into colors may be a prism or, more commonly, a *diffraction grating*. This is usually a piece of glass with a reflective coating (commonly aluminum) with fine lines ruled on the surface. These fine lines disperse the light. A crude example of a diffraction grating is a microgroove record, which shows colors if held at the correct angle to a beam of light. A grating with as many as 1200 lines per millimeter is not uncommon. The spectrum produced by the grating or prism is then photographed, and the entire device is called a spectrograph.

In the case of the spectrograph, and for ordinary pictures, photography is useful not only because it can detect objects not visible to the eye, but also because it is an objective record of the observation which can be studied at leisure in the daytime. Also, anyone can examine the photograph and extract his own data from it. Nighttime observing is thus for gathering data, and the daytime can be spent analyzing these. Usually much more time is needed to prepare for a night's observing, and to analyze the results, than to do the actual observing. Thus photography makes the use of a telescope more efficient. Especially for the largest telescopes, which are in great demand by astronomers in order to make observations of faint sources, this is an important point.

FIGURE 3–3
Photographs of an area of the sky taken with a Schmidt-type telescope. In the top photograph, a region of the sky is seen. In the bottom photo-graph, the same region has been photo-graphed with a thin prism placed over the front of the telescope. Each star image has been spread out into a spectrum. In this way many stellar spectra may be obtained at once, and analyzed later. Such techniques are very useful for surveying large areas of the sky. (Warner-Swasey Observatory)

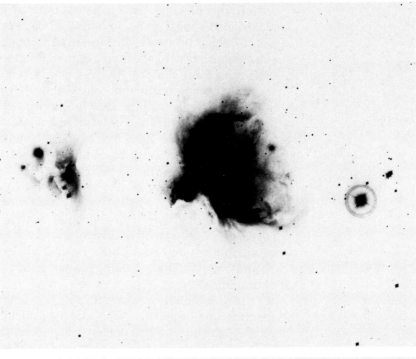

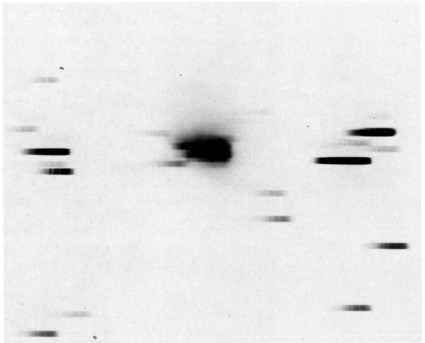

One type of telescope uses both a lens and a mirror, technically known as a *catadioptric system*. This telescope, called a *Schmidt telescope* after its inventor, Bernhard Schmidt, an Estonian optician, is really a specialized kind of camera, capable of photographing wide areas of the sky at one time (Figure 3-3). A Schmidt telescope is usually not used for viewing since the focal point is inside the telescope itself. One very useful application of a Schmidt telescope is to place a thin prism over the objective lens—an *objective prism*. Then each star image on the photographic plate will form a spectrum. This technique allows wide areas of the sky to be surveyed, and the stars classified by their spectra. These spectra are not, however, as detailed as the one-at-a-time spectra obtained with a spectrograph on an ordinary telescope.

Radio Telescopes

Detectors of radio radiation are essentially *antennas*. They may have some sort of structure to focus the radio waves to a point where the antenna can pick them up and pass the waves on to the receiver. Because radio waves are much longer than light waves, and of lower energy, radio telescopes are often much larger than optical telescopes, and are constructed in a greater variety of shapes and sizes (Figure 3-4). In order to improve the resolution of a radio observation, it is not uncommon for radio telescopes thousands of kilometers apart to be connected together directly or indirectly. This improves the ability to resolve close objects.

Radio telescopes may be steerable to any point in the sky, or partially steerable, or fixed. The latter two types must rely on the rotation of Earth to bring objects into their fields of view.

Radio astronomers do not, of course, look through their instruments. We do colloquially say, however, that a radio telescope "sees" an object; that is, the radio telescope detects radiation from that object. The observations are not displayed in pictures; rather, they are recorded as meter readings, as lines on oscilloscopes and chart recorders, as well as directly on computer tapes. Much analysis is needed to turn raw radio observations into usable data.

Astronomers working in the radio region of the spectrum also have the ability to send radio waves to nearby objects and receive the echoes. This is the field of *radar astronomy*. *Radar* is an acronym for Radio Direction and Ranging. A high-intensity beam of radio waves is directed into space from a powerful transmitter connected to a large antenna. Since the speed of the waves is known, by measuring the time it takes to receive the echo, the distance to an object can be obtained. We have been able to bounce radio waves

FIGURE 3–4
The 300-foot radio telescope at the National Radio Astronomy Observatory in Greenbank, West Virginia. Above the radio telescope is a radio map of a portion of the sky made with this telescope. This is made by assembling many scans of the radio telescope along adjacent lines in the sky. An upward deflection indicates that a source of radio waves is being received from that direction. All the radio sources seen here are found outside our galaxy. (National Radio Astronomy Observatory)

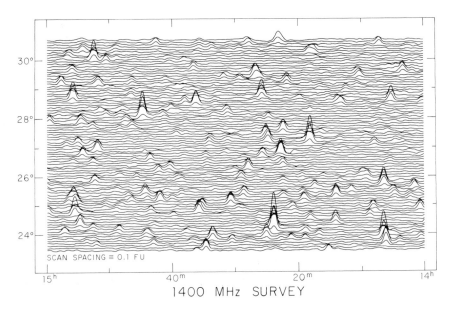

off the Moon, Venus, Mercury, Mars, some asteroids, and even the atmosphere of the Sun. This is an important technique for obtaining distances to these members of the solar system. Further, if the object is rotating, the Doppler effect will change the received wave slightly, and we can thus determine rotation speeds. This was done for Mercury in 1965 and showed, much to our surprise, that Mercury does not rotate on its axis in the same time it takes to go around the Sun, as had been thought. Radar also can "see" through the clouds of Venus, allowing us for the first time to determine its rotation (243 days per rotation, in an east to west direction) and to obtain a rough map of the surface, revealing craters and mountains. These features cannot be seen with optical telescopes.

Within the past few years, a sort of optical radar has been used to determine the distance to the Moon. The Apollo astronauts left special reflectors on the Moon's surface. A powerful laser beam directed at the Moon through a telescope is reflected back to Earth, and the time of travel is measured carefully. By this means, we have been able to obtain the distance to the Moon to within a few centimeters.

Other Wavelength Regions

Observations in the near infrared and ultraviolet regions may be made with photography. Beyond these regions, measurements are made in a manner similar to the techniques of photometry described above. Special equipment and techniques are needed for each spectral range, and for several areas the observations are still crude. Astronomy is dependent upon technology to supply it with the means of making desired observations (Figures 3–5 and 3–6).

To circumvent the effects of the atmosphere, instruments have been flown in aircraft, balloons, and rockets. Recently orbiting astronomical observatories have yielded much important information about spectral regions not accessible from the surface. Astronomers look forward to the time when they can man a permanent observatory on the Moon.

EFFECTS OF THE ATMOSPHERE

The observations we make from the surface of Earth are affected by the atmosphere. The air above us absorbs and emits light, and disperses and refracts the light from astronomical objects.

Various gases in the atmosphere, notably ozone and water vapor, absorb much energy. At the same time, ozone is beneficial to

FIGURE 3–5
An observatory in space. The Skylab space station was the first manned observatory in space. Observations of the Sun and other celestial objects were made in regions of the spectrum not accessible to ground-based astronomers because of atmospheric absorption. Analysis is still continuing of the great wealth of observational data obtained in the Skylab flights. (NASA)

FIGURE 3–6
Artists' drawing of the Orbiting Astronomical Observatory (OAO), one of the most advanced unmanned satellites launched. This series of satellites, in orbit 800 km above Earth, allowed astronomers their first long-term views of objects not observable from the ground. Such special purpose astronomical instruments greatly expanded our knowledge of the universe. (NASA)

us because it absorbs harmful ultraviolet rays from the Sun. If the ozone should disappear, life might not be possible on the surface of our planet.

At times of great solar activity, high-energy particles from the Sun cause the upper atmosphere to glow, creating the *aurora*, the northern or southern lights (Figure 3-7). Even when there is no aurora, the sky emits a faint glow detectable by astronomers. This glow sets the lower limit of the faintest objects observable. In addition to the aurora and night glow, scattered light from man-made illumination is a persistent problem for astronomers; this is why observatories are usually placed in remote, dark locations. In the United States it is practically impossible today to put an observatory in a location where artificial lighting cannot be observed, or where man-made air conditions do not adversely affect observations (Figure 3-8).

The atmosphere refracts, or bends, light. This refraction is responsible for the fact that the setting Sun appears flattened, and also causes an object to appear slightly higher in the sky than it really is. At the horizon this change in position amounts to only about half a degree, a small enough amount for most purposes, but

FIGURE 3–7
The aurora borealis, or northern lights, is a fluorescence process of gases high in the atmosphere. High-energy particles from the Sun excite the atoms and cause them to emit some of this energy in visible colors. Aurorae are most common near the north and south magnetic poles, but are occasionally seen at lower latitudes. (Allen Seltzer)

FIGURE 3–8
Artificial light is a serious and increasing problem for astronomers. This night view from Mt. Wilson Observatory overlooks Los Angeles and 40 other towns in the area. Scattered lights from urban areas plus polluted air make the sky bright and thus make photographs of faint astronomical objects difficult or impossible. There are very few dark-sky locations left in the world. (Hale Observatories)

it is extremely large when careful position measurements are desired. For this reason, measurements of star positions are made when the stars are high above the horizon (Figure 3–9).

FIGURE 3–9
The atmosphere refracts, or bends, light. This makes a star appear to be higher in the sky than it really is. The solid arrow shows the light path to the observer. The dashed arrow indicates where the observer sees the star. Because of refraction the setting Sun appears flattened.

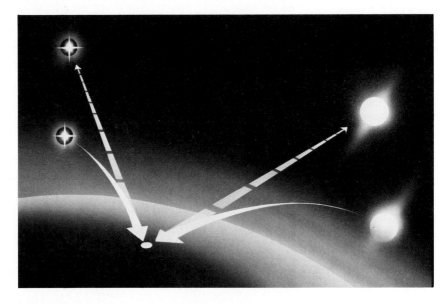

The atmosphere disperses light; that is, it does not treat all wavelengths alike. The air scatters blue light all around the sky, making the sky blue. The setting sun appears to be red because most of the blue light has been scattered out of the ray of sunlight reaching the observer. Scattering makes precise color measurements difficult, and the position of the object in the sky must be known in order to allow for the effects of scattering.

The astronomer must contend with all these effects in making his observations. Since he has no choice but to observe through the atmosphere, unless he is in space, he must take care to remove the effects from his data.

THE SYSTEM OF MEASURING BRIGHTNESS

Because of the large range of brightness of astronomical objects, the astronomer uses a system of brightness measurement which enables him to encompass a great range in a small range of numbers.

About A.D. 150 the Alexandrian astronomer Claudius Ptolemaeus, or Ptolemy, published his catalog of stars, giving their positions and indicating their brightness in an approximate way (Figure 3–10). The brightest stars he called "first magnitude," using the word in the same way we use it in conversation, meaning "importance." Those stars a bit fainter were "second magnitude," and so on. The faintest stars visible to the unaided eye were of the sixth magnitude.

This system was carried on through the Renaissance, through the invention of the telescope, and into the middle of the nineteenth century. It caused confusion, because two astronomers were likely to disagree on the boundary between, say, the second and third magnitude ranges. What was needed was to put the system on a mathematical basis, that is, to set up a scale of brightness measurement. This is the scale of *magnitudes*.

It was noticed that stars of the first magnitude are about 100 times brighter than those of the sixth magnitude. Around 1856, the English astronomer Norman Pogson suggested that the brightness factor be *defined* as exactly equal to 100. This sets up a definite scale of brightness. Since there are five magnitudes between 6 and 1, each step of one magnitude corresponds to a brightness factor of the fifth root of 100, or 2.512 . . . , or approximately 2½. Thus, differences in magnitudes correspond to *factors* in intensity. Here is how they relate:

1 magnitude $\;= $ a factor of 2½ times

2 magnitudes = a factor of $2\frac{1}{2} \times 2\frac{1}{2} = (2\frac{1}{2})^2 = 6\frac{1}{4}$ times

3 magnitudes = a factor of $2\frac{1}{2} \times 2\frac{1}{2} \times 2\frac{1}{2} = (2\frac{1}{2})^3 = 16$ times

4 magnitudes = a factor of $(2\frac{1}{2})^4 = 40$ times

5 magnitudes = a factor of $(2\frac{1}{2})^5 = 100$ times

If the difference in magnitudes of two objects is greater than 5, take out multiples of five magnitudes, and get the remainder from the table above.

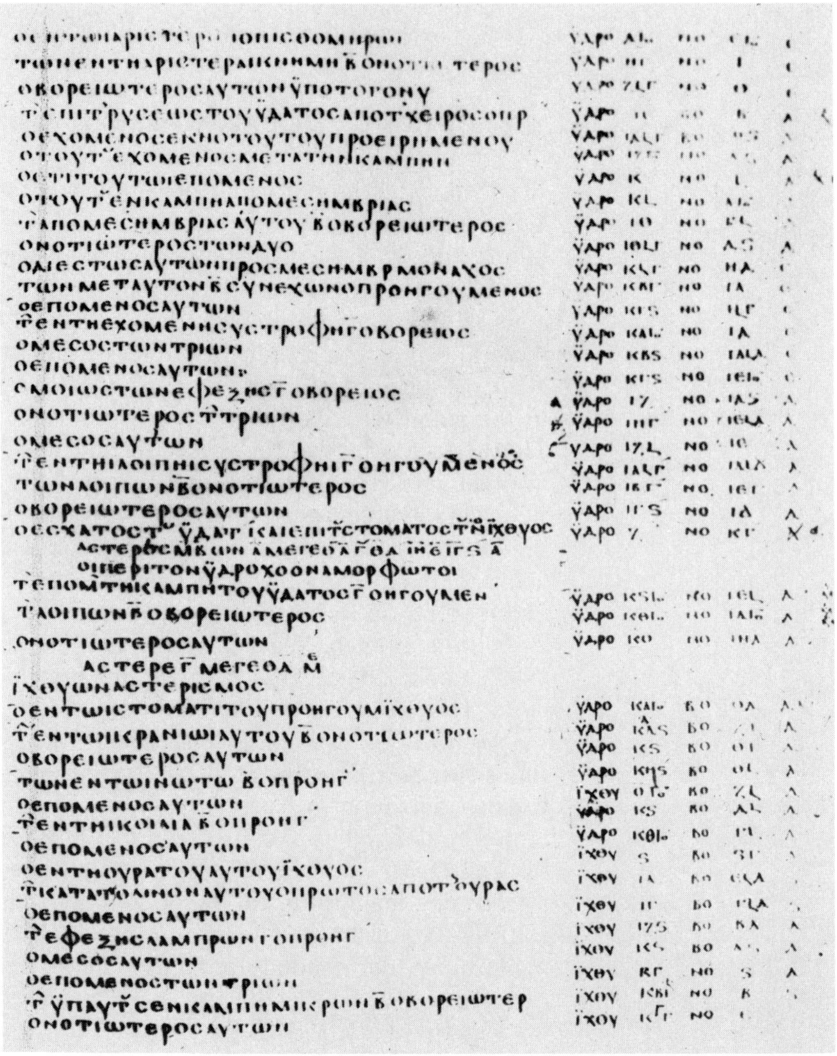

FIGURE 3–10
A page from the star catalogue of Ptolemy. This copy, written in uncial Greek, was made in the ninth century and is one of the oldest manuscripts of the Almagest known

Example: The magnitude of the faintest star that can be photographed by the 200-inch telescope is 23. As we have noted, the magnitude of the faintest star visible to the unaided eye is 6. The difference is 17 = (3 × 5) + 2. For each of the five magnitude differences, there is a factor of 100, so this accounts for a factor of $(100)^3$ = 1 000 000. The remaining two magnitudes represent a factor of 6¼, so the faintest star the eye can see is 6¼ × 1 000 000, or 6 250 000 times brighter than the faintest star photographed with the 200-inch telescope.

The advantage of this system is twofold: (1) one can express a large range of brightness—like 6 250 000, for example—with a small number like 17, and (2) it corresponds very closely to the way that the human eye responds to light.

In order for an object to look twice as bright, it must be actually more than twice as intense, because the eye does not respond linearly.

A linear relationship is one in which, if the input doubles then the output also doubles. The eye's response to light is *logarithmic*. (The same is true of the ear's response to sound; the scale of loudness is called the *decibel scale*.) If we remember what logarithms are (see Appendix A), we recall that we add logarithms in order to multiply numbers. This is exactly the same as saying that differences in magnitudes (a logarithm) are equivalent to ratios (multipliers) in brightness. For example, a magnitude difference of 2 is equivalent to a ratio of 6¼ in actual intensity.

The magnitude system must have a zero point, that is, something defined to be magnitude 1, or 6, or whatever is convenient. Attempts were made to use one star as a standard, defining its brightness as magnitude 0. This is not a very good idea, since individual stars may vary in brightness very slightly, giving us a variable standard. So astronomers chose a large number of stars and defined their average brightness to be exactly 6. The magnitude of other stars is measured relative to this average brightness.

To denote the magnitude of the brightest objects in the sky, it is necessary to use *negative* numbers, since there are objects brighter than magnitude 0 based on the standard scale (Figure 3–11). For example, the Sun has a magnitude of −27, the full Moon is of magnitude −12, and Venus at its brightest is of magnitude −4. The brightest star in the sky, Sirius, is −1.4. As mentioned before, the faintest star visible with the unaided eye is about magnitude 6.

Color Index

If we measure the brightness of a star in a particular range of color

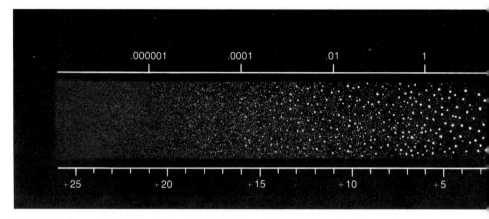

FIGURE 3–11   The magnitude scale allows us to express a large range of brightness in a convenient range of numbers. A scale of apparent magnitude runs along the bottom of the illustration, and a scale of actual intensity along the top. An object of apparent magnitude 6 is taken as the unit of intensity. This is the magnitude of the faintest objects visible to the unaided eye. Several representative objects are shown, ranging in brightness from the Sun to the faintest stars and galaxies photographable with the largest telescopes.

—say, yellow, or blue—and subtract one reading from the other, we will get an estimate of the *temperature* of the star. This magnitude difference is called the *color index*: the most common one is $B-V$, the magnitude in Blue light minus the magnitude in yellow (Visual) light. Since a brighter object will have a numerically smaller magnitude, the larger the color index $B-V$, the yellower or redder the star. Another common color index for stars is $U-B$, Ultraviolet minus Blue. Often data for a star are given in terms of a visual magnitude and one or two color indices. For example, the two brightest stars in the sky would be described this way:

Sun $V = -26.73$ $B-V = +0.63$ $U-B = +0.13$
Sirius $V = -1.42$ $B-V = +0.01$ $U-B = -0.4$

Earlier we discussed the curve of the spectrum of a hot object and found that the shape of the curve depends on the temperature. When we measure the intensity of light at two colors, we are just measuring the height of the curve at two points (Figure 3–12). Subtracting one measurement from the other gives the *slope* of the line. The slope depends on the shape of the curve, which in turn depends on the temperature. Thus, the color index can tell us the temperature of the star, which is an important piece of information from a com-

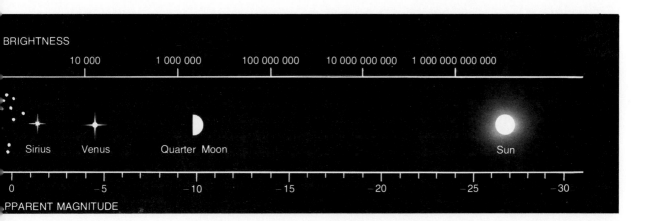

paratively simple observation. The color indices given above, for instance, tell us that Sirius is about 10 000°K, and the Sun is about 6000°K.

FIGURE 3–12   *Continuous spectra (Planck curves) for stars with surface temperatures of 24 000°K (upper curve) and 6000°K (lower curve). If we measure the brightness of one star in a small part of the spectrum around 5500 A and 6500 A, and subtract the first measurement from the second, we obtain the slope of the line. This is called the color index. Note that the slopes of the curves for the hot star and cooler star are different. In this way, a measurement of the color index of a star gives us its temperature.*

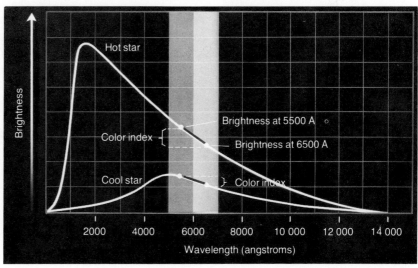

Brightness and Distance

There are two reasons why a star can look faint in a telescope: it can be very distant, or it can actually be intrinsically faint. There are thus three variables tied together: (1) the *apparent magnitude*, or brightness we measure with our instruments; (2) the intrinsic or actual brightness of the star; and (3) the distance of the star (Figure 3–13). To find any one of these, we must know the other two. Although it is relatively easy to measure the apparent magnitude of a star, knowledge of intrinsic brightness most often depends on spectral analysis, to be discussed in Chapter 5.

For reasons which will become apparent in Chapter 10, astronomers measure distances in *parsecs*. One parsec is equal to about 3.2 light-years, or 3.1×10^{18} centimeters. Measuring distances in parsecs makes our computations simpler.

If we wish to specify the intrinsic brightness of a lamp in a laboratory, we measure it in *foot-candles*, a unit incorporating a standard candle and a distance of one foot. For stars, we define the intrinsic brightness in terms of our magnitude scale and use the term *absolute magnitude*. The absolute magnitude of a star is how bright it would appear if it were at a distance of 10 parsecs. Thus, a star more distant than 10 parsecs will have an apparent magnitude greater (that is, it will appear fainter) than its absolute magnitude.

For example, the Sun (which is only 0.000005 parsecs from us) has an apparent magnitude of -26.73. But if it were 10 parsecs away,

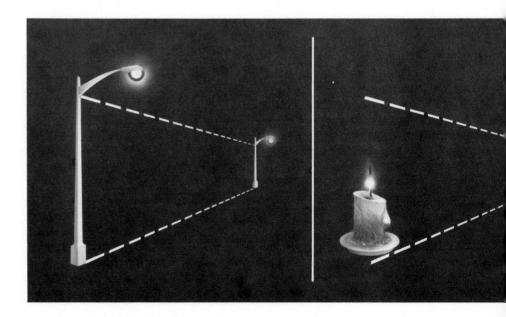

it would appear to be apparent magnitude +4.84. The star Betelgeuse, on the other hand, is 200 parsecs from us and has an apparent magnitude of +0.41. If Betelgeuse were 10 parsecs away, it would appear to be −5.6.

We give the symbols m to apparent magnitude and M to absolute magnitude. If d is the distance of the star in parsecs, the relation between these three important numbers is

$$m - M = 5 \log d - 5$$

The quantity $(m - M)$ is a measure of the distance of the star and is called the *distance modulus*. In using this equation, we must be careful, for there is the possibility that some of the light from the star has been absorbed along the way by the thin matter between the stars and us. If so, we will think it is farther away than it really is. Astronomers can sometimes get estimates of this absorption and take it into account.

FIGURE 3–13   *The relationship between brightness and distance. In the left panel, two objects of the same intrinsic brightness will appear to be of different brightnesses if they are at different distances. Or an intrinsically faint object may appear brighter if it is closer than a more distant, more intense source. The photograph is of M67, an open star cluster. Many of the stars in the photograph are in the cluster, some are foreground or background objects. In order to obtain a star's distance, we must measure its apparent magnitude and know its absolute magnitude.*

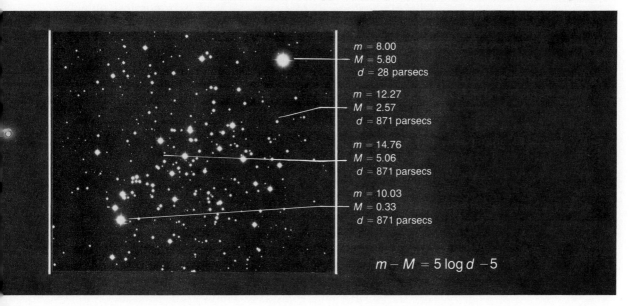

$m = 8.00$
$M = 5.80$
$d = 28$ parsecs

$m = 12.27$
$M = 2.57$
$d = 871$ parsecs

$m = 14.76$
$M = 5.06$
$d = 871$ parsecs

$m = 10.03$
$M = 0.33$
$d = 871$ parsecs

$$m - M = 5 \log d - 5$$

We have seen that the radiation from space is produced by many processes, affected by many things, and received with a diversity of equipment. Each photon carries information about the physical properties of the source and intervening space, and it is thus that we are able to deduce answers to our questions about the universe. As our knowledge of the electromagnetic spectrum grows, so too will our knowledge of the universe. Undoubtedly much is yet undiscovered about the spectrum, and about the universe. It is said that those who study the large-scale structure of the universe are finding out less and less about more and more, and those studying the atom are finding out more and more about less and less. Both groups have a long way to go before we understand nothing about everything and everything about nothing!

REVIEW

1   The main function of a telescope is to collect light and bring it to a focus.

2   Radio telescopes do not form images of objects in the sky. They record intensities of radiation.

3   The atmosphere distorts our view of astronomical objects by (1) absorbing some of the incoming radiation, (2) emitting some light of its own, (3) refracting the incoming light, and (4) transmitting different amounts of different colors.

4   The system of measuring brightness used by the astronomer is the magnitude system. A difference of five magnitudes corresponds to a ratio of intensities of 100. Thus one magnitude equals a factor of 2.512 . . . in brightness. The brighter object will have a numerically smaller magnitude (that is, as magnitude increases, brightness decreases).

5   The color index is the difference in brightness (expressed in magnitudes) of an object measured at two different colors (or wavelengths).

6   The apparent magnitude, m; the absolute magnitude, M; and the distance, d, of an object are related by

$$m - M = 5 \log d - 5$$

7 The absolute magnitude of a star is a measure of its intrinsic brightness, and is the magnitude it would appear if it were 10 parsecs away.

QUESTIONS

1 You are looking through a refractor telescope and are focused on a star. A fly lands on the lens. What would you see, and why?

2 If a pulse of light or radio waves is sent to the Moon, how long will it take to return? If we wish to measure the distance to the Moon to an accuracy of one meter, how accurately must we time the pulse?

3 Can we expect the observed color index of an object to stay the same, increase, or decrease as the object sets toward the horizon? Why? How might we correct for this?

4 Give some common examples of linear relationships.

5 Suppose you were observing a distant star through an invisible cloud of dust which absorbed a part of the starlight. How would this affect the apparent magnitude you would measure compared to what you would measure if the cloud were not there? What error would you make in calculating the distance of that star?

6 How can we correct the formula relating m, M, and d to take interstellar absorption into account?

READINGS

Hyde, F. W. *Radio Astronomy for Amateurs.* New York: W. W. Norton & Company, Inc., 1963.

King, H. C. *The History of the Telescope.* Cambridge, Mass.: Sky Publishing Corp., 1955.

Page, T., and L. W. Page, eds. *Telescopes.* New York: The Macmillan Company, 1966.

Sidgewick, J. B. *Amateur Astronomers Handbook.* London: Faber and Faber Ltd., 1971.

Wood, F. B. *Photoelectric Astronomy for Amateurs.* New York: The Macmillan Company, 1963.

Woodbury, D. O. *The Glass Giant of Palomar.* New York: Dodd, Mead & Company, 1970.

MATTER
IN SPACE

CHAPTER 4

The matter which makes up the universe is distributed not only in the form of stars and galaxies, but also in large tenuous clouds of material, and in an even more tenuous haze throughout space. The stars and galaxies, as well as some of the luminous clouds of material, are easily visible, but it has only been in this century that the existence and nature of some of the more diffuse material have become known.

In 1610, shortly after Galileo turned the first astronomical telescope to the sky, another astronomer, Fabri de Peiresc, discovered the Orion nebula, a faint patch of hazy light located in the "sword" of Orion and visible even to the unaided eye on a clear winter night (Figure 4-1). The word nebula is Latin for

The Trifid nebula in the constellation Sagittarius. This complex cloud of interstellar material is typical of the many clouds of gas and dust throughout the Milky Way Galaxy. The cloud's many small, dark globules may be stars and planetary systems in the making. (Lick Observatory)

FIGURE 4–1
The region of the constellation Orion, visible in the winter sky. In this photograph the stars are very slightly out of focus to make them more visible. The fuzzy 'star" in the middle of the "sword of Orion" is really a nebula first noticed in 1610. Compare the photograph in Figure 4–20. (Yerkes Observatory)

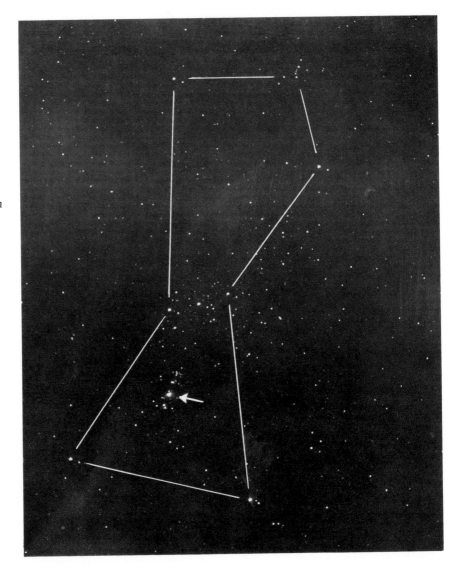

"cloud," and is appropriate for these fuzzy collections of gas and dust in space. At the time, of course, de Peiresc did not know the nature of the object he had discovered. In 1612, the German astronomer Simon Marius (1570–1624) discovered the Andromeda "nebula." Over the next 250 years, several more faint, fuzzy objects were discovered, and they added to the mystery (Figure 4–2).

In 1864, William Huggins (1824–1910), an English pioneer in astronomical spectroscopy, found that the Orion nebula showed a

FIGURE 4–2
The "nebula" in the constellation Andromeda. This area of hazy, faint light is just barely visible to the unaided eye on very clear nights in autumn. Its appearance is somewhat similar to clouds of gas in our galaxy, and for a long time it was thought to be just such another gas cloud. Today we know it is really a galaxy itself, very similar to our own and about 2.5 million light-years away. Compare Figure 10–7. (Yerkes Observatory)

spectrum of bright lines, indicating that the nebula is a collection of hot gases (Figure 4–3a). In 1899, it was found that the Andromeda "nebula" had a dark-line spectrum similar to the spectrum of a star, presumably indicating it was made of stars. This thoroughly confused matters, and the confusion was added to in 1918 when V. M. Slipher, the American astronomer, found that the nebulosity surrounding the Pleiades star cluster also showed a dark-line spectrum, even though the nebulosity was obviously gaseous (Figure 4–3b).

FIGURE 4–3
A comparison of the spectra of an emission nebula (left) and a reflection nebula (right) The emission nebula is in the region of Orion. The reflection nebula is associated with the Pleiades star cluster (see Figure 4–17). (Yerkes Observatory)

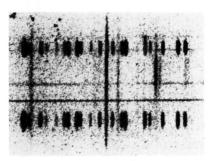

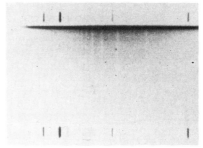

Emission nebula

Reflection nebula

Today we know that the observers were looking at different types of objects, and that the Andromeda "nebula" should not be so called, for it is a galaxy, not a cloud of gas. Later we will go into detail about the observations which showed this to be so. For now, we will concentrate on the nebulae proper.

THE CHEMICAL MAKE-UP OF THE UNIVERSE

As indicated in Chapter 3, all matter is composed of one or more of the chemical elements, of which 105 are known at this writing. Of these, 92 are so-called "natural elements"—those which occur in nature. Several are naturally radioactive, including the heaviest, uranium, which is number 92. The elements heavier than uranium are man-made in nuclear accelerators. It is likely that more elements will be produced in the future.

In stars, and in space itself, the most abundant element by far is hydrogen, followed by helium. Together these two elements make up about 99 percent of the matter in the universe.

We find different relative proportions of the elements when we look at different kinds of astronomical objects. For instance, Earth, a planet, has very little hydrogen and helium, but abundant oxygen, silicon, aluminum, and iron. The Sun is mainly hydrogen and helium, but with many more heavy elements than the average for the universe as a whole. In the chapter on stellar evolution we will see that richness in the heavier elements provides clues about the age of the Sun compared with that of other stars.

Because of the overwhelming preponderance of hydrogen and helium in the universe, astronomers for convenience label everything other than these two elements *metals*. Thus, when we mention metals in an astronomical sense (chemists define metals in a different way), we mean all elements heavier than helium.

The periodic chart in Figure 4–4 indicates the elements identified in various types of astronomical objects. About two-thirds of the known elements have been found in the spectrum of the Sun. Presumably the others are there as well, but in amounts so small that they have not yet been detected. No elements above uranium, the heaviest natural element, have been found in astronomical objects. Of the natural elements, a few, such as radium and technetium, are radioactive, and their presence is not expected since they would have decayed in the time since their formation. However, in rare cases, stars show traces of some radioactive elements, for reasons not yet explained adequately.

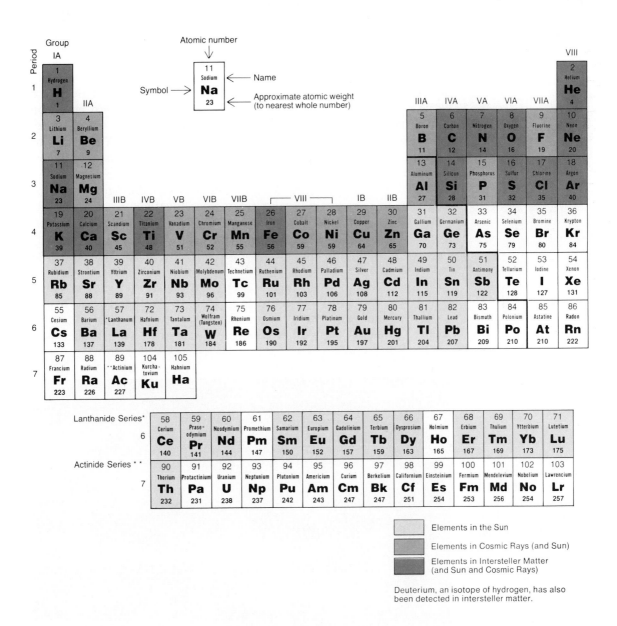

FIGURE 4–4   A comparison of the elements detected in the Sun, in cosmic rays, and in interstellar matter. Observe, in the periodic chart, that all elements detected in interstellar matter also have been detected in cosmic rays, and that all elements detected in interstellar matter and cosmic rays also have been detected in the Sun.

Intergalactic Matter

We do not have a very good estimate of the amount of diffuse matter between the galaxies. If the galaxies condensed out of a large cloud of material produced by the Big Bang, there might be some material left over which did not go into making the galaxies. There is some rather sketchy evidence for this belief. Tentative observations reveal haze in the middle of clusters of galaxies (Figure 4–5). Also, a line in the spectrum of a distant galaxy has been detected, possibly produced by matter between the galaxy and us. Occasionally we see two galaxies so close together that they distort one another, perhaps tearing gas out of each other to be left behind when the galaxies have receded (Figure 4–6).

More recently, the astronomical satellite *Uhuru* has made observations in the X-ray part of the spectrum of distant clusters of galaxies. These observations reveal that there may be as much material or more in the form of thin gas between the galaxies of the cluster as there is in the galaxies themselves. More observations are needed to ascertain if other clusters of galaxies have as much gas.

FIGURE 4–5
A cluster of distant galaxies located in the constellation Hercules. Some observers have reported observing a faint haze within such clusters, and recently satellite-borne ultraviolet telescopes have detected a suggestion of intergalactic matter. (Hale Observatories)

If so, this would increase by many times the average density of the universe.

Astronomers have speculated on the possibility of "rogue" stars in intergalactic space. These stars, because of their high speeds, would have escaped from their parent galaxies to wander alone in the darkness of intergalactic space. None have been seen, but they may exist.

The question of the amount of intergalactic matter has important consequences for the large-scale structure and evolution of the universe. If the overall average density, taking into account all the galaxies and any intergalactic matter, is high enough, then the mutual gravitational pull will be strong enough to stop the expansion of the universe and reverse it. This would produce a contracting universe, and perhaps eventually, many billions of years in the future, another Big Bang. Since observation of intergalactic material is very difficult, we still do not have a reliable figure for the amount.

Interstellar Matter

The diffuse matter which occurs between the stars in our galaxy

FIGURE 4–6
Two galaxies so close together that they distort one another. Some of the interstellar matter from the galaxies may be left behind in intergalactic space when they have receded. (Lick Observatory)

has been studied in detail in the past half century (Figure 4–7). It is important not only because of the effects it has upon our observations of distant stars, but also because the stars themselves presumably came from the interstellar material when knots of gas contracted under the pull of gravity until they became hot enough for nuclear reactions to occur. In distant galaxies, interstellar material is easily seen, as in Figure 4–8, and our galaxy is similar to many other galaxies. It has been estimated that roughly 20 percent of the mass in our galaxy is in the form of diffuse material.

Interstellar material is of two types: gas and dust. "Dust" is a general term meaning any small particles composed of many atoms or molecules. The composition of the dust is not known, and the discovery of its nature and properties is one of the most actively pursued topics in this field. It is known that the dust grains are very small, about the size of the wavelength of visible light, a few thousand angstroms, or half a micrometer. Compositions of ice, iron, graphite, frozen hydrogen, and combinations of these have been suggested, but none meets all the requirements needed to explain the observations.

FIGURE 4–7
A portion of the Milky Way Galaxy toward the constellations Cygnus and Cassiopeia. Note the complicated structure of bright and dark areas, produced by variations in the distribution of interstellar material. Also note that most of the absorbing material lies close to the plane of the Galaxy. (Hale Observatories)

The gas is mostly hydrogen and helium. The presence of helium is inferred, since it cannot be observed directly in most cases. Recently, a number of molecules have been discovered, and we will discuss them after we have seen how the interstellar matter is detected.

It may be helpful here to have an outline of the types of nebulae in the Galaxy. The two main types of galactic nebulae, luminous and dark, are classified by whether we see them glowing, or whether they absorb light from stars behind them. Often nebulae show both effects. The luminous nebulae may be further subdivided into diffuse nebulae, planetary nebulae, and supernova remnants. The light from a diffuse nebula may come from emission by the gas, or by reflection of starlight.

Dark nebulae. Even a casual glance into the sky along the band of the Milky Way on a clear night reveals the complicated distribution of dark material in the Galaxy. The band of light is not smooth but is interrupted by dark streaks and lanes. These were at one time thought to be holes in the Galaxy, devoid of stars, through which

FIGURE 4–8
A distant galaxy which happens to be edge-on to our line of sight. Note that the dark absorbing material lies along the plane of this galaxy. Compare Figure 4–7. Our Milky Way Galaxy would look similar to this if seen edge-on from a distance of several million light-years. (Lick Observatory)

FIGURE 4–9
A portion of the
Milky Way Galaxy
looking toward the
constellation
Sagittarius.
(Yerkes Observatory)

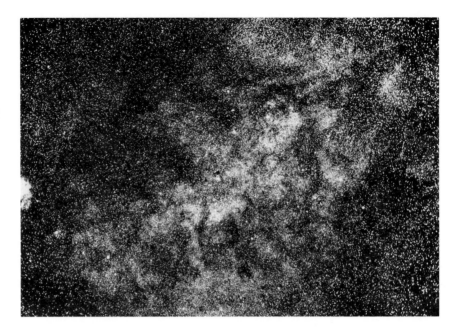

we could see into the dark depths of space. It was later discovered that these areas are actually regions where the light of distant stars is absorbed. These clouds of dark, absorbing material occur particularly near the band of the Milky Way, along the plane of the Galaxy.

FIGURE 4–10
A portion of the
Milky Way Galaxy
looking toward the
constellation
Ophiuchus.
(Yerkes Observatory)

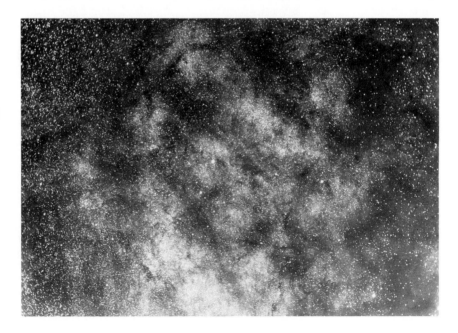

FIGURE 4-11
A portion of the
Milky Way Galaxy
looking toward the
direction of
Cassiopeia. (Yerkes
Observatory)

This is similar to the distribution of dark material seen in other galaxies, particularly those seen edge-on. The dark clouds in our galaxy take up larger or smaller areas of the sky, depending on their actual size and on their distance from us (Figures 4–9 to 4–12).

FIGURE 4–12
A dark nebula in the
constellation Serpens.
Note the bright rims
to some of the
clouds, caused by hot
stars hidden from
view. Note also the
small, dark globules.
(Lick Observatory)

FIGURE 4–13
The "Horsehead"
nebula in Orion. (Hale
Observatories)

FIGURE 4–13
The "Horsehead"
nebula in Orion. (Hale
Observatories)

A close look at the photograph of the "Horsehead" nebula in Figure 4–13 will show how a dark cloud obscures our view. Note that there are many times more stars in the upper part of the photo than in the lower part. The dark cloud covers the lower part, so the only stars we see are those closer to us than the cloud. The cloud is not very apparent except for the large blob of material which juts out into a brighter area, producing a silhouette which gives this beautiful nebula its name. Other photographs of other nebulae show that occasionally the dark clouds have bright rims. Small, very dense globules are sometimes seen; some people suspect these are stars and planetary systems in the process of formation.

The actual density of the dust in dark nebulae is very low, about 10^{-26} grams per cubic centimeter (compare the density of Earth's atmosphere at sea level of 10^{-3} g/cm^3). The density of the gas in the same cloud is about a thousand to a million times greater. It is the grains of dust in the nebulae which remove the light, but the process is one of scattering, not absorption. The grains deviate the starlight from its original path, or scatter it, somewhat similar to the way that air molecules scatter sunlight. Just as air scatters more blue light than red light, so too does the interstellar material scatter more blue light. This reddens the light of stars. The amount of reddening is proportional to the amount of total dimming, and gives us a clue as to the thickness of the cloud. In this way we can detect very thin clouds we otherwise might not know were dimming the light of stars behind them (Figure 4–14).

In addition to discrete clouds, dark material is diffused throughout all of space. We detect its presence by the absorption and reddening just mentioned, and by the effect that the gas has on the starlight. The gas which exists along with the dust does not absorb much light, but may produce faint dark lines in the light we receive from stars behind the cloud (Figure 4–15). Sometimes several lines appear in the spectrum, lines which do not belong in stars of that temperature. This means that some of the starlight has been absorbed by the cool gas in the clouds. The clouds are moving, each at a different radial velocity. Thus, each cloud will produce a dark line shifted slightly by the Doppler effect. In this way we can get an idea of how thick the clouds are, how many of them there are, what

FIGURE 4–14 *Starlight passing through a cloud of dark material is scattered in a way somewhat similar to that in which sunlight is scattered by the air. Blue light is scattered more than red light, so the starlight emerges from the cloud redder than it went in. Thus, we will observe the star to be redder, and fainter, than it really is. The dust component of the cloud is responsible for the scattering.*

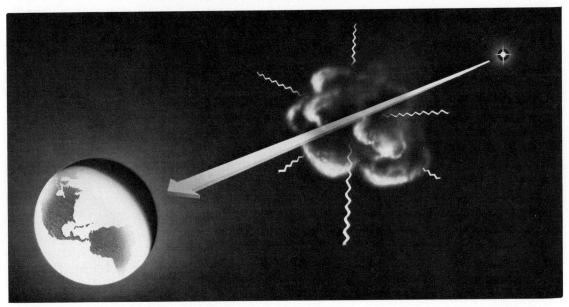

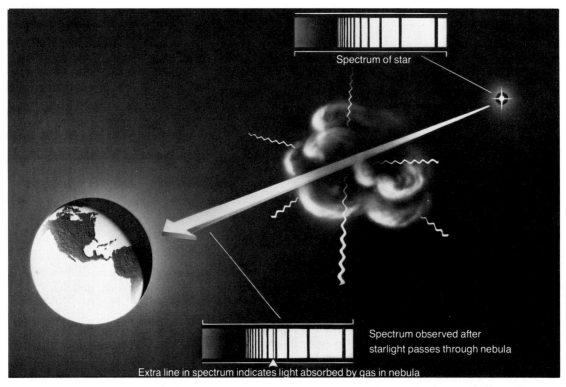

Spectrum of star

Spectrum observed after
starlight passes through nebula

Extra line in spectrum indicates light absorbed by gas in nebula

FIGURE 4–15 *Occasionally the spectrum of a star observed through a cloud of interstellar material will not only be reddened and dimmer, but also one or more spectral lines may appear that should not appear in a star of that type. The gas in the cloud has absorbed some light characteristic of the composition of the gas.*

they are made of, and how they are moving. Most of the light is blocked by the dust, not the gas. On the average, the diffuse interstellar material absorbs about three-quarters of a magnitude for each kiloparsec (1000 parsecs) of dark cloud the starlight passes through (0.75 magnitude is a drop in intensity by a factor of 2). (Figure 4–16.)

A further effect of the diffuse interstellar medium is that it sometimes polarizes the light from stars. The light from the star, originally unpolarized, is scattered by the cloud. Since the scattering is different for different planes of vibration of the light waves, only those waves in certain planes reach the observer. This is similar to the way in which polarized sunglasses absorb light polarized in a horizontal plane but let through vertically polarized light. The result is that the light comes through slightly polarized. Polarization gives

FIGURE 4–16
The spectra
of several stars,
showing spectral lines
of calcium and
sodium which have
been absorbed by
interstellar material
between the star and
Earth. Notice that
there are several
components to each
line, indicating that
there are actually
several clouds, each
with a different
radial velocity. (Lick
Observatory)

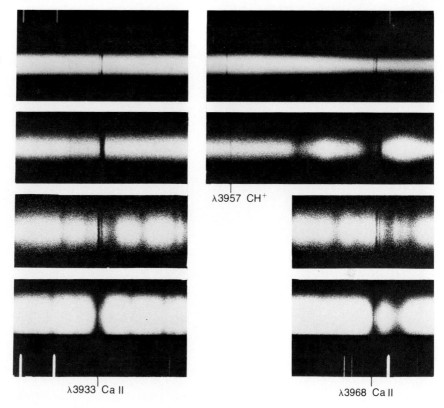

λ3957 CH⁺

λ3933 Ca II

λ3968 Ca II

important information about the size and composition of the dust, and about the magnetic fields of the galaxy which align the dust to produce this effect.

Luminous nebulae. Sometimes a cloud of gas and dust will be near a star, either by accident or because the star has recently formed from part of the cloud. The star may illuminate the cloud and produce a luminous nebula. If the star is not too hot, less than about 25 000° K, the starlight will simply be reflected by the dust component of the cloud, giving us a *reflection nebula.* The spectrum of the nebula will of course be similar to that of the nearby star, but the overall color of the cloud will be somewhat bluer than the star owing to the preferential scattering of blue light by the cloud. The most famous reflection nebula is that surrounding the Pleiades star cluster seen in Figure 4–17. This explains why Slipher found a dark-line spec-

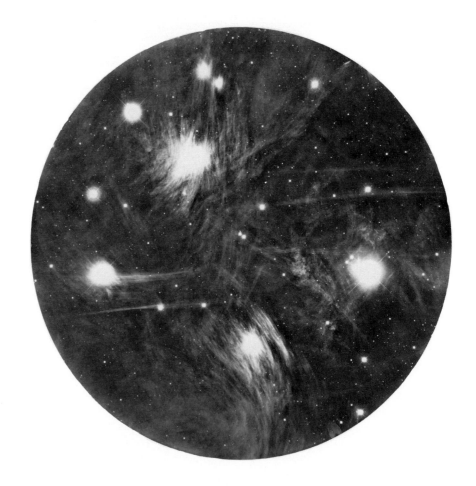

FIGURE 4–17   *The reflection nebula around the Pleiades star cluster.*
The light from the stars is reflected by the dust component of the cloud.
The stars are too cool to cause the gas in the cloud to glow. (Kitt Peak
National Observatory)

trum when he observed this nebula, even though it was believed to
be entirely gaseous. Now we have an explanation: the nebula contains dust as well as gas, and the dust part reflects the starlight to us.

Should the nearby star be very hot, the ultraviolet radiation of
the star will excite the hydrogen gas in the cloud and cause it to
glow by a process of fluorescence. This produces an *emission nebula,*

FIGURE 4–18
The "North
American" nebula in
the constellation
Cygnus. This is an
emission nebula, but
the shape is caused
by a dark nebula in
front of the glowing
cloud. (Lick
Observatory)

such as that shown in Figure 4–18. *Fluorescence* is the process of converting radiant energy of one wavelength into another. (A more familiar example of this is a fluorescent mineral, which glows under ultraviolet light.) From the discussion of radiation in Chapter 2, we recall that a very hot object will have a great deal of ultraviolet radiation. This excites or ionizes the hydrogen, and when the electrons de-excite or recombine, visible light is given off (Figure 4–19). A cloud of ionized hydrogen is referred to as an "HII" region. (HII stands for ionized hydrogen; HI is neutral hydrogen.) Since the gas in the cloud is thin, the spectrum will be one of many bright lines, sometimes with a faint continuous spectrum added by reflection from the dust in the cloud. From the bright lines, we detect not only hydrogen and helium, but also oxygen and nitrogen, plus other elements. Often these gases emit "*forbidden lines.*" These are spectral lines from energy levels of the atoms not possible in dense gases in

FIGURE 4–19   *In an emission nebula, the ultraviolet radiation from the star ionizes the hydrogen gas. For this reason, an emission nebula is often called an HII region. When the electrons recombine with the hydrogen, light is produced, some of it in the visible part of the spectrum. The amount of ultraviolet radiation from the star determines the size of the nebula.*

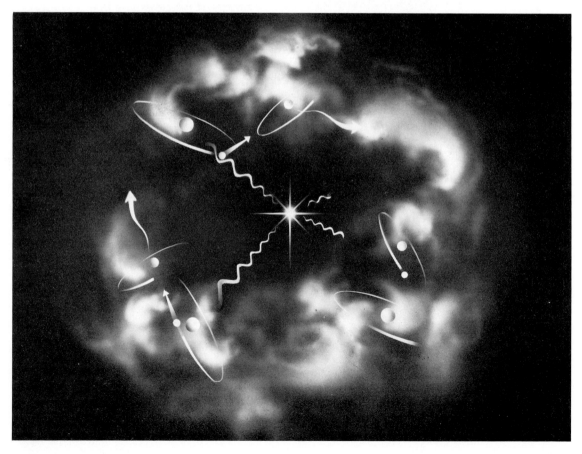

laboratories on Earth. Despite the fact that these beautiful glowing clouds are easily seen in photographs, where they appear dense, they are really many times less dense than the best vacuum that can be produced in a laboratory. The most famous emission nebula is the Great Nebula in Orion (Figure 4–20).

The exciting star of an emission nebula is very hot and blue. Because of the tremendous amounts of energy these stars expend every second, they cannot last for very long and are therefore young in an astronomical sense. The Orion nebula is excited by four of

FIGURE 4–20
The Orion nebula, one of the most beautiful of the nebulae, and one of the brightest. Four extremely bright and hot stars embedded in the center of the cloud cause the cloud to glow. They cannot be seen in this photograph because the central region of the nebula was overexposed to bring out the outer portions of the cloud. (Lick Observatory)

these hot blue stars, called the *Trapezium,* which are among the youngest stars in the Galaxy. A couple of million years ago, these stars would not have been seen—they would not yet have been formed!

The gas in an emission nebula is at a temperature of about 10 000°K, and has a density of perhaps 10 or 100 atoms per cubic centimeter, along with some dust. Recently, molecules have been discovered in some clouds, and also dark, cool globules which may be protostars forming. Much current research is devoted to explaining the many curious physical processes taking place inside emission nebulae (Figures 4–21, 4–22, and 4–23).

FIGURE 4–21
A nebula in the region near the star 12 Monocerotis. (Yerkes Observatory)

FIGURE 4–22
The emission nebula
in the region of the
star Eta Carina.
(Cerro Tololo
Interamerican
Observatory)

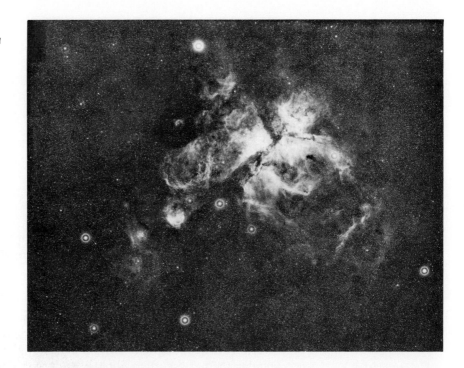

FIGURE 4–23
A section of the
"Rosette" nebula.
Some of the dark
globules may be stars
in the process of
forming. Perhaps
some of them will
form planetary
systems. (Hale
Observatories)

FIGURE 4–24
The "Ring" nebula in
the constellation
Lyra. This famous
planetary nebula
looks like a smoke
ring since along the
edges we are looking
through a greater
depth of material
than we are through
the center. The star
in the exact center
threw off this cloud
of gas long ago. (Hale
Observatories)

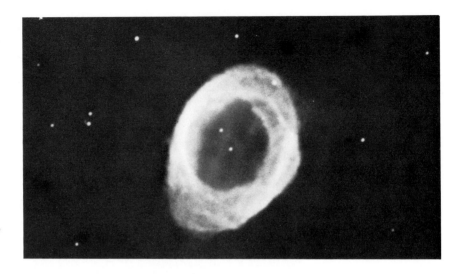

Planetary nebulae. In the late 1700s, Uranus became the first planet to be discovered telescopically. In a telescope it looks like a faint greenish disk. When other objects of this appearance were sighted, they were at first thought to be planets, but the fact that they did not wander among the stars showed they were related to the stars. They were called planetary nebulae. Some, such as the famous Ring nebula in Lyra, shown in Figure 4–24, appear as smoke rings. A careful look, however, shows that they are not really rings but spherical shells of gas surrounding very hot stars. The edges

FIGURE 4–25
Two large planetary
nebulae. On the left
is the "Dumbbell"
nebula. The one on
the right is known as
NGC 7293. (Lick
Observatory)

are thicker than the center, so they look like rings. The spectroscope confirms this, and reveals that planetary nebulae must be enormous shells of gas thrown off by stars near the end of their lives (Figure 4–25).

The material in a planetary nebula, which may be about the mass of our Sun, glows by a fluorescence process similar to other emission nebulae. The exciting star in the center is much hotter than ordinary stars, up to 100 000°K. They are the hottest stars we have seen.

We know from Doppler shifts of the spectral lines that the gas is expanding away from the star at tens of kilometers per second. Since the stars are quite small, and have a high escape velocity, we believe that the gas was thrown off at an earlier stage in the star's life, when it was bigger and thus had a smaller escape velocity.

Shortly after a star throws off a cloud, the gas is quite dense, and the ultraviolet light from the star cannot penetrate all the way to the edges of the cloud. Thus, a "young" planetary nebula will look bright near the center and dark around the edges. As it expands and thins out, the whole shell of gas can receive the ultraviolet radiation, so all of it glows. We know that about 100 000 years after the cloud is thrown off it will have dispersed so much that it will be hard to see (Figure 4–26). Since we notice many planetary nebulae in

FIGURE 4–26
A very dispersed nebula in the constellation Cygnus. Often called the "Veil" nebula, this is what remains of a stellar explosion thousands of years ago. Such clouds can exist only for about 100 000 years before becoming too thin to be seen. Some of the luminosity of this nebula may be produced by collisions between the atoms and grains of the expanding cloud and the surrounding interstellar medium. (Hale Observatories)

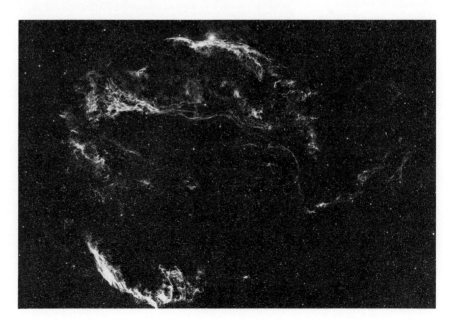

our galaxy (about 1000 known so far just in the few kiloparsecs near the Sun), we suspect that a planetary nebula is a common stage of life for some types of stars. The cause of planetary nebulae is a challenging area. We will meet them again in later chapters.

Supernova remnants. On what would correspond to July 4, 1054, in our calendar, Oriental astronomers noticed a "new" star in the sky which outshone even Venus, and could in fact be seen in the daytime for several weeks. They called it a "guest star," and today we would call it a *supernova*. A supernova is a tremendous stellar explosion, almost destroying the star, and increasing the star's brightness by millions of times for a space of a few weeks or months.

The astronomers of 1054 located the new star between the horns of Taurus the Bull. In just the same location today we see a vast energetic cloud of gas known as the Crab nebula, seen in Figure 4–27. We believe it is the remains of that explosion, the matter that was once part of a star thrown off into space at great speed. Since the Crab nebula was identified as a supernova remnant, several other similar clouds have been identified with other stellar explosions,

FIGURE 4–27
The Crab nebula in the constellation Taurus. One of the central stars was seen to explode on July 4, 1054, and was recently discovered to be a pulsar. (Hale Observatories)

and still others are suspected of being the remnants of supernovae which exploded too long ago to have been recorded (Figure 4–28).

Supernova remnants are highly energetic clouds of gas, expanding at thousands of kilometers per second. They produce light by the synchrotron process, not by a thermal process or by fluorescence as in other types of nebulae. Synchrotron radiation implies energetic particles and strong magnetic fields, as mentioned in Chapter 2. Our knowledge of supernova remnants is far from complete, but much work is being done on them. They, and the Crab nebula in particular, are among the most investigated phenomena in the branch of astronomy known as high-energy astrophysics. One astronomer has gone so far as to say that all astronomy can be divided into two parts: the study of the Crab nebula, and everything else!

Many supernova remnants have strong radio emission, and this is one of the ways they can be identified (Figure 4–29). Radio telescopes have detected some objects which do not correspond to any optical object. These may be very old or very distant supernovae remnants. The recent discovery that the star in the center of the Crab nebula, the star left behind after the explosion, is a pulsar, adds

FIGURE 4–28
A part of the "Veil" nebula, seen in its entirety in Figure 4–26. This may be the remnant of a supernova explosion. (Hale Observatories)

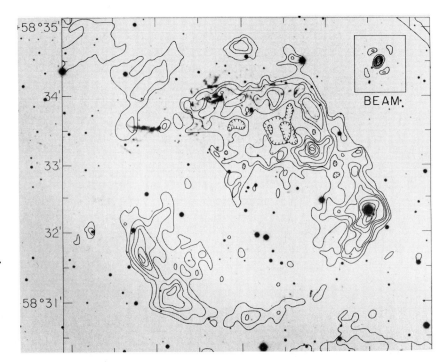

FIGURE 4–29
A supernova remnant observed in visible light, with the intensity levels of radio emission superimposed. This is known as Cassiopeia A to radio astronomers. The supernova probably occurred about 300 years ago and is expanding at the rate of 5000 kilometers per second. It is about 3500 parsecs away. (National Radio Astronomy Observatory)

to the importance of these strange objects (Figure 4–30). Much more work, both theoretical and observational, needs to be done to understand them.

Cosmic Rays

In the last century, experimenters noticed that a charged electroscope would lose its charge if left alone. The reason for the loss was not apparent, but it was certainly through the air. The only time that air conducts electricity is when it is ionized, and ionization occurs only when the air has been subjected to some energetic process. One researcher found that the electroscope discharged faster if it were taken up in a balloon to high altitude. He concluded that the ionizing radiation was coming from outside Earth's atmosphere. For this reason the radiation was called *cosmic rays* (Figure 4–31).

Today we know that cosmic rays are not radiant energy but are streams of high-energy particles which speed through space. When one strikes a molecule high in the atmosphere, the molecule is broken up and the pieces shower down to the surface. The incoming particle is called the *cosmic ray primary*, and the pieces are called *secondary rays*. We get some information about the energy and abundance of the primary particles from the surface since the number and energy of secondaries is proportional to the number and energy of primaries.

Primary cosmic rays are the nuclei of atoms, completely

FIGURE 4–30
The pulsar in the Crab Nebula turns on (left) and off 30 times a second. The nearby stars are constant. This pair of photographs was taken by placing a rotating shutter in front of the film to act as a stroboscope. (Lick Observatory)

FIGURE 4–31
When a cosmic ray particle, the primary, strikes a molecule in the upper atmosphere of Earth, other particles, called secondary cosmic rays, are produced and shower down to the surface. Thus one cosmic ray particle produces a number of detectable events. From analysis of the secondary particles, we can deduce the nature of the primary.

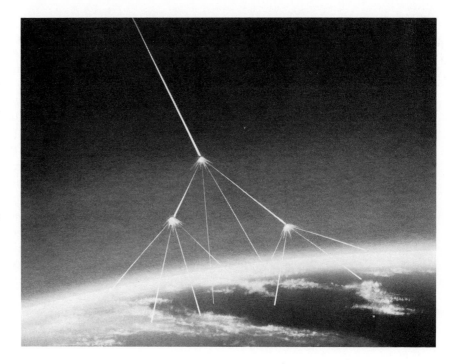

stripped of their electrons. We can, if we wish, think of cosmic rays as a very thin, very hot gas pervading all of space. The particles have a wide range of energies, some extremely high owing to speeds very close to the speed of light. Along with the nuclei are a few electrons, and even some positrons. The distribution of the nuclei among the chemical elements is very different from the average cosmic abundance of the elements. Table 4–1 shows some of the relative proportions of the elements in cosmic rays.

Table 4–1 *Relative abundance of elements in cosmic rays as compared with their cosmic abundance*

Element	Cosmic abundance	In cosmic rays
Hydrogen	10 000	10 000
Helium	1 500	700
Lithium, boron, beryllium	10^{-5}	15
Carbon, nitrogen, oxygen, fluorine	$1\frac{1}{2}$	40
Neon through potassium	$\frac{1}{5}$	14
Elements with atomic number greater than 20	$\frac{1}{10}$	5

Notice that the abundance of the metals (everything heavier than hydrogen and helium) is much higher than we would expect. We can explain this in two ways. First, the only place we find large numbers of heavy elements is in old stars, stars which have gone through most of their lives. These must be very massive stars, for reasons we will discuss in the section on stellar evolution, and they will probably become supernovae. Since cosmic rays must have been produced in some very energetic event, we think they come from the supernova explosions or possibly from exploding galaxies such as the one shown in Figure 4–32. This explains the heavy elements.

The light metals—lithium, boron, and beryllium—are very rare in the universe because they are not produced in stars. To explain their abundance in cosmic rays we must assume they are the result of collisions of cosmic rays with other nuclei. This also gives us a way of measuring the amount of material through which the cosmic rays have passed. The more light metals there are, the greater the amount of material the cosmic rays have encountered.

The Sun has an important influence on the cosmic rays we detect. The Sun throws off into space vast amounts of both light and particles. Further, there is a large magnetic field in the solar system extending far out among the planets. The magnetic field and particles shield Earth from some of the cosmic rays, and only the ones above a certain energy get through.

Astronauts in space have reported seeing flashes of light when their eyes are closed. This is the result of primary rays interacting with molecules inside their eyes, releasing light. More data are needed before the health threat of cosmic rays to astronauts is known.

Cosmic rays and the Galaxy. Because of the thinness of the cosmic ray "gas," we can treat it as another component of the interstellar medium. The particles, or streams of gas, interact with the magnetic field in our Milky Way Galaxy, and most cosmic rays will be confined to the Galaxy, bottled in by the magnetic field. Rarely a very high energy particle will not be trapped by the field of the Galaxy, and so must be on its way out of the Galaxy or just passing through, a messenger from the depths of intergalactic space.

Many physicists have worked on the problem of how cosmic rays interact with the other interstellar material, perhaps heating it or exerting some small pressure. It now seems that cosmic rays do contribute to the small pressure in space. Also, the amount of energy due to cosmic rays in any given volume of space is about equal to the energy of the starlight passing through the same volume.

FIGURE 4–32
The exploding galaxy M82, photographed in blue light. The filament extends 25 000 light-years from the center of the galaxy. Violent events such as those occurring in the nucleus of this galaxy may be producing cosmic rays. (Hale Observatories)

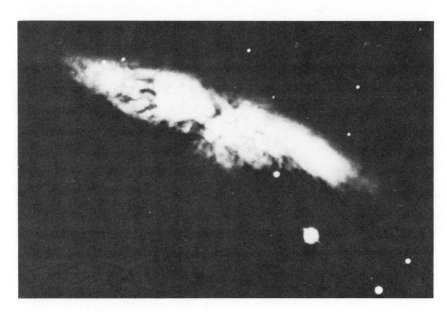

We know that very energetic phenomena are going on in other galaxies, and presumably cosmic rays are being produced. They appear to play a role in the structure and evolution of those galaxies. So far we do not know just what role they play in these galaxies or in our galaxy, but in the next few years, using observations from spacecraft, we hope to learn more.

TRACING THE HYDROGEN IN SPACE

The Milky Way Galaxy does not rotate as a solid disk, but instead each star or interstellar cloud has its own period of revolution about the center. Just as the inner planets of the solar system move faster around the Sun, so too do objects close to the center of the Galaxy move faster about it than objects further out. This produces a shearing effect between objects at different distances.

In particular, we observe that gas clouds have different radial velocities when we look in different directions around the Galaxy. We can use this fact to determine the distribution of hydrogen in space.

Usually hydrogen is noticeable only when it is illuminated by a nearby hot star and caused to glow. Neutral (unionized) hydrogen does not show up in the visible portion of the spectrum. It does appear, however, in the radio region of the spectrum. The radiation is caused not by the change of the orbit of the electron about the nucleus, but by a more subtle effect: a change in the "spin" of the electron relative to the proton.

Both the proton of the hydrogen nucleus and the electron orbiting it are spinning. If the two have spins in the same direction, the atom has a slightly higher energy than if they spin in opposite directions. If the atom is left alone with the two spins in the same direction, after a time the electron will spontaneously flip over, reducing the energy of the atom, and this small amount of energy is converted into a photon having radio wavelengths. The average time for the flip-over is about 10^{11} years. However, there is so much hydrogen in space that at any given time enough atoms are flipping to radiate a detectable amount of radiation. The wavelength is 21 centimeters, or the frequency is 1.420 gigahertz (1 gHz is 10^9 Hz)—about ten times the frequency of the TV band. This frequency is so important for the study of the universe that it has been declared a protected region of the electromagnetic spectrum by international agreement.

Careful observations of stars near the Sun reveal that objects at different distances from the center of our galaxy are moving around the Galaxy at different speeds. By knowing how the speeds differ at different distances from the center of the Galaxy, we can turn the problem around and find the distance of a cloud of gas by measuring its speed. The method is shown in Figure 4–33.

FIGURE 4–33   *When we observe with a radio telescope in a certain direction in the Galaxy, we use knowledge of how objects move around the Galaxy to find out where the clouds of hydrogen are. On the right is seen the beam (field of view) of a radio telescope intersecting several clouds. The intensity of the received signal from each cloud will depend on the amount of hydrogen in the cloud and how far away it is. The wavelength of the received signal would be 21 centimeters if the cloud were not moving. The clouds shown here are moving toward the Sun, and the more distant cloud is moving faster. The Doppler effect will increase the frequency (decrease the wavelength) of the signal from each cloud. The "oscilloscope" images on the left show the signal that would be received from the distant cloud alone, the signal from the near cloud alone, and the actual combined signal as received. This is a powerful technique for studying the Galaxy.*

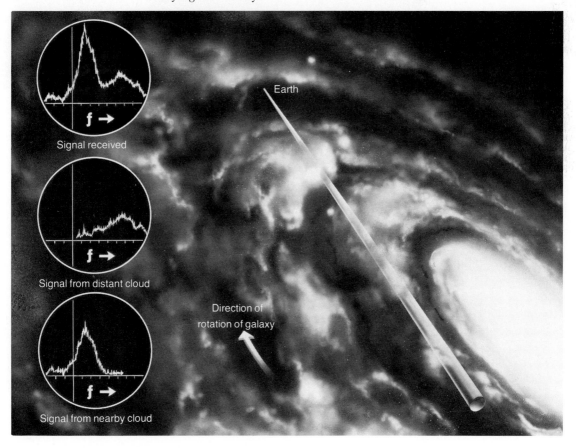

Signal received

Signal from distant cloud

Signal from nearby cloud

Earth

Direction of rotation of galaxy

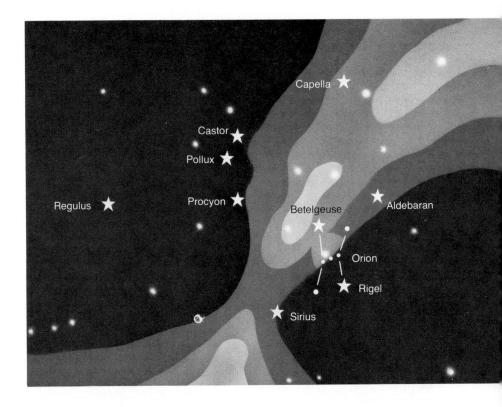

FIGURE 4–34 *If our eyes could see radio radiation with a wavelength of 120 centimeters, this is what the sky would look like. The brighter areas are regions of more intense radiation. Some stars are shown for reference. Actually they would be invisible at this wavelength.*

We assume that a radio telescope is "looking" in a certain direction. The intensity of the signal received will depend on the amount of hydrogen in the cloud, and the frequency will depend on the radial velocity of the cloud. Thus we can trace the structure of our galaxy over large distances and obtain a map of our galaxy such as Figure 4–34. The details of how this can be done will be explained

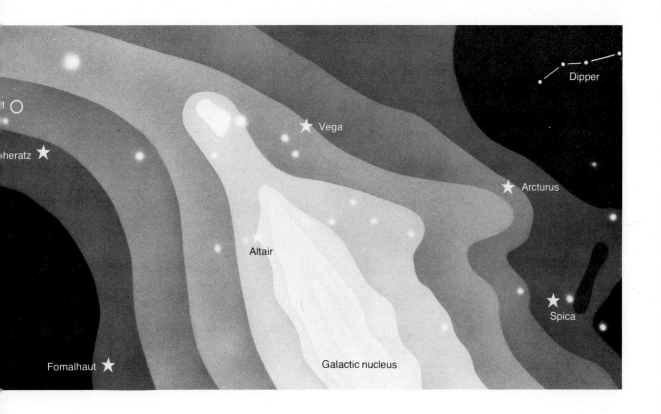

more fully in Chapter 9, where we discuss the structure of the Milky Way.

MOLECULES IN SPACE

Before 1963, only a few molecules were known to exist in interstellar space, and these all showed up as extra lines in stellar spectra. In 1963, another molecule was discovered. From 1968 to 1972, almost two dozen more had been found, all by radio astronomy. Many are rather complicated molecules, and at present there is no adequate explanation of why they occur in the amounts they do. Several are seen only in the direction toward the center of the Galaxy, while others seem to be associated with star formation. Table 4–2 shows the molecules discovered as of late 1974.

Table 4–2   *Molecules discovered in interstellar clouds*

Molecule	Chemical Formula	Wavelengths
——	CH	4300A
Cyanogen	CN	3875A
—— (ionized)	CH^+	3745-4233A
Hydroxyl	OH	18, 6.3, 5.0 and 2.2 cm
Ammonia	NH_3	1.3 cm
Water	H_2O	1.4 cm
Formaldehyde	H_2CO	6.2, 2.1 and 1 cm; 2.1 and 2.0 mm
Carbon monoxide	CO	2.6 mm
Cyanogen	CN	2.6 mm
Hydrogen	H_2	1100
Hydrogen cyanide	HCN	3.4 mm
"X-ogen"	?	3.4 mm
Cyano-acetylene	HC_3N	3.3 cm
Methyl alcohol	CH_3OH	36 and 1 cm; 3 mm
Formic acid	CHOOH	18 cm
Carbon monosulphide	CS	2.0 mm
Formamide	NH_2CHO	6.5 cm
Silicon oxide	SiO	2.3 mm
Carbonyl sulphide	OCS	2.7 mm
Acetonitrile	CH_3CN	2.7 mm
Isocyanic acid	HNCO	3.4 mm; 1.4 cm
Hydrogen isocyanide	HNC	3.3 mm
Methyl-acetylene	CH_3C_2H	3.5 mm
Acetaldehyde	CH_3CHO	28 cm
Thioformaldehyde	H_2CS	9.5 cm
Hydrogen sulphide	H_2S	1.8 mm
Methyleneimine	H_2CNH	5.7 cm
Deuterated hydrogen cyanide	DCN	2.1 mm
Ethynyl	C_2H	3.4 mm
Sulfur monoxide	SO	3.0 mm
Dimethylether	$(CH_3)_2O$	3.3 mm
Silicon sulfide	SiS	2.8 mm
—— (ionized)	N_2H^+	3.2 mm
Heavy water	HDO	3.7 mm
Ethyl alcohol	C_2H_5OH	3.3 mm

As you can see from Table 4–2, almost all the "organic" elements such as carbon, sulfur, oxygen, and nitrogen have been found in interstellar space. Some odoriferous compounds, such as ammonia, are present in enough abundance to be smelled if they were in air. Methyl alcohol has been found, and astronomers have recently discovered ethyl alcohol!

It has been suggested that the molecules are formed on the surfaces of dust grains, rather than by collisions in free space, since a chance collision between atoms is very unlikely. The mechanism by which molecules might form on dust grains is not known in enough detail to say whether it could produce the amounts we see. It is safe to say that more molecules will be discovered, as they are currently of much theoretical and observational interest. Not only do the molecules tell us about the conditions in the interstellar medium, but they also have implications for the existence of life elsewhere in the universe.

REVIEW

1 Space is not empty, but is thinly filled with gas and dust. The density of the material is much less than that in the best vacuum we can make in a laboratory.

2 Discrete clouds of interstellar matter are called nebulae. The interstellar material is composed of gas, including atoms and molecules, and dust of presently unknown composition.

3 The matter in interstellar space, as in the rest of the universe, is mostly hydrogen, with helium next in abundance. Together they make up 99 percent of all matter.

4 Interstellar matter scatters starlight passing through it. More blue light is scattered than red light, so a star appears redder (the color index is higher). In some cases the light is partially polarized.

5 If a cloud of interstellar matter is near a very hot star, it may be excited to fluorescence, producing an emission nebula. If the nearby star is not hot enough to excite the gas, a reflection nebula, in which the starlight is reflected from the dust component of the cloud, may appear.

QUESTIONS

1 Explain how the existence of intergalactic matter would affect theories of the evolution of the universe.

2 It was noticed early in the twentieth century that no distant galaxies were seen in the direction near the band of the Milky Way. This was called the "Zone of Avoidance." Can you suggest an explanation for this effect?

3   How can we use the Doppler effect to find out if a planetary nebula was really an expanding shell of gas? Can we determine how long ago it exploded?

4   What effect does the gas component of an interstellar cloud have on starlight passing through it? What data can be obtained from this?

5   Would "space" have a temperature? Would a grain of interstellar dust? What would heat it?

6   A typical emission nebula has a temperature of about 10 000°K. What would be the effect on a spaceship that penetrated the cloud? Why?

READINGS

Greenberg, J. M. "Interstellar Grains." *Scientific American*, October 1967.

Middlehurst, B., and L. Aller. *Nebulae and Interstellar Matter.* Chicago: University of Chicago Press, 1968.

Pasachoff, J. M., and W. A. Fowler. "Deuterium in the Universe." *Scientific American*, May 1974.

Robinson, B. J. "Hydroxyl Radicals in Space." *Scientific American*, July 1965.

Rossi, B. *Cosmic Rays.* New York: McGraw-Hill Book Company, 1964

ORDINARY STARS

CHAPTER 5

Stars, luminous spheres of gas which get their energy from nuclear reactions, make up the bulk of the mass of galaxies. They occur singly, in pairs, triples, and higher multiples, and in clusters of a few dozen to a few hundred thousand. The Sun is typical among the hundred billion stars in our galaxy. Study of the Sun provides information about the properties and processes in more distant stars, and observations of other stars yield data on the probable origin and destiny of the Sun, and with it Earth and the solar system. In this chapter we will discuss the principal characteristics of ordinary stars.

A star can be described by its radius, mass, density, chemical composition, pressure, temperature, and the distribution of these

Each star in the sky is a sun. Except for our own Sun we see every star as a point of light. From theory and from observations of the light emitted by the outer layers of the stars, we know they are nuclear furnaces, converting matter into energy.

conditions inside the star. Since we can directly observe only the outer layers of a star, the study of stellar interiors consists of making models, which are numerical estimates of the properties of the star as a function of distance from the center of the star. With the aid of powerful computers, we attempt to trace both the structure of a star at one instant and the course of the star's aging. The most important information we need in order to predict the details of the life of a star is its mass, that is, the amount of material in it. Later, we will see how this determines the life story of the star.

THE SOURCE OF ENERGY

For thousands of years men have wondered about the seemingly unlimited energy coming from the Sun. Here on Earth (outside the filtering effects of the atmosphere), we receive 2 calories per square centimeter every minute. This is called the *solar constant*. It is equivalent to 1.4×10^6 ergs per square centimeter per second. Knowing how far Earth is from the Sun and the radius of the Sun, we can calculate that each square centimeter of the surface of the Sun radiates 6.4×10^{10} ergs/sec. Knowing the size of the Sun, we find that the total radiation is 4×10^{33} ergs/sec. On Earth we receive one two-billionth of this amount. What is the source of this energy?

Early astronomers thought the Sun must be burning, perhaps a lump of wood or coal. But a fire would not give off enough energy for long enough. It was later suggested that the Sun was a ball of burning hydrogen and oxygen. But even this energetic reaction could not account for all the energy, and, besides, it would eventually leave us with a big ball of water at the center of the solar system!

Approximately one hundred years ago Lord Kelvin (1824–1907) and H. L. von Helmholtz (1821–1894) proposed that the Sun might be contracting. When any ball of gas shrinks, it converts some of its gravitational potential energy into heat and radiation. The important question was how much the Sun would have to shrink in order to supply the amount of energy we observe. The surprising answer was that a shrinkage of only a few hundred kilometers a year would suffice. This amount of shrinkage would be detectable from Earth only after many years. Nevertheless, if it had been shrinking at this rate, the Sun would have been as big as the orbit of Neptune only a few million years ago.

Thus, astronomers thought that the age of Earth was a few million years. Geologists, however, had fossil evidence that life had been around for longer than that, and maintained that Earth must be much older, certainly several hundred million years, possibly even *several billion* years.

A few other sources of energy were suggested in attempts to lengthen the age of Earth and the Sun. Some proposed a great infall of meteoroids onto the Sun. Others suggested that some sort of radioactive decay such as had been recently observed by the Curies was producing heat. By that time, however, spectroscopy had advanced to the point where we could be sure that no spectral features of uranium or radium were present in the spectrum of the Sun.

In the early 1900s, with the advent of the theory of relativity,

FIGURE 5–1   *Earth receives about one two-billionth of the energy radiated by the Sun. By knowing how much energy crosses each square centimeter each second at Earth, we can calculate how much is radiated in all directions if we know the distance to the Sun. The energy is produced in the center of the Sun and slowly diffuses outward, interacting with the opaque material in the Sun's interior. By the time it reaches the "surface," some 20 million years later, the spectrum of the energy resembles that of a black body.*

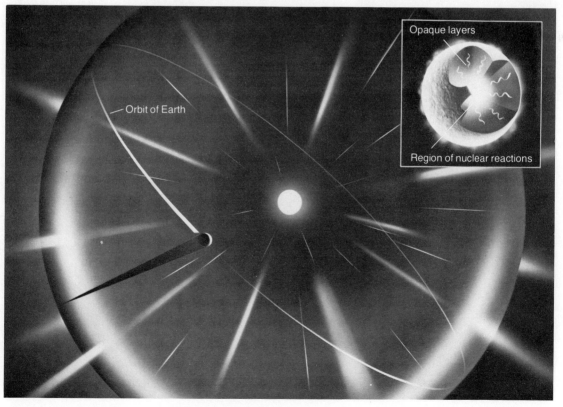

and advances in the theory of atomic structure, some astrophysicists began to think that a subatomic process was responsible. Not until 1938 was the exact process explained. Hans Bethe (1906–   ) worked out the details of how hydrogen atoms could, with the help of some carbon, combine (fuse) into helium, with a release of energy. This became known as the *carbon cycle*, or sometimes the carbon-

FIGURE 5–2   *The carbon-nitrogen-oxygen cycle. This series of nuclear reactions converts 4 hydrogen atoms into 1 helium atom plus energy. This is the principal source of energy in stars with central temperatures greater than about 15 million degrees Kelvin.*

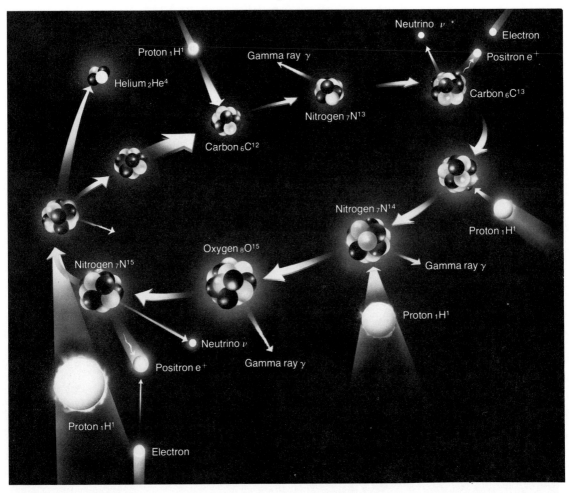

nitrogen-oxygen cycle. It is important for stars whose central temperatures are above 15 million degrees Kelvin. For this important breakthrough Bethe was awarded the Nobel Prize.

The carbon cycle can be written in chemical symbols as this series of nuclear reactions:

$$_6C^{12} + {}_1H^1 \rightarrow {}_7N^{13} + \text{a photon of energy}$$
$$_7N^{13} \rightarrow {}_6C^{13} + e^+ + \nu$$
$$_6C^{13} + {}_1H^1 \rightarrow {}_7N^{14} + \text{a photon of energy}$$
$$_7N^{14} + {}_1H^1 \rightarrow {}_8O^{15} + \text{a photon of energy}$$
$$_8O^{15} \rightarrow {}_7N^{15} + e^+ + \nu$$
$$_7N^{15} + {}_1H^1 \rightarrow {}_6C^{12} + {}_2He^4$$

H is hydrogen, He is helium, N is nitrogen, O is oxygen, e^+ is a positive electron, or *positron*, and ν is a neutrino (see page 140). The e^+ will quickly combine with an e^- (a normal negative electron) to provide more energy. The *subscript* before a symbol tells the atomic number, which is the number of protons in the nucleus. The *superscript* after the symbol is the atomic mass, the total number of protons and neutrons in the nucleus. For example, $_8O^{15}$ is an isotope of oxygen, with atomic number 8. There are a total of 15 protons and neutrons in the nucleus. Since we know there are 8 protons, there must be 7 neutrons. The CNO cycle is illustrated in Figure 5–2.

There is another process of nuclear fusion which can turn hydrogen into helium: the *proton-proton process* which is important in stars as cool as a few million degrees (Figure 5–3). Below a few million degrees nuclear reactions do not readily occur. In chemical symbols, the proton-proton process would be written in the following steps:

(1) $_1H^1 + {}_1H^1 \rightarrow {}_2He^2 + e^+ + \text{a photon of energy}$
(2) $_2He^2 \rightarrow {}_1H^2 + e^+ + \nu$
(3) $_1H^2 + {}_1H^1 \rightarrow {}_2He^3 + \text{a photon of energy}$

The previous two steps take place twice, then

(4) $_2He^3 + {}_2He^3 \rightarrow {}_2He^4 + {}_1H^1 + {}_1H^1$

Notice that two hydrogens go into the first step and one into the third step. Since these three steps happen twice each in order for the fourth step to occur, the total number of hydrogens going into the reaction is six. In the fourth step, the output is one helium atom and two hydrogens. Since we get two hydrogens back, the input is only four hydrogens, as in the CNO cycle. The net reaction is thus

$$_1H^1 + {}_1H^1 + {}_1H^1 + {}_1H^1 \rightarrow {}_2He^4 + \text{four photons}$$

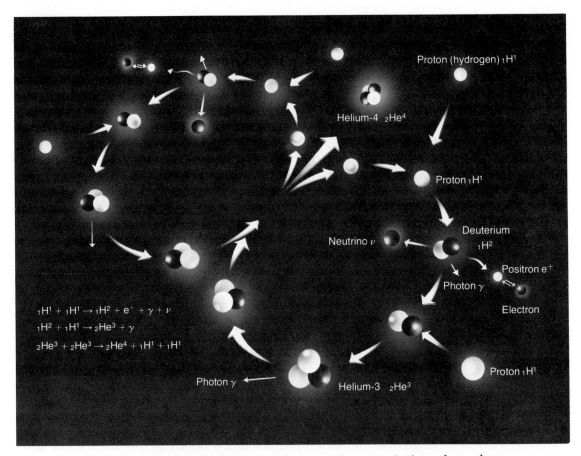

FIGURE 5–3 *The proton-proton process, which produces the energy inside the Sun. This reaction is the principal source of energy for stars with central temperatures less than about 15 million degrees K.*

The energy to produce the photons, usually at gamma-ray energies, comes from the important fact that one helium nucleus is slightly lighter than the sum of the four hydrogen nuclei. This small difference in mass is converted into energy by the famous equation $E = mc^2$, energy equals mass times the square of the speed of light. The amount of mass lost in the reaction is very small, less than 1 percent of the mass of a hydrogen nucleus, which is itself very small. However, c^2 is a very, very large number, and so each gram of hydrogen turned into helium releases about 6×10^{18} ergs. That is enough energy to accelerate a 60-kilogram person to a speed of 140 kilometers per second!

Each second the Sun converts about 600 million tons of hydrogen into helium, and of this, about 4 million tons become energy. A minuscule fraction of this reaches Earth. We should note, however, that the energy does not escape into space directly from the core of the Sun where the reactions take place. The internal material of a star is quite opaque to the gamma rays given off in the reaction. These photons work their way out slowly, by a process of diffusion, and gradually, after many, many interactions with the gas inside the Sun, the radiation is smoothed out into a black-body distribution, with most of the energy coming out in yellow wavelengths. The high opacity of the gas is like a blanket which maintains the very high internal temperature of the Sun necessary for the nuclear reactions.

Most stars produce energy by the CNO cycle or the proton-proton reaction or both of these processes. A few older stars may utilize fusion of heavier elements such as carbon, sulfur, and silicon to produce elements as heavy as iron. Beyond this element ordinary fusion cannot go. But we will consider the source of the heavy elements in the next chapter.

For now, we should emphasize that all stars are mostly hydrogen and helium, the two most abundant elements in the universe. The proportion of hydrogen to helium varies widely and is imperfectly known because of the scarcity of spectral features of helium in the visible range of the spectrum. Observations from satellites outside the atmosphere promise to add to our knowledge of helium in the universe. Despite the preponderance of these two elements, many other kinds of atoms have been detected in the spectra of stars. About two-thirds of the known elements have been identified in the Sun, and the others (except the radioactive elements) are probably there in amounts too small to be detected. Despite their low abundance, these "metals" play an important role in the structure of a star and in its eventual fate. They are especially important in establishing the opacity of the stellar material which in turn determines the flow of energy outward from the center of the star.

We should keep in mind that we actually observe only the outside layer of a star, the layer of gas thin enough to allow the escape of visible radiation. The spectrum we observe from a star is produced in its lower atmosphere. Thus, knowledge of the star's interior comes only from models. We often assume, for all but a few exotic stars, that the chemical composition is the same throughout the star, except in the core. There the nuclear reactions take place in a volume only about a thousandth of the whole star, and involving less than 20 percent of the mass of the star. It is only in the very central region that the temperatures are high enough for nuclear fusion!

FIGURE 5–4
The Size of the Sun compared with that of other stars. Supergiant stars such as VV Cephei and Betelgeuse are almost as large as our entire solar system. Giants like Capella and Arcturus would engulf the orbit of Earth if placed where our Sun is. The diagram changes scale to show a comparison between the Sun and Earth, and then changes again to show the relative sizes of Earth and a neutron star.

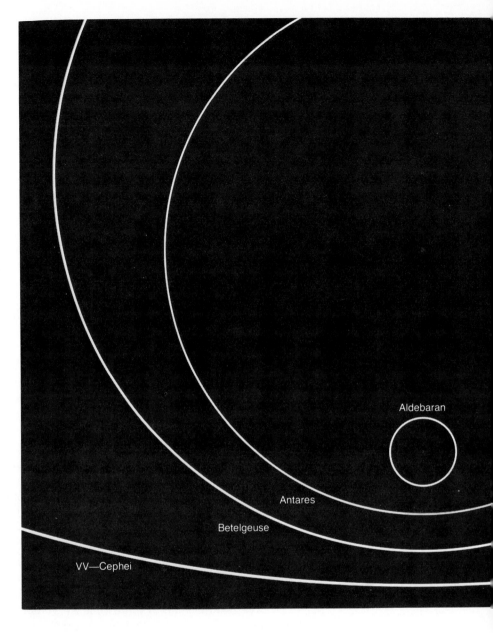

DESCRIBING A STAR

In trying to describe other stars, it is convenient to use our Sun as a standard. This is useful because it gives us something familiar as a reference, and because our Sun turns out to be quite average in its properties. Some of these properties are the following:

Mass $= 2 \times 10^{33}$ g $= 2 \times 10^{27}$ tons $= 333\ 000$ times the **mass of Earth**

Radius $= 7 \times 10^5$ km $= 432{,}000$ miles $= 110$ times
    the radius of Earth $= 0.005$ times the
    mean Sun-Earth distance

Luminosity $= 4 \times 10^{33}$ ergs/sec $= 5 \times 10^{23}$ hp

Surface temperature $= 5800°$K (effective temperature)

Absolute magnitude $= +4.7$

Apparent magnitude $= -26.78$

We will see in this and later chapters how these properties are determined for the Sun and for other stars. We will call the data for the Sun "1 solar mass," "1 solar radius," "1 solar luminosity," and so on. In this way we can better appreciate the ranges in stellar properties.

For other stars we have measured masses ranging from about $\frac{1}{100}$ that of the Sun to somewhat less than 100 solar masses. Stellar sizes range from slightly less than 0.1 solar radii up to about 30 solar radii for what we might call "normal" stars. For "abnormal" stars, the range is from less than 0.01 to over 1000 solar radii (Figure 5–4).

Luminosity, or the total amount of energy radiated by a star, varies widely. Although many normal stars emit more than a million times more energy than the Sun, some normal stars emit less than $\frac{1}{1000}$ the energy of the Sun. The extreme range is from several million times more down to less than a millionth the energy output of the Sun. In terms of absolute magnitudes most stars are within 15 magnitudes on either side of the Sun, that is, M ranges from −10 to +20.

Surface temperatures range from around 2000°K to 40 000°K. Since many physical laws depend on the logarithm of the temperature, the range here is from log $T = 4.6$, to 3.6 for the Sun, down to about 3.3 for the coolest stars.

STELLAR SPECTRA

Information about the stars comes from an analysis of their spectra, the detailed distribution of light from the star's surface. The *photosphere* of the star, the outer layer which is opaque to visible light coming from greater depths, emits light very close to the spectrum emitted by an ideal black body. Thus, the shape of the graph of energy versus wavelength will tell the temperature of the photosphere. The continuous spectrum from the photosphere is broken by dark lines produced by elements in the lower atmosphere of the star absorbing certain wavelengths. Since the state of excitation and ionization of a given element depends on temperature, the dark lines in the spectrum also give an indication of temperature and are much easier to measure than the continuous radiation (Figure 5–5).

When astronomers first observed stellar spectra such as those in Figure 5–6, over a century and a half ago, they could not explain the differences among the spectra of different stars. Thus, the observers merely tried to find patterns and similarities among the data

FIGURE 5–5
Spectral lines of different elements become prominent at different temperatures. The elements which appear define the spectral type of the star. Thus, by determining the spectral type of a star, we have determined its temperature.

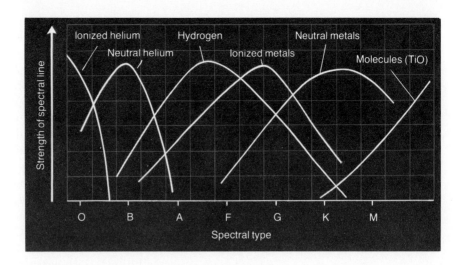

in order to establish some sort of classification scheme. This exercise in taxonomy, scientific name-calling, began with just four "pigeonholes" suggested by the Italian astronomer Angelo Secchi (1818–1878). The types were classified by the amount of the spectrum that was blocked by dark lines.

FIGURE 5–6
A photograph made by placing a prism in front of a Schmidt telescope. The light from each star is spread into a spectrum which may be classified. In this way many stars may be studied at once. Compare these spectra with the standard spectral types in Figure 5–7 to try to classify some of the stars. (Warner and Swasey Observatory)

It was soon realized that greater detail was needed. Astronomers at Harvard Observatory, compiling a catalog of over 200 000 stars, the *Henry Draper Catalog*, devised another classification scheme. This also depended on the number of lines in the spectrum, and the types were given the labels A, B, C, and so on. Later it was realized that this was not a physically meaningful order: that is, the order was not related to any useful physical property of the stars. With the birth of modern physics at about the same time, it was apparent that within the series there was a variation of temperature, but the sequence was out of order. Some of the letters were dropped, others were combined, and the remaining ones were rearranged to produce a sequence of *decreasing* temperature: O, B, A, F, G, K, M. These are the *spectral types* in use today. The standard mnemonic for remembering the order of the letters is "Oh, *be a fine girl kiss me.*"

The O-type stars are hottest, with surface temperatures around 40 000°K. The Sun is a G-type star, with a surface temperature around 6000°K. The M-type stars are coolest. It was found that this classification was not fine enough, so the types were subdivided into tenths. Thus, one may classify the Sun as a G2 star, two-tenths of the way from G0 (G-zero) to K0. Vega is an A0 star, and Arcturus is spectral type K2.

Figure 5–7 shows spectra of several stars. Careful inspection will reveal that different lines show up with different strengths (darknesses) at different types or temperatures. To classify a spectrum, the astronomer compares the strengths of some of the lines—say, those of hydrogen with those of others, for example, calcium or helium. Comparison with established "standard" stars allows precise typing. With a little practice it is easy to classify stellar spectra.

Note that hydrogen is most prominent in A-type stars. In somewhat cooler stars, such as the Sun, many metals show up; among them are calcium, iron, and magnesium. The very cool M stars show strong bands produced by molecules of titanium oxide. Recall that these dark lines do not directly reflect different abundances of these elements, but primarily only temperature differences. The analysis of the abundances of elements from a spectrum is a much more difficult problem.

Astronomers refer to spectra of types O and B as *early-type* stars and to spectra of types K and M as *late-type* stars. Or, an F star might be said to be "earlier" than a G star. While this is convenient terminology, it must not be thought to carry any implication of age. At one time it was believed that stars started out as O stars and

Spectral type   Star name

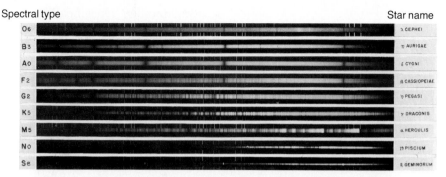

O6	λ CEPHEI
B3	η AURIGAE
Ao	δ CYGNI
F2	β CASSIOPEIAE
G2	η PEGASI
K5	γ DRACONIS
M5	α HERCULIS
No	19 PISCIUM
Se	R GEMINORUM

FIGURE 5–7   *The principal spectral types of stars. O-type stars are hottest, M-type are coolest. Most types have about the same chemical composition. Type N and a few other unusual types not shown have different compositions. (Hale Observatories)*

eventually became M stars, but this is known today to be far from the truth. Late-type stars are not *necessarily* older.

STELLAR LUMINOSITIES

The luminosity of a star, as we mentioned previously, is the total amount of energy radiated by the star. The luminosity of a star will depend on two things: (1) the amount of energy being radiated by each square centimeter of the star and (2) the number of square centimeters of star surface, that is, the size of the star.

Soon after the development of the concept of stellar spectral classification, it became apparent that there were stars of the same temperature but different absolute magnitude, or different luminosity. This could only be explained by the brighter star being bigger than the less luminous one; that is, the brighter star had more surface area from which to radiate. Astronomers at Mt. Wilson Observatory invented a system of adding lower-case letters before the spectral type of a star to indicate the star's *luminosity class*. The symbols used were the following:

c   supergiant

g   giant

d   dwarf

wd   white dwarf

Later, a more detailed classification was devised by W. W. Morgan (1906–) of the Yerkes Observatory. In this scheme the luminosity classes are the following:

Ia   bright supergiants

Ib   less bright supergiants

II   bright giants

III   giants

IV   subgiants

V   dwarfs (ordinary stars)

Thus, the complete spectral type of a star consists of both the spectral type proper and the luminosity class. The Sun is G2V (read "G-two-five"), since it is a dwarf, an ordinary star. Arcturus is K2III, since it is a giant, and Antares is M1Ib, a red supergiant.

ORGANIZING THE DATA

Statistics about stars become significant and useful when they are organized in a meaningful way. Early investigators knew that there are physical laws which relate one property of a star to another property. For example, a high-temperature star will emit more energy per square centimeter than a star of lower temperature. Or, a star of high mass is likely to be bigger than one of lower mass.

Two astronomers independently decided that the two most important properties of stars are spectral type (temperature) and absolute magnitude (luminosity). They plotted a graph of these two properties for all stars for which data were available at the time (about 1912). The two astronomers were Ejnar Hertzsprung (1873–1968) in Denmark and Henry Norris Russell (1877–1957) in the United States. The diagram which resulted, known as the *Hertzsprung-Russell diagram* (or H-R diagram, for short), is shown in Figure 5–8.

Note that by far the majority of stars lie within a narrow band from the upper left to the lower right. This is known as the *main sequence*: About 95 percent of all stars fall in this region. These are the ordinary stars of luminosity class V. Since on the graph luminosity increases upward, and temperature increases to the left, we see the expected correlation between these two properties, with the hotter star being brighter. There are a few other regions in which some stars lie off the main sequence. These are not ordinary stars; we will consider them in the next chapter.

Note that the Sun lies about in the middle of the diagram. It is a yellow star, spectral type G2, and absolute magnitude +4.7. Stars on the main sequence are called "dwarfs," not to be confused with the "white dwarfs" we will encounter in the following chapter.

For many years the physical significance of the main sequence was not known. Today we can say that the main sequence is the location of stars which are mature, stars producing energy by nuclear fusion of hydrogen to helium. They are "hydrogen-burning" stars.

It is hard to overemphasize the importance of the H-R diagram to astronomy. It is here that observation meets theory, that data are organized and displayed in a manner that is physically significant and easily evaluated. Deviations from the main sequence give important information about the properties of a star. Calculated models of stars and their paths through the diagram as they age may

FIGURE 5–8   A Hertzsprung-Russell diagram for many stars. The principal features are easily seen: the main sequence, red giant branch, and white dwarfs. The almost empty region between the main sequence and red giants is called the Hertzsprung gap. (Yerkes Observatory)

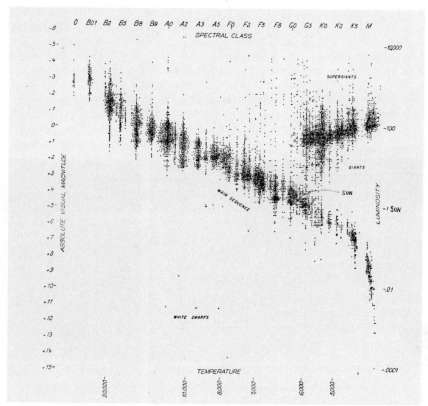

be plotted and compared with observation. Because of the importance of the H-R diagram, we will return to it again and again. One should become quite familiar with it and be able to sketch it schematically, indicating the stellar properties it displays.

Figures 5–9a and 5–9b show H-R diagrams for the nearest stars and for the apparently brightest stars, respectively. Note that among the nearby stars, only a few are brighter than the Sun, intrinsically. Also, among the apparently bright stars, only a few are nearby. This tells us that intrinsically bright stars are rare and occur only at great distances from the Sun. If we consider the "solar neighborhood" to be a typical region in space, then we see that the intrinsically faint stars are common. Compared with the number of faint red dwarfs, even such middling stars as the Sun are uncommon. Figure 5–10 shows how the stellar sizes vary along the main sequence.

We thus have a great deal of data on ordinary stars, those stars along the main sequence. But there is wide variation in the properties of even these ordinary stars. How do they come to be, and how do they reach the main sequence? What determines their lives? To answer these questions, we must go back to the interstellar material from which they came.

FIGURE 5–9   (a) An H-R diagram of the 100 nearest stars. Note that most are intrinsically less bright than the Sun, indicating that they are numerous, whereas intrinsically very bright stars are rare. Also note the white dwarfs below the main sequence. (b) An H-R diagram of the 100 apparently brightest stars in the sky. Most of these are intrinsically brighter than the Sun. Although they are far away, their great luminosities make them the brightest objects in our sky. Compare this carefully with (a).

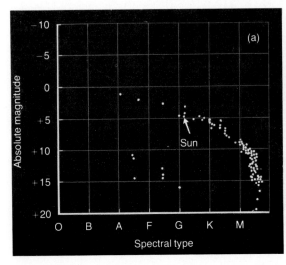
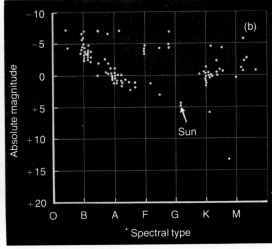

EARLY STELLAR EVOLUTION

The clouds of gas and dust in the Galaxy are in constant motion. Each atom and each grain of dust is attracted to every other atom or grain by gravity. Each attraction is very small, but the enormous number of atoms of gas and grains of dust produces a significant gravitational field. If the cloud is dense enough, it will begin to contract under its own gravity. Many things might break up the cloud: the rotation of the Galaxy, turbulence within the cloud, hot stars imbedded in it. If the cloud survives all these dangers, it can become an independent body. As it shrinks, gradually its temperature builds because gravitational energy is turned into heat and radiation. Much of this energy is radiated away into space, but as the contraction continues, the density of the cloud increases, and so too does the

FIGURE 5–10   An H-R diagram showing how star sizes vary along the main sequence.

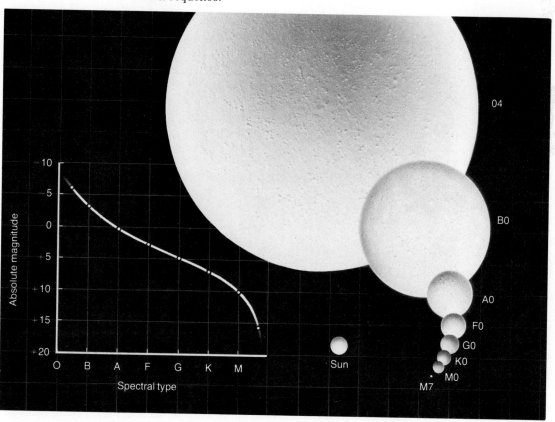

opacity to radiation. More and more of the radiation is trapped, and the temperature rises still further. Eventually all the dust grains are vaporized, and at still higher temperatures the atoms are ionized. Pressure within the cloud builds up. The contraction speeds up, and finally the cloud, which we may now call a protostar, reaches a size at which the amount of pressure trying to force the material outward almost equals the force of gravity. At this point contraction slows down, and the star is almost in equilibrium. It may now be about the size of our solar system. From here on, it slowly contracts and radiates, and the inside of the protostar heats up more than the outside.

The slow shrinkage and heating continue until the internal temperature is high enough to trigger the nuclear reactions, which then supply the energy that is being radiated away. The nuclear reactions provide enough added pressure to exactly balance gravity. The cloud of dust and gas has become a star, a self-sustaining

Table 5-1   *The nearest stars*

Star	Distance light-years	Spectral type	Apparent magnitude	Absolute magnitude
Proxima Cen	4·3	M5	+10·7	+15·1
α Cen A	4·3	G2	0·0	4·4
α Cen B	4·3	K5	1·4	5·8
Barnard's star	6·0	M5	9·5	13·2
Wolf 359	8·1	M8	13·5	16·5
Lal 21185	8·2	M2	7·5	10·5
Sirius A	8·7	A1	−1·5	1·4
Sirius B	8·7	wA5	+8·5	11·4
UV Cet A	9·0	M6	12·5	15·3
UV Cet B	9·0	M6	13·0	15·8
Ross 154	9·3	M6	10·6	13·3
Ross 248	10·3	M6	12·2	14·7
ε Eri	10·8	K2	3·7	6·1
L 789-6	11·1	M7	12·2	14·6
Ross 128	11·1	M5	11·1	13·5
61 Cyg A	11·2	K5	5·2	7·5
61 Cyg B	11·2	K7	6·0	8·4
Procyon A	11·3	F5	0·3	2·6
Procyon B	11·3	wF	10·8	13·1
ε Ind	11·4	K3	4·7	7·0
Σ 2398 A	11·6	M4	8·9	11·1
Σ 2398 B	11·6	M5	9·7	11·9
Grb 34 A	11·7	M1	8·1	10·3
Grb 34 B	11·7	M6	11·0	13·3
Lac 9352	11·9	M2	7·4	9·6

nuclear furnace. The star settles down to become a mature star for a period lasting from several million years for very massive stars, up to several trillion years for stars of very low mass.

The best way to illustrate the pre-main-sequence evolution of a star is on the H-R diagram as in Figure 5–11. The cold, dark cloud starts off at the far-right bottom portion of the graph, actually far off the graph as normally drawn. During its early rapid contraction, its track is upward and slightly to the left; that is, it becomes brighter and slightly hotter on the surface. When it reaches the almost-equilibrium stage, it turns downward and proceeds almost vertically for a while until the inside of the star heats up enough to change the process by which energy is carried out of the star from one of convection to one of radiation. Then the track turns horizontally to the left and proceeds until the internal temperature sets off nuclear fusion. A slight turn downward brings the star to the main sequence, which we now see is the location of stars deriving their energy from

Table 5–2 *The brightest stars*

Star	Proper name	Distance light-years	Spectral type	Apparent magnitude	Absolute magnitude
α CMa	Sirius	8·7	A1 V	−1·47	+0·7
α Car	Canopus	300	F0 Ib	−0·71	−5·5
α Cen	Rigil Kent	4·0	G2 V	−0·27	+4·6
α Boo	Arcturus	36	K2 III	−0·06	−0·3
α Lyr	Vega	26	A0 V	+0·03	+0·3
β Ori	Rigel	850	B8 Ia	0·08	−7·0
α Aur	Capella	45	G8 III	0·09	+0·12
	(binary)		G0 III		+0·37
α CMi	Procyon	11	F5 IV–V	0·34	+2·8
α Eri	Achernar	75	B5 V	0·49	−1·3
β Cen	Hadar	300	B1 III	0·61	−4·3
α Aql	Altair	16	A7 IV–V	0·75	+2·1
α Tau	Aldebaran	65	K5 III	0·78	−0·2
α Cru	Acrux	270	B1 IV	0·80	−3·8
α Ori	Betelgeuse	650	M2 Ib	0·85	−5·5
α Sco	Antares	400	M1 Ib	0·92	−4·5
α Vir	Spica	220	B1 V	0·98	−3·2
β Gem	Pollux	35	K0 III	1·15	+0·7
α PsA	Fomalhaut	23	A3 V	1·16	+1·8
α Cyg	Deneb	1500	A2 Ia	1·26	−7·0
β Cru	Mimosa	370	B0 IV	1·28	−4·0
α Leo	Regulus	85	B7 V	1·33	−1·0
ε CMa	Adhara	620	B2 II	1·42	−5·0
γ Ori	Bellatrix	450	B2 III	1·61	−4·1
λ Sco	Shaula	300	B2 IV	1·61	−3·3
β Tau	El Nath	270	B7 III	1·64	−3·0

conversion of hydrogen to helium.

Remember that the points and tracks plotted on the H-R diagram are observed properties of the stars' surfaces only. We use these to interpret the structure of the interior. Thus, when we say that the track moves to the left, for example, we are saying that the star's surface temperature is increasing while its brightness remains constant.

A problem is that we cannot follow the path of a single star from cloud to main sequence, and beyond. The time period is too long. Even the rapid contraction to reach equilibrium takes thousands of years. For the Sun, the time from this point to the main sequence is many millions of years, and the Sun has been on the main sequence for about 5 billion years and will remain there for about another 5 billion years. The astronomer thus has a problem in trying to discover the outline of stellar evolution. The times involved are much longer than the life of an astronomer, and are indeed much longer than the entire history of the science.

An analogy might be the following assignment: go to a forest and take several snapshots of the forest with a camera. Then deduce the life story of a tree.

On each photograph we will see trees of all sizes, ages, and

FIGURE 5–11
An H-R diagram can be used to plot the successive stages in the evolution of a star. The path of the Sun is shown from the condensation of a cloud of interstellar matter to the main sequence. The rapidity of evolution is not uniform along the track. Evolution is rapid at first and slows down as the Sun approaches the main sequence. Once on the main sequence, the Sun remains fairly constant for about 10 billion years.

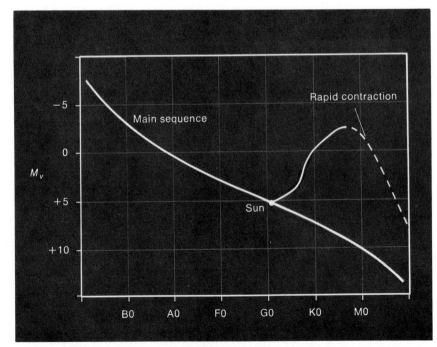

conditions. We will see different types of trees, each with different properties. From this we must determine which are the young trees, which are the old ones, where the trees come from, and what happens to them after they die—all this without seeing them do anything.

The way around this problem is to look at many stars, particularly to look at stars of different types which were formed about the same time, so that differences seen today reflect differences in the aging processes of various kinds of stars. The place we find such conditions is in star clusters. We will consider them in Chapter 8.

By making theoretical models of stellar evolution, astronomers quickly found out that the single most important property of the star is its mass. Massive stars, since they reach higher internal temperatures, produce more energy from each gram of material. They thus "burn out" faster. Low-mass stars produce less energy from each gram and so live longer. Every step of the aging process of a star is determined primarily by its mass: in more massive stars, each step is of shorter duration. The paths on the H-R diagram of stars of different mass are different, as seen in Figure 5–12.

As mentioned above, the Sun will remain on the main sequence for about 10 billion years. Small red dwarfs may last a trillion years

FIGURE 5–12
Evolutionary paths showing the formation of stars of different masses. Very massive stars evolve much more quickly than low-mass stars, but the percentage of time spent in each stage of evolution is about the same for all stars.

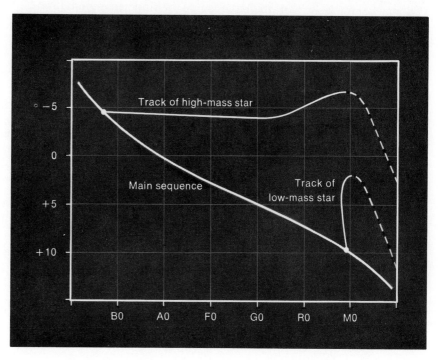

as mature stars. Other stars, such as the very hot ones in the Orion nebula, expend energy so fast that they were not mature stars when man became a species a few million years ago. Such are the extremes of stellar life.

SETTING LIMITS ON STARS

Since the main sequence has a top and bottom, there must be something which restricts stars to this range. At both ends, it turns out, the factor which limits the stars is temperature. At the upper end of the main sequence the central temperatures are so high that the force of the energy flowing outward is about equal to the gravitational force inward. A star more massive would be unstable and blow itself apart.

At the lower end, the mass is so low that the internal temperature never reaches the point at which nuclear reactions take over the generation of energy. The star just keeps on contracting and eventually winds up as a white dwarf.

ALONG THE WAY . . .

In the final stages of contraction some stars seem to throw off gas, or to leave some behind. In some (maybe many) cases, this gas becomes planets and satellites. Thus, the planets, too, are made of "star-stuff," and the materials in our bodies were once inside a star. In this way, material that began as part of a cold cloud in interstellar space ends up as intelligent creatures capable of tracing their ancestry back to cosmic debris.

REVIEW

1 Stars are self-luminous spheres of gas deriving their energy from nuclear fusion.

2 The energy-producing process in stars involves the conversion of hydrogen to helium, with a slight loss of mass. This mass is converted into energy. The specific chain of reactions may be the carbon-nitrogen-oxygen cycle, or the proton-proton process.

3 The luminosity of a star is the amount of energy it radiates. Another measure of the same quantity is the absolute magnitude of the star.

4   Stellar spectra are classified into types O, B, A, F, G, K, and M, in order of decreasing temperature. A few other types are recognized for unusual stars.

5   Stellar temperature and luminosity are described on a Hertzsprung-Russell diagram by one point for each star. The actual data plotted are spectral type and absolute magnitude.

6   About 95 percent of the stars on the H-R diagram lie on the main sequence. Further, there are many, many more faint, red stars than bright, hot stars. The Sun is intermediate in temperature and luminosity.

QUESTIONS

1   In another five billion years or so, the Sun will have converted all the hydrogen in its center into helium. Will the Sun be much less massive when it has ended its main-sequence life?

2   Draw an H-R diagram, similar to Figure 5–9. Include along the horizontal axis a scale of temperature as well as spectral type, and along the vertical axis a scale of luminosity as well as absolute magnitude. Let the radius and luminosity of the Sun be 1. The luminosity (L) of a star is related to its radius (R) and temperature (T) by the equation $L=kR^2T^4$, where k is a constant. Draw lines on the H-R diagram giving the location of stars with radius $\frac{1}{100}$, $\frac{1}{10}$, 1, 10, and 100 times the Sun.

3   Draw an H-R diagram and plot these stars on it:
    (a)   Sun   G2V   (f)   Procyon   F5V
    (b)   Sirius   A1V   (g)   Arcturus   K2III
    (c)   Antares   M1I   (h)   Aldebaran   K5III
    (d)   Betelgeuse   M2I   (i)   Barnard's Star   M5V
    (e)   Rigel   B8I   (j)   Alpha Centauri   G2V

4   Explain why "late-type" stars are not necessarily older than "early-type" stars.

5   On an H-R diagram, plot the following steps in the life a star, beginning at the lower right-hand corner:
    (1)   increases greatly in brightness but becomes only a little hotter;
    (2)   decreases somewhat in brightness and heats up a little;
    (3)   heats up a good deal but remains at constant luminosity;
    (4)   decreases in brightness and remains at a constant temperature;
    (5)   moves along a line of constant radius (see question 2).

6   Eventually the Galaxy will run out of interstellar matter from which more stars can form. Ten billion years after this happens, what would you estimate the stellar population of the Galaxy would be like?

READINGS

Bahcall, J. N. "Neutrinos from the Sun." *Scientific American*, July 1969.

Herbig, H. "The Youngest Stars." *Scientific American*, August 1967.

Johnson, M. *Astronomy of Stellar Energy and Decay.* New York: Dover Publications, Inc., 1959.

Page, T., and L. W. Page, eds. *The Evolution of Stars.* New York: The Macmillan Company, 1968.

Tayler, R. J. *The Stars: Their Structure and Development.* New York: Springer-Verlag New York, Inc., 1970.

EXTRAORDINARY STARS

In Chapter 5, we left the life story of the stars just as they reached the main sequence on the Hertzsprung-Russell diagram. They had become mature stars, obtaining energy from the conversion of hydrogen to helium by nuclear fusion. The location of stars which have just reached this stage on the H-R diagram is called the zero-age main sequence, or ZAMS. Just as the star's mass determines where it will be on the main sequence, so too the evolution, or aging, after the main sequence proceeds differently depending on the mass of the star.

In order to understand the subsequent evolution of stars, we must first briefly explore the laws of physics which govern the structure of a star. Using these laws, astrophysicists can make

The star Beta Lyrae is really two stars very close together. Material is pulled out of one by the other, and some escapes to stream into a giant pinwheel around the system. Such close pairs of stars have a life history very different from that of ordinary stars.

computer models of a stellar interior from which we can predict the surface-temperature properties (spectral type and luminosity) of a star and compare these with observation.

PHYSICS OF STELLAR INTERIORS

In order to study the inside of a star, we usually consider a small cube of matter somewhere in the interior at a distance r from the center of the star. We assume that the star is a perfect sphere, and that every cube at the same distance from the center has the same properties. See Figure 6–1. The pressure pushing the cube down from above due to the weight of overlying layers, and the weight of material in the cube, must be equaled by the pressure from below holding up the cube. This is called the law of pressure balance, or *hydrostatic equilibrium.*

Another important law relates the pressure, volume, amount, and temperature of a gas. This law is called the *ideal gas law* and is correct for gases in which the atoms are far apart. This would *not* be

FIGURE 6–1 *A mathematical model of the inside of a star is made by considering small regions inside the star and applying the law of physics. In the cube on the right we consider the pressure forces acting on a typical cube. If the cube is not moving, these must balance out. In the cube on the left, we consider the energy flowing in and out of a typical cube. The energy out must equal the difference between the energy in and any energy produced in the cube.*

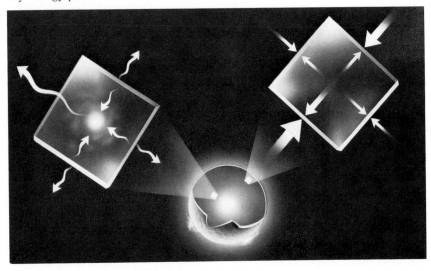

true for a gas at the very high densities that occur within a star, except for the fact that the very high temperatures completely ionize the atoms, removing all their electrons. The remaining nuclei are very small compared to the cloud of electrons they had, and so are quite far from one another. Thus, the ideal gas law is usually valid inside a star.

Another physical law is simply the conservation of energy: the amount of radiation coming out of the cube must equal what went in, plus any energy produced within the cube itself.

Further, we must consider the *opacity* of the material in the cube. Opacity is the resistance of material to the passage of radiation. The higher the opacity, the more it tends to keep the energy inside a star, like a blanket. The amount of opacity is dependent on the temperature, density, and other physical properties of the material in the cube.

Another important piece of information which goes into calculations of stellar structure and evolution is chemical composition. There will be slight differences in spectral type and luminosity between stars having slightly different proportions of helium and heavier elements. Also, the paths of stars of different chemical composition will be different after the stars leave the main sequence. We can compare predicted paths with those observed to try to determine the chemical composition of stars. Usually we will be considering stars with chemical compositions similar to that of the Sun: about 75 percent hydrogen, about 25 percent helium, and about 1 percent everything else.

STARS ON THE MAIN SEQUENCE

The aging of a star depends on its structure, which depends largely on its temperature, which in turn depends on its mass. It is useful to divide the main sequence into three regions. The upper main sequence is the locus of stars of more than 1.75 solar masses ($M_\odot$) and the lower main sequence those below 1.1 $M_\odot$. In between is a transition zone with stars having some of the properties of both groups.

A high-mass star, on the upper main sequence, will have a very high central temperature. This means that the source of energy is the carbon cycle. Because this cycle is critically dependent on the temperature, the energy generation inside the star will take place in only a very small volume at the star's center. The energy output of this core is so great that radiation, the slow outward diffusion of photons discussed in Chapter 5, is not sufficient to carry all the

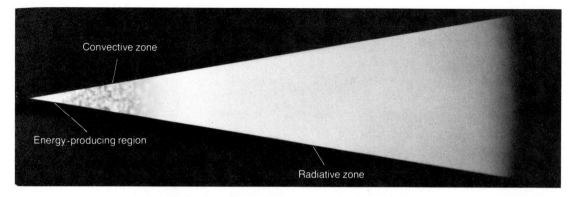

FIGURE 6–2   The internal structure of a star of high mass. Because of the very high central temperature and flow of energy, radiation alone is not enough to carry the energy away from the center. Convection occurs which transports energy by bulk motion of material. This brings a fresh supply of hydrogen to the core. Thus, eventually hydrogen will be depleted over a volume larger than just the energy-generating core.

energy. So *convection* occurs, the transfer of energy by the circulation of material up to a cooler region. In this upper region radiation takes over to carry the energy outward to the surface and then into space (Figure 6–2).

In a lower main-sequence star, such as our Sun, the central temperature is lower, and the proton-proton process provides the energy. This reaction is much less dependent on temperature, so the conversion of hydrogen to helium takes place over a large volume in the center of the star. The energy is carried outward by radiation toward the outer parts of the star where convection completes the job of carrying the energy to the surface. As the star ages in the main sequence, the hydrogen decreases gradually over a larger fraction of the star than is the case for a star on the upper main sequence (Figure 6–3).

The time a star will spend "burning" hydrogen depends on how much hydrogen is available and how fast it is converted. The amount available is obviously related to the mass of the star. However, not all the material in a star is available for fusion, but only that portion near the center. The *rate of conversion* is the luminosity of the star, which depends on the mass. Thus, we can get a relation between mass and lifetime. The relation obtained states that the dependence of lifetime lies between M^{-3} and M^{-5}, that is, on the inverse third to inverse fifth power of the mass of the star compared with the mass of the Sun. For example, a star twice the mass of the Sun will stay on the main sequence between $\frac{1}{8}$ ($= 2^{-3}$) and

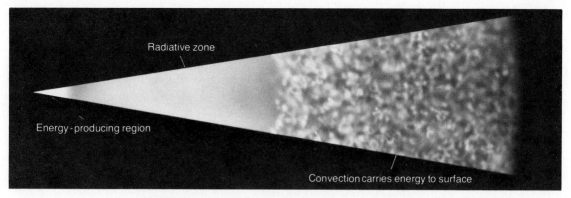

Radiative zone

Energy-producing region

Convection carries energy to surface

FIGURE 6–3   The internal structure of a star of low mass. Radiation
carries energy away from the core. Part of the way toward the surface,
the rate of temperature decrease becomes so great that convection occurs
and carries the energy to the surface.

$\frac{1}{32}$ (= 2^{-5}) the length of time the Sun will. Thus, very massive stars
will have very short lives compared with the Sun's, and the Sun
itself will live a short time compared with the low-mass stars at the
bottom of the main sequence. Figure 6–4 shows estimates of the
lifetimes of stars of different masses.

FIGURE 6–4
A graph of the rela-
tionship between
stellar mass and
lifetime on the main
sequence.

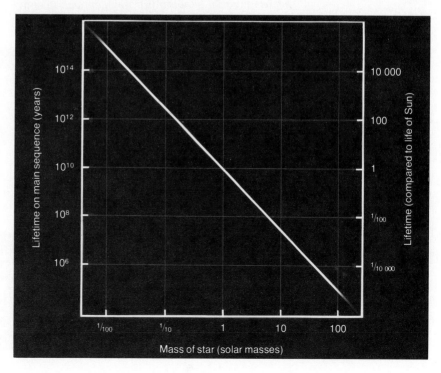

From Figure 6–4 we see that there is a great range in the time a star will remain on the main sequence. Thus, whenever we observe a very luminous main-sequence star—say, an O or B star—we know it must be young compared with the Sun. On the other hand, a small red star will have a lifetime comparable to the age of the Galaxy. When we observe one of these red dwarfs, it may be young—that is, just formed—or it may be old. There is no easy way of telling which. It is thus proper to speak of early-type stars as being young, but not correct to call late-type stars old.

STELLAR GERONTOLOGY

Eventually every star runs out of available hydrogen fuel to con-

FIGURE 6–5   A comparison of the post-main-sequence evolution of a low-mass star with one of high mass. The high-mass star reaches the red giant stage and begins to convert helium to carbon in its core. Surrounding this core is a region in which remaining hydrogen is being converted to helium. After using all its energy sources, the star evolves to the white dwarf stage. The low-mass star also becomes a red giant, but never reaches internal temperatures high enough to convert helium. The star then evolves to the white dwarf stage. The low-mass star evolves much more slowly than the high-mass star.

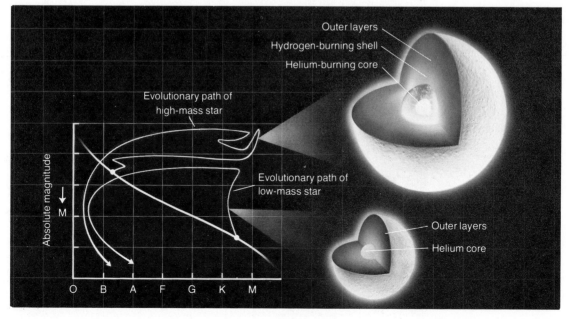

vert to helium in its core. As the hydrogen runs out, an upper main-sequence star will move slightly upward and to the right, becoming brighter and cooler. A star on the lower main sequence will move slightly up and to the left, almost along the main sequence. Remember that the H-R diagram is a graph of only the surface properties of the stars. The surface properties reflect the internal properties. The advantage of describing stellar evolution by a track on the H-R diagram is readily apparent, since it shows at a glance the life story of a star which would otherwise take many words of description. Figure 6–5 shows the paths of two stars of different mass.

After the core of hydrogen is gone, a high-mass star moves rapidly to the right, remaining at about the same luminosity. Since it is getting cooler, it becomes redder. The inside, however, becomes hotter as the core of the star contracts in order to maintain the energy output it had before. Heating of the core also heats the region surrounding the core in which hydrogen is still unburned, starting fusion in a thin shell. This shell contributes much less energy than the earlier period of burning in the core did. The star expands to compensate for the changed sources of energy. Thus, it becomes a *red giant,* luminosity class III.

Lower mass stars also become red giants, expanding and cooling on the surface as their cores contract and heat. This seemingly contradictory situation produces a star which is of low average density, but very high central density. Whereas the massive star moved to the right horizontally across the diagram, a low-mass star moves mainly upward and slightly to the right. *All stars eventually become red giants.* At some point in or before the red giant branch is reached, the central temperatures become hot enough to trigger the nuclear fusion of helium to heavier elements, with a consequent release of energy. This reaction is mildly explosive and causes another readjustment in the structure of the star, causing it to move slightly down and to the left on the H-R diagram.

We can thus explain the red giant branch as a sort of *helium main sequence* of stars obtaining energy from helium fusion reactions, just as the ordinary main sequence was the location of stars obtaining energy from hydrogen fusion. Just as there was a lower limit to the main sequence, there is also a limit to the red giant branch. Stars from the lower main sequence, stars of low mass, will expand toward the red giant branch, but their small mass will never cause the internal temperatures to rise high enough to burn helium. From this point on, their only sources of energy are gravitational and leftover heat energy. From here on they are dying stars.

The lower limit of star mass which will just begin to burn helium is uncertain. We can calculate how much the interior will heat up as a result of contraction, but we are still uncertain as to how much energy will be lost, cooling off the star. The energy is lost by the escape of neutrinos from the core.

Astronomy in a Gold Mine

In the nuclear reaction listed in Chapter 5, we remember that a couple of the reactions release a neutrino, which escapes the star altogether. Neutrinos have no mass or electrical charge. They are very elusive and can pass through ordinary matter as if it weren't there. In contrast to photons, neutrinos do not often react with anything. Thus, unlike the photons released in fusion, neutrinos can escape through the material of the star into space, carrying away energy that would otherwise go into heating up the interior.

Neutrinos interact very little with ordinary matter; it has been estimated a neutrino would have only a 50-percent chance of being stopped while passing through 10 million light-years of lead. In fact, billions of billions of neutrinos pass through our bodies each second, and we never know it.

Thus, neutrinos act to cool a star's interior, and this cooling is critical to the evolution of the star. In order to try to find out about neutrino losses from the Sun, physicists have tried to detect neutrinos, attempting to catch them in a large tank of cleaning fluid. Of the billions upon billions of neutrinos passing through the tank, the prediction is that two or three should strike the chlorine atoms every few days. After several months, the physicists filter the fluid to look for chlorine atoms that have been transformed into radioactive argon atoms by the collisions. Figure 6–6 show one such neutrino "telescope."

They place the tank in a mine deep in Earth. The layers of rock above the tank shield it from cosmic rays which could also turn chlorine into radioactive argon, confusing the measurements.

The surprising result of the experiments is that fewer neutrinos than expected have been detected. This means one of three things, all of which scientists are now eagerly investigating. Either our predictions of how often a neutrino will hit a chlorine atom are in error; or our models of how many neutrinos are produced in the nuclear reactions inside the Sun are in error; or the Sun's core is cooler than we thought. The rate of the nuclear reactions is very sensitive to temperature. We are uncertain at the present time as to the amount of cooling of other stars as well. This is an important piece of data we hope to obtain within the next few years.

Advanced Stages of Stellar Aging

The structure of those stars which get hot enough to convert helium consists of several layers. In the core, helium is converted into carbon, oxygen, neon, and other elements. In a thin shell surrounding the core, hydrogen is being converted into helium. From this layer outward to the star's surface, no nuclear reactions take place. And, just as a star on the main sequence will eventually run out of hydrogen, so too will a red giant run out of available helium.

FIGURE 6–6
A neutrino "telescope" of the Brookhaven National Laboratory. This tank contains 100 000 gallons of perchloroethylene and is 1½ kilometers below ground level in the Homestake Gold Mine at Lead, South Dakota. Neutrinos from the Sun are captured by the chlorine atoms and changed into argon atoms. The argon is removed from the tank and measured, indicating the rate of solar neutrinos. (Brookhaven National Laboratory)

Once again internal contraction and external expansion begin, until the core is hot enough to fuse the next element. Then helium will begin to fuse in a thin shell around the core. In the more advanced stages of nuclear fusion, internal temperatures are in billions of degrees. And, just as some stars are not massive enough to reach the helium fusion stage, so, too, some stars which did manage to convert helium will not be massive enough to reach temperatures required to fuse heavier elements; they, too, drop by the wayside and begin to evolve toward the white dwarf stage. This sequence of internal contraction with external expansion, development of a new energy source with consequent slight external shrinking, exhaustion of the new source, and so on, takes place over and over again for sufficiently massive stars. It is important to keep in mind that all this happens over a span of time which is much less than the time the star spends on the main sequence. The exact number of years a star spends in each stage varies with the star's mass but the fraction in each stage relative to the main-sequence life of the star is roughly the same.

For the most massive stars, somewhat less than 100 $M_\odot$, the end product is a core of iron. Such a star, shown in Figure 6–7, would have an almost pure iron core, surrounded by a shell burning silicon to iron, surrounded by a shell burning carbon and oxygen to silicon, surrounded by a shell burning helium to carbon and oxygen, surrounded by a shell burning hydrogen to helium, surrounded by the outer layers of the star. By this time, the star is a supergiant several thousand times the size of our Sun, perhaps as big as the orbit of Saturn or Uranus in our solar system!

Recall that the reason we can get energy from nuclear fusion reactions is that the mass of the atom which has been produced is less than the sum of the masses of the atoms which went into its production. Thus, a helium atom is less massive than four hydrogen atoms, and the mass difference is converted into energy. But this is true only for elements lighter than iron. An element heavier than iron (atomic weight greater than 56) would have more mass than the sum of the atoms that produced it. This means that iron and elements heavier than iron cannot produce energy by fusion. On the contrary, energy is needed to make them fuse. Thus, iron is the limit, and only a very few, very massive stars reach this stage. The less massive stars have dropped by the wayside. After the last nuclear fuel has been burned, the star has only gravity and leftover heat energy. What happens now again depends on the mass of the star. The details of these later stages of evolution are still uncertain, owing to the great number of varieties and incomplete laboratory data on nuclear fusion. Despite this, we do know that the end product of all this evolution for most stars is the formation of a white dwarf.

FIGURE 6–7   *The internal structure of a highly evolved supergiant star. Iron has been produced in the core. Surrounding this are a number of shells in which remaining nuclear fuels are being converted. Although more than 50 times more massive than the Sun, the star is so large that it is little more than a "hot vacuum" with an indistinct surface. The orbit of Saturn (about 3 billion kilometers across) is shown to scale.*

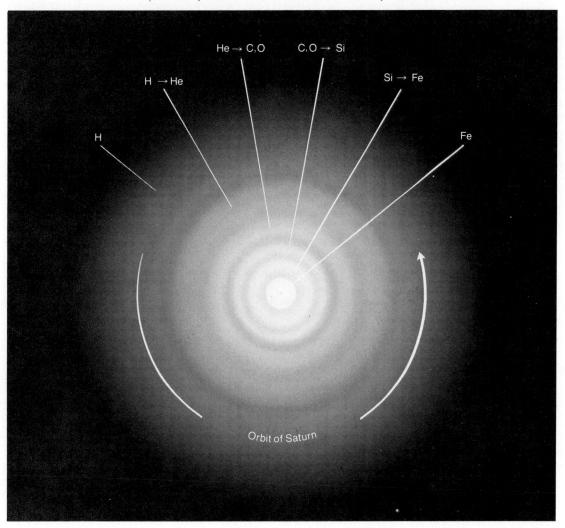

WHITE DWARFS

About a hundred and fifty years ago it was noticed that Sirius was not moving among the stars in a straight line. It had a wavy path, and it was suggested that Sirius might have an unseen companion in orbit about it, or, more exactly, that both stars were in orbit about a common center of mass. In the 1850s Alvan Clarke (1804–1887), a telescope maker, while testing a new lens made for an observatory, discovered Sirius's companion. It was white in color, but much, much fainter than Sirius, which is also white. The luminosity of a star depends on temperature and size. Since the color, hence temperature, of the two stars was about the same, the faint star known as Sirius B was deduced to be quite small. In fact, it is a star with a mass comparable with that of the Sun, but with a size comparable with that of Earth! Thus, its density must be much higher than the Sun's. Earth's radius is about one one-hundredth that of the

FIGURE 6–8　*The double nature of Sirius (Alpha Canis Majoris) was detected by the wavy motion of the star through the sky. At first only the brighter star (A) was visible. In the upper left is shown the apparent orbit of Sirius B around Sirius A. In the lower right are shown the true orbits of each star around the center of mass of the two.*

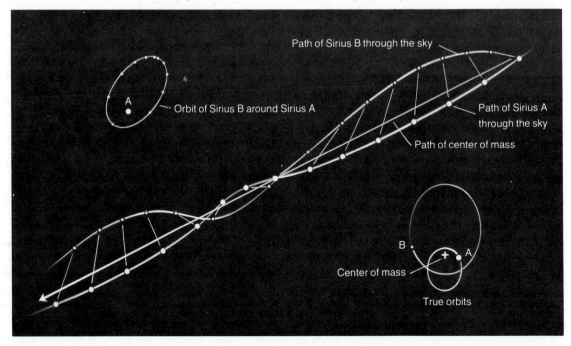

Sun, so the density of Sirius B must be about $(100)^3$, or 1 000 000 times the Sun's density, that is, about 1 000 000 grams per cubic centimeter. Sirius B was the first white dwarf known (Figure 6–8).

A white dwarf is so extremely dense that matter is no longer describable in ordinary terms. Although the nuclei of the atoms are still close to a perfect gas, the electrons are a strange kind of "degenerate" gas. In a degenerate gas, the electrons do not behave independently of each other, but in concert. This is predicted by relativity and quantum mechanics, which restrict the positions and velocities of the electrons in a way not comparable to everyday experience. Despite the unusual nature of such stars, the theory of white dwarfs is rather well understood—in fact, better than some other stages of stellar evolution.

We know, for instance, that white dwarfs cannot have just any mass. They have masses between about 0.2 $M_\odot$, in which case they will be about the size of Earth, up to about 1.4 $M_\odot$, which will make them about half the size of Earth. (Note that the larger the mass, the smaller the star!) It has been computed that a white dwarf cannot have a mass greater than 1.44 times the mass of the Sun; if it did, its volume would be zero. Since we know of many stars with main-sequence masses greater than this, we must look for either another end product of stellar evolution, or a way of reducing the mass of a star below this upper limit. We think we have found both.

The process of mass loss for a star may be either gradual or abrupt. If it happens slowly, it is similar to the *solar wind.* This is a stream of high-energy particles ejected from the Sun, to be discussed in Chapter 11. A *"stellar wind"* is a very slow way of losing mass; for the Sun it is only about 10^{-13} $M_\odot$ per year, and completely negligible as far as any effect on the Sun's evolution is concerned. However, if distant stars have much higher loss rates, they are undetectable, so perhaps this is one way a star reduces its mass below the upper limit for white dwarfs. A star can also lose mass from pulsations which would eject material into space.

The star may also explode in the form of a *nova,* ejecting several tenths of the star's total mass. There is evidence that some stars may explode several times, and this would be a very good way of reducing their mass. The result of a nova may be a *planetary nebula,* a roughly spherical shell of gas surrounding the star. Figure 6–9 shows a series of photographs of Nova Lacerta 1910. NGC 7293, a typical planetary nebula, is seen in Figure 6–10.

Even more powerful is a *supernova* explosion, a stellar catastrophe, practically destroying the star. The result of such an explosion may be a *neutron star,* plus a huge cloud of gas, a *supernova*

FIGURE 6–9
Three photographs of Nova Lacerta 1910. In the first photograph (left), taken in 1907, the star is magnitude 13. The second photograph shows the star near maximum brightness of magnitude 7 on December 31, 1910. By September 29, 1911, the star had faded to magnitude 11, seen in the third photograph. (Yerkes Observatory)

remnant, similar to the Crab nebula. We will consider these further (Figure 6–11).

The stages of stellar evolution after the star leaves the main sequence occur rapidly compared to the time the star spent on the main sequence. After the star's fuel has been used up, the star's path from the red giant or supergiant branch is extremely rapid, perhaps punctuated with a nova or supernova explosion along the

FIGURE 6–10
A planetary nebula, NGC 7293, in the constellation Aquarius, photographed in red light. This is the result of a stellar explosion. (Hale Observatories)

FIGURE 6–11
The Crab nebula, a remnant of a supernova explosion seen in A.D. 1054. These photographs in different regions of the spectrum reveal information about the source of the light. Most of the radiation is produced by the synchrotron process, which requires strong magnetic fields and energetic particles. (Hale Observatories)

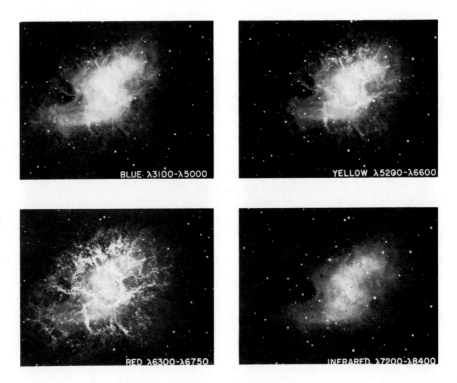

way. The star moves almost horizontally to·the left across the H-R diagram, crosses the main sequence, turns downward, and moves along a line of constant size as it becomes a white dwarf. From the giant branch to the white dwarf region the star has been contracting, converting its remaining gravitational energy into heat. When the white dwarf stage is reached, when all the material in the star (except perhaps for a thin surface layer) has become degenerate, it can no longer contract if it lies in the mass range for a white dwarf. From this point on it remains the same size. It lives on stored heat energy, and only the outermost layers resemble an ordinary star. The outermost layer is quite opaque, so energy leaks out slowly. For this reason a white dwarf takes a long time to cool down to a "black dwarf," or whatever you might call the cinder of a once-brilliant star.

BEYOND THE WHITE DWARF

Stars whose final masses are greater than the limit for white dwarfs do not stop contracting even when the material in them has become degenerate.

The contraction continues, and the increased pressure reduces the distances between the atomic particles until the electrons are forced into the nucleus, combining with the protons to produce neutrons. The stars contract until the neutrons are degenerate, that is, do not behave independently, and the star's interior is a mixture of solid and gas, composed of neutrons. The density is on the order of a billion tons per cubic centimeter. This we call a *neutron star*. It is now believed that stars with final masses between about 1.4 and 4 solar masses will become neutron stars. Try to imagine a star with a mass several times that of the Sun, in a volume only 10 kilometers across (Figure 6–12).

These strange objects were a theoretical prediction made by astrophysicists long before one was discovered. In 1967, radio astronomers in England observed short, regular pulses of radio waves coming from several sources in space. More were discovered, and they were called *pulsars*. An early suggestion was that they might be "lighthouses" set up for interstellar navigation by superior extraterrestrial species, and were explained by the "LGM" theory. LGM stood for "little green men." Soon, however, it was realized that these were natural objects. Now they are believed to be neutron stars, rapidly rotating and emitting synchrotron radiation in a very narrow beam, like a searchlight. When the beam passes over us, we receive a pulse of radio waves.

Confirmation of this theory came from observation of the

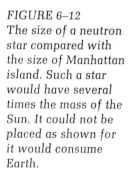

FIGURE 6–12
The size of a neutron star compared with the size of Manhattan island. Such a star would have several times the mass of the Sun. It could not be placed as shown for it would consume Earth.

pulsar in the Crab nebula. We know that the Crab nebula is a remnant of a supernova explosion that was seen in A.D. 1054, and
theory says that such an explosion should leave behind it a neutron
star. Astronomers finally observed that the suspected star sends
out pulses of light, flashing at the same rate as the radio pulses,
about 30 times a second. This suggests that a common source was
responsible for both types of radiation (Figure 6–13).

*FIGURE 6–13   The Crab Nebula. In the upper right are contours
of equal radio intensity. The graph below shows the pulses
received from the pulsar located within the nebula. (Hale Observatories;
graphs by National Radio Astronomy Observatory)*

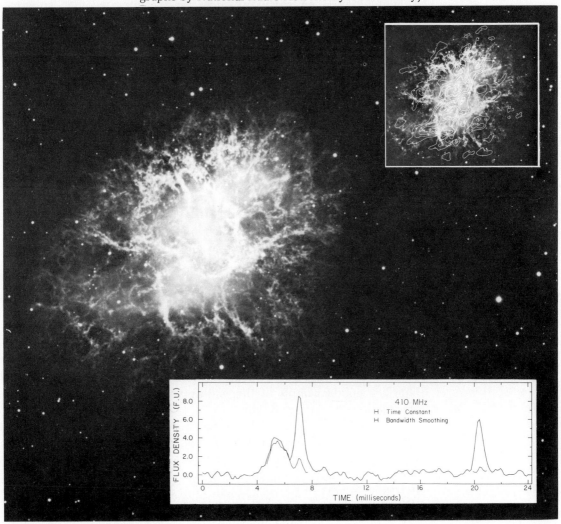

The pulsars are slowing down, very slightly, at a uniform rate, except for a few abrupt slight speed-ups which have been explained as "starquakes," or sudden, slight readjustments of the star. If a neutron star shrank slightly, its rotation rate would increase slightly. Thus, pulsars fit quite neatly into our theories of stellar evolution.

PREDICTING THE INVISIBLE

For more massive objects than the neutron stars, the final contraction will not stop even at the terrifically dense neutron-star stage. They will keep on shrinking until the density becomes so great nothing can escape; the object is then called a *"black hole."* See the cover of this book for an artist's conception of a black hole.

One might think of a black hole as an object which dug a hole, jumped in, and pulled the hole in after it! Consider a dense object. It will have an *escape velocity,* the speed necessary to get completely away from it. The escape velocity will be higher the more massive the object is, and the denser it is. That is, a small object will have a higher escape velocity than an object of the same mass but a larger size. Now imagine an object with several times the mass of the Sun, but extremely small. As it shrinks, its escape velocity will eventually become greater than the speed of light. Thus, even light will not be able to escape from it, much less matter. So the term "black hole" is appropriate for these strange objects. Interestingly enough, such objects were theorized decades ago, but no one believed they existed. Now some astrophysicists believe that black holes might be the answer to the problem of the density of the universe mentioned in Chapter 1. Perhaps a significant part of the mass of the universe is in the form of black holes.

Anything approaching within a certain distance of a black hole would be drawn in, irretrievably. A black hole can be detected only by looking for radiation from clouds of matter which might surround it. Material being pulled into a black hole would radiate intensely; we may be able to detect some of this radiation. A black hole is a good example of the real universe being stranger than the speculations of science fiction.

Recent observations from Earth and from satellite observatories have revealed a star which might be a black hole. It is called Cygnus X-1, since it was the first X-ray source discovered in the constellation Cygnus. Cygnus X-1 is a double star, and one of the two stars seems to be very small. Periodically, the big star eclipses the other. (For more on eclipsing double stars, see Chapter 7.) From the length of the eclipse we can obtain an idea of the sizes of the stars. The X rays we receive are about what we predict should be radiated by a gas cloud surrounding a black hole. Much more observation is needed to confirm this.

VARIABLE STARS

At some (maybe several) stages in the lifetime of a star, it may be unstable. For most stars this instability is brief in duration, and slight, but for a few the instability leads to the destruction or drastic alteration of the stellar structure. Most variable stars pulsate, alternately swelling and contracting, heating and cooling, and consequently changing in brightness. Others flare up suddenly. (There are also stars that vary in brightness because one star eclipses another. These eclipsing variable stars will be discussed in Chapter 7.) Figure 6–14 shows two variable stars over a period of time.

In the study of variable stars, amateur astronomers have provided important data. Many of the stars can be seen and measured in small telescopes, and in this way we can keep track of the light variations of thousands of stars for which it would be inefficient to

FIGURE 6–14
The variable stars R and S Scorpii, showing their changes in brightness. (Yerkes Observatory)

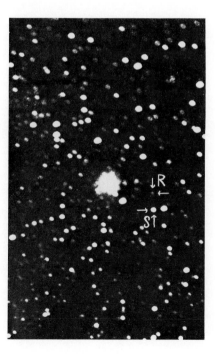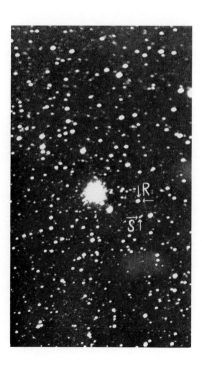

use the large, professional telescopes. In a year, amateur astronomers report over 100 000 observations to the American Association of Variable Star Observers. These data are then available to anyone who needs them for research.

Aside from the eclipsing stars, variable stars can be classified as either *pulsating* or *eruptive* types. About 80 percent of the known variable stars are of the pulsating type. The variable stages of a star's evolution play an important role in the life of a star. Some types of variable stars are extremely useful indicators of distance due to a known connection between their variability and their absolute magnitude. Figure 6–15 shows where many of the types of variable stars lie in the H-R diagram.

Pulsating Stars

Long-period variables. The most common type of pulsating star is the so-called *long-period variable* (LPV). Such stars are also known as Mira types, or Omicron Ceti types, after the first one known. Mira, in the constellation Cetus, comes from the same Arabic word as our word "miracle." Mira is a typical member of its class. Most of the time it is invisible to the unaided eye. Once every 300 days it increases by five or six magnitudes, to become about magnitude 2.5. Like all other LPVs, Mira is a cool, red giant star. During its varia-

FIGURE 6–15
The locations of some of the major types of variable stars on the H-R diagram. Note that most are in regions of advanced stellar evolution. The notable exceptions are the T Tauri variables which are believed to be stars still in the process of contracting to the main sequence.

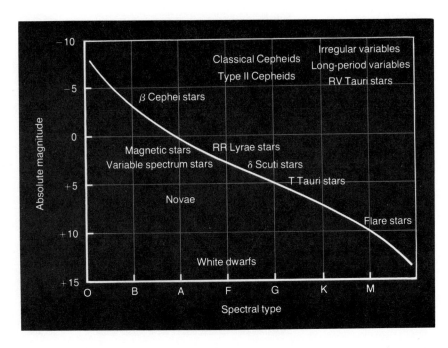

tion, the surface temperature changes from about 2600°K (spectral type M6) to about 1900°K (spectral type M9). It is estimated that Mira is more than 400 times the size of the Sun and has an average density less than one one-millionth the Sun's. This is so tenuous that some people have called such a star a "hot vacuum." Almost 4000 Mira-type stars have been identified.

RR Lyrae stars. The next most numerous class is comprised of the *RR Lyrae stars*, named after the prototype star found in the constellation Lyra. They are white or yellow-white giant stars with very regular pulsations. They have short periods, usually between about a quarter and three-quarters of a day. The important fact about them is that they all seem to have the same absolute magnitude, about +0.5. Thus, if we observe an RR Lyrae star, we can determine its distance by the formula relating m, M, and r discussed on page 65. RR Lyrae stars are commonly found in globular clusters (see Chapter 8), and a few isolated ones occur in the outer parts of the Galaxy. They have played an important role in our determination of the distance scale of the universe.

Irregular variables. Next most numerous are the *irregular* and *semiregular variable stars*. These have no periodic variation, or only a hint of a period. The increases and decreases in brightness are almost at random. The irregular variables are cool giants and supergiants, but the semiregular variables may be of any spectral type.
Other types of variables stars, such as RV Tauri stars, Beta Cephei stars, Delta Scuti stars, and so on, are very rare. Their existence is quite overshadowed by the best known of all the variable stars, the Cepheids.

Cepheids. These extremely important stars are named for Delta Cephei. They are spectral class F and G supergiants, among the largest stars in the Galaxy. They have a very regular pulsation with a range from a few tenths of a magnitude to a couple of magnitudes. Somewhat less than a thousand are known so far.
Most of the Cepheids have periods of pulsation that range from 50 days down to a few days. They are intrinsically very bright, and the important thing about them is that there is a relation between their period of pulsation and their average brightness. In the first two decades of this century, a study of Cepheids in the Magellanic clouds (two small satellite galaxies of the Milky Way which are not visible from northern temperate latitudes) showed that a graph on which was plotted the magnitude of the Cepheids against the

logarithm of their periods was almost a straight line. This is known as the *period-luminosity relation* and is graphed in Figure 6–16.

Since the original discovery, it has been found that there are really two types of Cepheids, called types I and II, which differ by about 1½ magnitudes. We can, however, tell the two types apart, and they both have a period-luminosity relation. Thus, whenever we see a Cepheid, we can tell its distance. If the variable is in a cluster or another galaxy, we get the distance to that cluster or galaxy. The importance of this distance-measuring tool to astronomers cannot be overestimated. It was this method which was used by Edwin Hubble in 1924 to show that the so-called Andromeda "nebula" was actually another galaxy, another island universe (Figure 6–17).

For any variable star, a graph of its light variation with time is called a *light curve*. When a star is observed to vary, a careful series of observations enable astronomers to classify the star as to type

FIGURE 6–16
The period-luminosity relation for Cepheid variables. The longer the period of pulsation, the greater the average luminosity of the star. The RR Lyrae variables are similar to the Cepheids but have very short periods and an absolute magnitude of +0.5. There are two types of Cepheids, differing by about 1½ magnitudes for a given period.

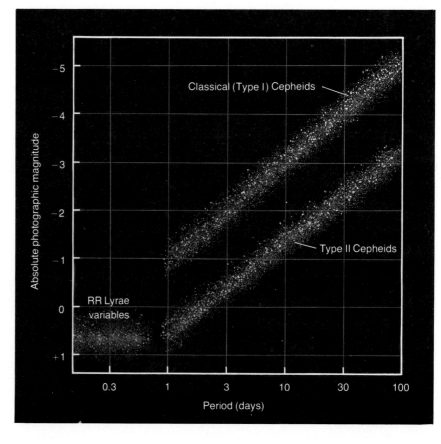

by the shape and amount of variation of the light curve, together with the spectral type and period. More variable stars are discovered each year.

Eruptive variables. Variable stars which change brightness abruptly are of many different types. The variations are almost always sudden increases in brightness. Some are believed to be flare-ups on the surface of a star, similar to but much larger in scale than the solar flares on the Sun. Others are large-scale variations in the structure of the star. One peculiar type, known as R Coronae Borealis stars, shows irregular drops in brightness. These stars have less hydrogen than most stars, and much more carbon. Like most other variables, they have not been adequately explained.

The *T Tauri stars* provide one very interesting type of eruptive variable. These are class G to M stars, found slightly above the main sequence in the H-R diagram. They show quick, irregular flashes of up to a few magnitudes. Further, they are always associated with interstellar gas and dust and are believed to be low-mass stars which have not quite completed their contraction to the main

FIGURE 6–17
A field of variable stars in the Andromeda galaxy. Two of the variables are indicated. These can serve as distance indicators. (Hale Observatories)

FIGURE 6–18
*Examples of a type
of variable object
called a Herbig-Haro
object, after its
discoverers. These
are believed to be
stars in the process
of formation and may
be predecessors of
the T Tauri stars.
(Lick Observatory)*

sequence. Hydrogen fusion is just beginning in these stars; they are stars in the process of creation (Figure 6–18).

NOVAE

The most common eruptive variables, aside from the T Tauri stars, are the *novae,* and they are considerably more spectacular. The word comes from the Latin *nova stella* or "new star," because the sudden increase in brightness may cause the star to appear in the sky where no star has been seen before, so it seems to be "new." The increase in brightness may be up to 10 or 15 magnitudes (that is a factor of 10 000 to 1 000 000 in luminosity).

Novae are actually old stars, not new ones. Before they brighten, they are subdwarfs, that is, somewhat below the main sequence, but not as small as white dwarfs. In a day or so, they increase their light by 10 to 15 magnitudes, then fade away, taking a few months to decline to their former brightness. Occasionally they become bright enough to be seen with the unaided eye. Several are usually discovered by astronomers each year.

The mechanism of the explosion is not known. We do know that a large shell of gas is thrown off into space. This may contain a small fraction of the mass of the star and is rapidly dissipated into space. Currently we believe that most, if not all, novae are members of close systems of double stars. It is hypothesized that one star ejects material onto the other, heating it suddenly and explosively. More data will have to be accumulated to confirm or refute this.

Supernovae

Among the most spectacular objects in the universe, a supernova is a star which increases suddenly by millions to hundreds of mil-

FIGURE 6–19
Two photographs of the galaxy NGC 7331. The top plate was taken before the explosion of the supernova seen in the bottom plate. At maximum, the light output from this one star may have equaled the combined output of the rest of the galaxy. (Lick Observatory)

lions of times in brightness, reaching an absolute magnitude of -20 or so, in just a day. Thus one star, a supernova, may for a short time outshine the entire rest of the galaxy, more than 100 billion stars! Figure 6–19 is a photograph of a nova in the galaxy NGC 7331. A supernova is a cataclysmic stellar explosion, virtually destroying the star and leaving behind a remnant which may be a neutron star, or perhaps a black hole, surrounded by a cloud of energetic gas.

There are two types of supernovae, each thought to be caused by a different process. Type I supernovae may be caused by the explosive beginning of the fusion of carbon to heavier elements in the star's core. Supernovae of type II are altogether different, and are thought to be the death throes of the most massive stars, stars which have finally produced an iron core. At a temperature of billions of degrees, iron is unstable. The core collapses, the outer parts follow, and the rapid contraction produces a sudden temperature rise. The unburned nuclear fuel then fuses, explosively, and disrupts the star, casting most of the star's material into space. Remaining behind will be the degenerate core of the star, by now probably a neutron star, which we might detect as a pulsar.

Since very massive stars are rare, supernovae are rare. Only three have been seen and well documented in our galaxy in the past thousand years. Several dozen have been seen in other galaxies, but telescopes are required to find them. The most famous supernova in our galaxy was seen by Oriental astronomers on what would correspond in our calendar to July 4, 1054. They called such apparitions "guest stars." So bright did this supernova become that it was visible for over a year, and it was bright enough for a few weeks to be seen in the daytime if one knew where to look.

The next supernova was seen in 1572 and is called "Tycho's star" after Tycho Brahe (1546–1601), the Danish astronomer who made careful observations and published a book about it. See Figure 6–20. This was an important factor in refuting the ancient idea of the universe which held that the sky was immutable. The next supernova appeared in 1604 and is known as "Kepler's star." None has been seen since then.

We can only theorize and try to imagine the appearance and effects of a nearby supernova. One group of scientists has even proposed that a nearby supernova far back in prehistory was responsible for the dinosaurs dying out, as a result of the intense burst of sterilizing radiation!

Although only three supernovae have been seen, another eleven which occurred long ago have been detected by radio telescopes. A supernova leaves behind it a large expanding cloud of energetic gas, like the Crab nebula. By looking at known supernova remnants with radio telescopes, and knowing what to look for, radio astronomers have found others. All are less than about 2000 years old, which gives us a rough idea of how often we might observe one. We might expect a supernova every 250 years or so. Since it has been much longer than that since 1604, we might see one any night now.

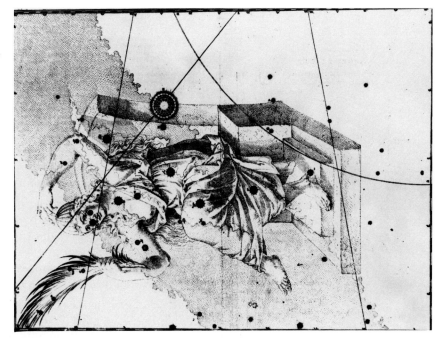

FIGURE 6–20
In 1572 the Danish astronomer Tycho Brahe studied a "new" star which appeared in the sky in the region of the constellation Cassiopeia. This was the first supernova recorded in modern times, and helped contradict the classical idea that the heavens were immutable. In 1603, Johannes Bayer compiled the first modern star atlas and included in it this map of the region of Cassiopeia, showing the supernova as the brightest object in the area, even though it was invisible by that time.

OTHER UNUSUAL STARS

In addition to the stars discussed above, most of which are in advanced stages of evolution (the T Tauri variables are the big exception), there are other stars with unusual properties. These are stars with huge extended shells of gas around them, others which emit X rays or gamma rays, and stars with chemical make-ups obviously different from most stars. Because of their strangeness, these unusual stars are both difficult and fascinating to study. More data need to be accumulated about them. Many of these facts will be obtained from observations made outside the atmosphere where we can get data from portions of the spectrum not accessible from Earth's surface.

Astronomers have uncovered a wide variety of stars, most of which fit into a general scheme of the formation, life, and death of stars. As more and better data are obtained, more pieces of the puzzle will fall into place, filling in some of the gaps that are still present. Nevertheless, we know in a general way the biography of a star.

REVIEW

1 The location on the Hertzsprung-Russell diagram of stars which have just begun to convert hydrogen into helium is called the zero-age main sequence (ZAMS).

2 An astronomer makes a mathematical model of the interior of a star, collecting observed data on mass, size, temperature, and composition, and applying the laws of physics.

3 The most important property of a star is its mass. This determines its life history.

4 More massive stars have higher luminosity and shorter lives than lower mass stars.

5 When a star has exhausted the hydrogen in its core, the core contracts and heats, and the outer layers expand and cool. The star becomes a red giant. The core will continue to contract until the central temperatures are high enough to cause the conversion of helium to carbon.

6 The nuclear reactions in the Sun give off neutrinos, which travel immediately out of the Sun. The rate at which neutrinos are detected on Earth should enable us to measure the internal temperature of the Sun, since the rate of the nuclear reactions depends on temperature.

7 After using all possible nuclear energy sources, a star contracts to become a white dwarf or other very dense object.

8 Most stars go through unstable portions of their later lives in which they vary their light output.

9 The most important type of variable stars is the Cepheids. There is a relationship between their average brightness and period of pulsation, called the period-luminosity relation. This enables us to use these stars as measuring devices.

QUESTIONS

1 Suppose you discover a Cepheid variable in a distant group of stars. Explain in detail how you would go about finding the distance to the group. Are there any possible problems you might encounter?

2 Explain why a more massive star will live a shorter time than the Sun.

3 When a Hertzsprung-Russell diagram of a large number of stars is plotted, the large majority of stars lie on the main sequence. Some are

in the red giant region, and some are white dwarfs. Between the main sequence and red giant branch is a region of the diagram almost devoid of stars, called the Hertzsprung Gap. Using what you know about stellar evolution, explain why we do not see many stars on this part of the diagram.

4 Based on what you know about stellar evolution, which types of stars would be candidates for extraterrestrial life, and why?

5 List the major stages in the life of a star, and the main source of energy of the star during each stage.

6 Explain how a model of a stellar interior is constructed.

READINGS

Glasby, J. S. *Variable Stars*. Cambridge, Mass.: Harvard University Press, 1969.

Greenberg, J. L. "Dying Stars." *Scientific American*, January 1959.

Hewish, A. "Pulsars." *Scientific American*, October 1968.

Jastrow, R. *Red Giants and White Dwarfs*. New York: The New American Library Inc., 1967.

Kraft, R. P. "Pulsating Stars and Cosmic Distances." *Scientific American*, July 1959.

MULTIPLE STARS

One can see many pairings of stars in the sky without the aid of instruments, and with the development of the astronomical telescope many more have become visible. The question naturally arises whether these pairs of stars are merely accidentally aligned from our point of view but in fact are very far removed from one another, or whether they are truly physically associated. Sir William Herschel in the late 1700s assumed that most of these "double stars" were accidental, and he tried to find the relative distances to them. Instead, he found that in most cases the stars are in orbit about one another, or, more exactly, in orbit about a common center of mass. These are the true physical binary stars, by which we mean that they are gravitationally

A "double-double" star system. A pair of relatively close stars orbit one another, and the pair orbits another similar, more distant pair. Multiple star systems are common in space.

FIGURE 7–1
A photograph of a starfield along the Milky Way. Many close pairs of stars are seen, but it is not immediately obvious which are merely random associations and which stars are actually physically connected. (Yerkes Observatory)

associated with each other. The chance coincidences in direction of unrelated stars are called *optical doubles* and have no particular importance in astronomy. Physical pairs, triples, and larger groupings are, however, extremely important, for they enable us to ascertain sizes and masses of stars, information available in no other way (Figure 7–1).

THE LAWS OF ORBITS

Johannes Kepler (1571–1630), around 1610, studied the data on the positions of Mars left as a legacy by Tycho Brahe. With painstaking longhand calculations, Kepler derived, purely empirically, the laws of the orbit of one object about another. Although the two objects are actually in orbit about a common center of mass, it is usually more convenient and informative to think of one object in orbit about another. Whether it be Earth and Sun, Moon and Earth, or two stars, the laws by which they move are the same.

Which object we choose to call the primary and which the secondary are arbitrary, but we usually choose the larger or brighter member as the primary. When there is a large difference in mass, the more massive component is the obvious choice. Although we could speak of the Sun as orbiting Earth, it would be a bit ridiculous since the center of mass of the Sun-Earth system is inside the body of the Sun.

Kepler's descriptions of orbital motions, which have come to be called *Kepler's laws*, are the following:

1   The shape of the orbit of one body about another is an ellipse. The primary body is at one focus of the ellipse (Figure 7–2).

FIGURE 7–2
Kepler's first law states that the orbit of one object about another is an ellipse with the primary body located at one focus of the ellipse. The other focus is empty. An ellipse can be drawn by putting a loop of string around two pins (the foci) and drawing a curve, keeping the string taut.

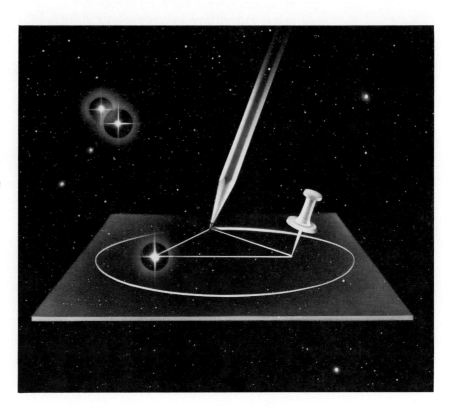

2 The line from the primary body to the orbiting body sweeps over equal areas in equal times (Figure 7–3).

3 The distance between the objects is related to the period of revolution by the equation $T^2 = kA^3$, in which T is the time for one revolution and A is the average distance between the bodies. k is a constant of proportionality. In the solar system, if T is in years and A in astronomical units, $k=1$. (An *astronomical unit* (a.u.) is the average distance of Earth from the Sun.)

These laws are strictly valid only for the case of two objects in orbit about one another. They are very good approximations in the case of the Sun and planets, since the Sun is so very much more massive than all the planets. In order to appreciate how these laws aid us in determining stellar properties, let us consider each one.

Kepler's First Law
The shape of the orbit is an ellipse with the primary body at one of the foci. An ellipse is a mathematical figure in which the sum of the distances from a point on the ellipse to the two "foci" is always the same. An ellipse may be easily drawn by placing two tacks in a

FIGURE 7–3
Kepler's second law states that the line from the primary body to the secondary sweeps over equal areas in equal times. Thus, if the object moves from 1 to 2 in the same time it takes to move from 3 to 4, the areas swept over are equal. This means that the object moves fastest when near the primary.

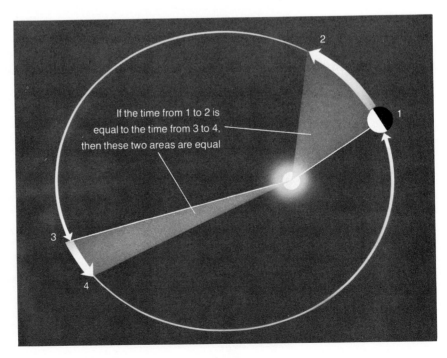

If the time from 1 to 2 is equal to the time from 3 to 4, then these two areas are equal

board, making a loop of string, placing it over the tacks, and stretching it out with a pencil. Now move the pencil about, keeping the string taut. The resulting figure is an ellipse, and the two tacks are at the two foci. In an orbit, the primary (more massive) object is at one focus, the other focus is empty, and the orbiting body travels around the curve. A circle is a special case of an ellipse in which the two foci coincide.

The long dimension of the ellipse is called the *major axis,* and half of it is the *semimajor axis,* designated as *A.* This is the average distance between the two objects. The short dimension is called the *minor axis.* If the tacks are both at the same place, the curve is a circle. The farther apart the two tacks are placed, the more the curve is flattened. This flattening is called the *eccentricity* of the orbit. Thus, the semimajor axis and the eccentricity, *e,* give the size and shape of the orbit.

Kepler's Second Law

The line from one object to another sweeps over equal areas in equal times. This law turns out to be the law of conservation of angular momentum, one of the basic laws of the universe. Conservation of angular momentum is what makes an ice skater spin faster when he pulls his arms in closer to his body.

By looking at Figure 7–3, you can see that Kepler's second law says an object must move slower when far from the star and faster when nearby. Thus, the speed in orbit is not constant. The effect of this law is most forcefully demonstrated by comets in our solar system. Their highly eccentric orbits carry them far away from the Sun, at which distances their speeds are low. Thus, they spend most of their time in the depths of the solar system, and only a short time near the Sun, where their speeds may be over 100 000 kilometers per hour.

Kepler's Third Law

The square of the period of revolution is proportional to the cube of the average distance. Recall that this average distance is the semimajor axis of the orbit, which we called *A.* The period, or time for one revolution, we will call *T.* It is easy to confuse the two and to forget which is squared and which cubed. (A mnemonic is the following: think of the intersection in New York City of Broadway, 42nd Street, and Seventh Avenue: Times Square.) In symbolic terms, this law is $T^2 = k \times A^3$, and k is the constant of proportionality. In the solar system, if A is in astronomical units and T in years, $k = 1$, so $T^2 = A^3$. For double stars, k is different, depending on the mass of the

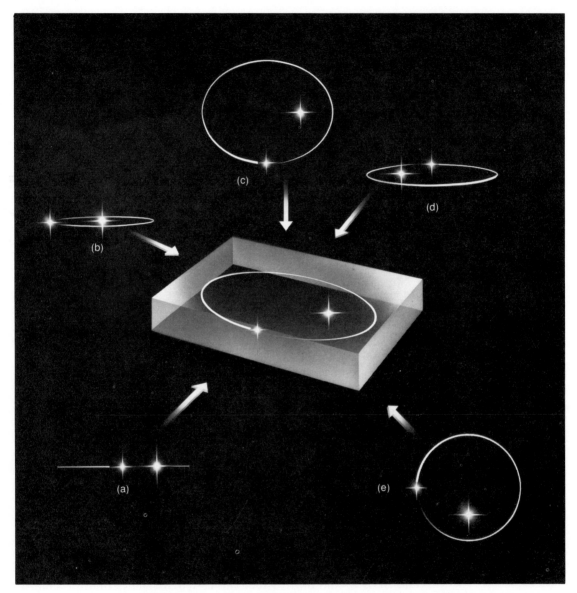

FIGURE 7–4   We usually do not observe the true orbit of a double star.
The observed orbit depends on our angle of view. The true orbit is seen
in the center. Only from (c) would we observe the true orbit. In other
cases we would observe an ellipse, but the primary star would not lie
at a focus. This is the clue that enables us to deduce the true orbit.

stars. If M_1 and M_2 are the masses of the two stars in a binary system, and G is the gravitational constant, then

$$k = \frac{4\pi}{G(M_1 + M_2)}$$

It is this law which allows us to determine the sizes and masses of binaries.

THE ORBIT

When two objects are in orbit about each other, both orbits are in the same plane. The plane of the orbit, as seen from our vantage point on Earth, may have any orientation. Some orbits we may see as if we were looking straight down on the pair of stars, some from right along the plane of the orbit, and most of them we see inclined at some intermediate angle. This means that we don't observe the true path of the companion about the primary, but only a projection. A tilted ellipse still looks like an ellipse, but an ellipse of different size and shape, and, importantly, the focus of the ellipse (that is, the primary star) looks out of place. It is this last feature which enables us, with a lot of mathematics, to determine the true orbit from what we observe (Figure 7–4).

Most double stars take years, sometimes hundreds of years, to complete one revolution. Thus, many observations over a period of time are needed to give enough information on the orbit. Each observation consists of two numbers: the *separation*, that is, the distance observed between the two stars, usually in seconds of arc, and the *position angle*, the direction of the companion from the primary. Many such observations are plotted to give the apparent orbit.

We may also use another piece of information if it is available —the radial velocity of each star. Here we use the Doppler principle, looking for shifts in the spectral lines. Each star participates in two motions: (1) the motion around the center of mass, which we will observe as a varying radial velocity as the star sometimes comes toward us, sometimes goes away, as it travels around in its orbit; and (2) the motion of the whole double star system through space, which is constant (Figure 7–5).

For example, if we observe that one star of a binary system has maximum and minimum radial velocities of 40 and 60 kilometers per second, we easily determine that the average of these two numbers, or 50 km/sec, is the radial velocity of the system as a whole.

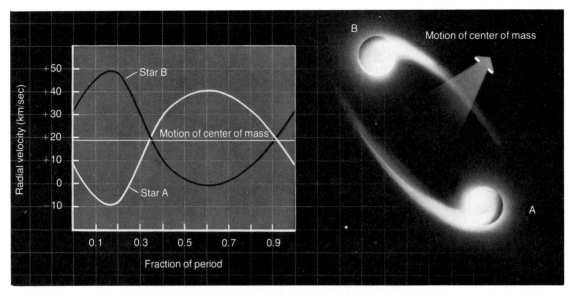

FIGURE 7–5   *On the right is a double star system. The two stars are in orbit about the center of mass, while the entire system moves through space. We cannot directly observe the motion of the center of mass, but we can deduce it by graphing the motions of the individual stars, as on the left. If the two stars are alike, the curves will be mirror images of each other.*

The difference between this average and the extreme values, which is 10 km/sec, is the orbital speed of the star around the center of mass. If the secondary star varies between 30 and 70 km/sec, we know its orbital speed is 20 km/sec, with, of course, the same center-of-mass speed as the system—50 km/sec. The laws of physics tell us that the speeds in orbit will be inversely proportional to the masses of the stars. Thus, the primary, which has an orbital speed half that of the companion, is twice the mass. Already we have obtained the ratio of the masses of the two stars. In actual practice, the situation is more complicated, but the principles are the same.

About 70 000 multiple star systems are known. Many of these are sufficiently separated to enable us to use analyses like the one just discussed to get some information about the properties of the system. Often, however, we are able to observe only one star, and its path is wavy in space. Recall the situation of Sirius A and Sirius B discussed earlier. Sirius A was found to have a wavy path, and no companion star was seen until much later. The reason for this was that the companion was much less luminous than Sirius A, and very close to it. In many double stars we have not seen the companion at all, and do not expect to. The double nature of these stars is deduced only from the path of the brighter star. Such stars are called *astrometric binaries* since the path of the star is determined by astrometry, the branch of astronomy dealing with the precise measurement of positions and motions of stars.

The second nearest star, Barnard's star, was observed to be an astrometric binary. Very careful analysis showed that the wavy motion of the primary could best be explained by the effects of two Jupiter-mass objects in orbit about it. If so, this would be the first detection of planets outside our solar system. Other researchers have also studied the system, and find no evidence for a wavy motion, and thus no planets. At the present time, the question has not been resolved, but it shows very well the possibility for disagreement when the effects being looked for are very small.

If in a pair of stars one is rather close to the other, their speeds in orbit will be high, and we may be able to detect this by studying the spectrogram of the stars. The stars will be so close together in the sky that they cannot be seen as two stars, only as a single point of light. If the radial velocity of the spectral lines varies, we deduce that this is indeed a double star. If the two stars are about the same luminosity, we will see spectral lines from both stars, and they will be displaced in opposite directions, one coming toward us and the other going away. Usually the two stars are of different luminosities, and the light from the brighter washes out the other. In this case we see only the lines of the brighter star, but the fact that they vary reveals the double nature of the star. In either case, these stars are called *spectroscopic binaries*. The terms "double-line" and "single-line" are used to indicate whether we see spectral lines from both stars or just one. Figure 7–6 shows the spectra of single- and double-lined spectroscopic binaries.

ECLIPSING STARS

In 1783, a young English amateur astronomer named John Goodricke noticed that the star Beta Persei, or Algol, varied its light output. Every 69 hours it dimmed for a few hours, then resumed its normal brightness. Although the variation of Algol had been casually noticed over a hundred years before Goodricke saw it, he was the first to study the variation systematically and to suggest an explanation. (*Algol* means "the demon" in Arabic, suggesting that Middle Eastern astronomers knew of the variability.)

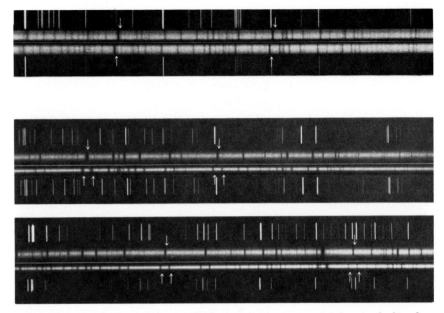

FIGURE 7–6   The top photograph shows the spectrum of the single-lined spectroscopic binary Alpha-1 Geminorum. Notice the shifts in the spectral lines. The bottom photographs show the spectrum of the doubled-lined spectroscopic binary Kappa Arietis at times when the lines from the two stars are together and when they appear as two lines. The two stars in this system must be about the same luminosity for both lines to appear.

Goodricke proposed the theory that Algol is really two stars, one much brighter than the other, and that we happen to lie nearly in the plane of the orbit of one around the other. Because of this, periodically one star gets in the way of the other, and the total light output from the two is dimmed for the duration of the eclipse. The two stars are too close together to be seen as two stars, but the light variation, which is unlike that of any other type of variable star, reveals the dual nature of what looks to us to be a single point of light. These are known as *eclipsing binary stars*.

A typical *light curve* (the variation of brightness with time) is shown in Figure 7–7. The biggest decrease in light occurs when the star of greater surface brightness is eclipsed. This is so because in each case the same amount of surface area is hidden from view. Actually, the star with the lesser surface brightness might be (and often is) brighter overall owing to its large size.

The detailed shape of the light curve enables us to deduce certain properties of the system. For instance, we can easily measure the period of revolution and the duration of the eclipses. This yields information on the sizes of the stars relative to the size of the orbit. This information, combined with information on the actual size of the orbit derived from applying Kepler's laws to the pair, leads to actual sizes for each of the stars. In this way we directly determine stellar radii, one of the important pieces of data needed for studies of stellar properties.

If we can observe the variations in radial velocities of the two stars from their spectra, we can determine the lower limit of the mass of each star. Combining these data with eclipsing data, we can find the actual mass. Thus, eclipsing binary stars are just about our only source of fundamental data on stellar properties.

Checking Theories of Stellar Evolution

Presumably the binary stars were formed at the same time from the same cloud of interstellar matter. The only thing which should cause a difference in their evolution is the difference in mass. We therefore have a useful comparison for our studies of stellar evolution. All this assumes that the two stars do not exchange matter, and in most cases this is true, since the separations are on the order of hundreds of astronomical units. In some cases, though, the two stars

FIGURE 7–7   The eclipsing binary Algol (Beta Persei). The light curve on the left gives information about the star's size and orbit. Primary eclipses occur when the star of greater surface brightness is behind the other.

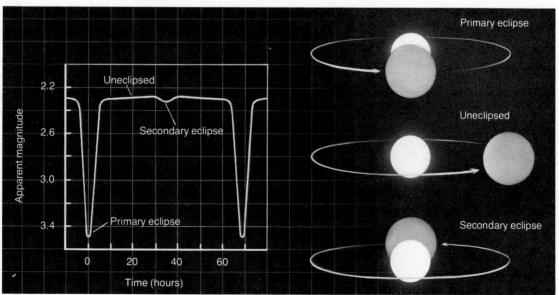

are so close together that they influence each other, each one raising significant tides on the other, distorting the shape from that of a sphere to an ellipsoid. They may even pull matter out of each other. These are known as *contact binaries.*

CONTACT BINARIES

Contact binaries are strange systems. The stars are quite distorted in shape, and this means that most of our assumptions about their evolution are not as valid as they were for spherical stars (see Figure 7–8). If these stars are sufficiently large, they may exchange matter, and some of it may escape into space, producing a spiral of gas around the system. This is detected by the fact that the spectrum shows bright lines from the gas, as well as the dark-line spectrum due to the stars.

Several such systems are known, but the most famous is Beta Lyrae. Refer back to the illustration on page 132. The brighter star is of spectral type B. Orbiting it is a cooler star, which is gaining mass at the expense of the B star. Surrounding the smaller star is a ring of gas, and some gas escapes into a large spiral surrounding the whole system. Many questions remain to be answered about this pair of stars, and research continues to perfect a model of the Beta Lyrae system. Otto Struve, a famous astrophysicist who died in 1963, spent

FIGURE 7–8 *Most binary stars are separate (a). If one expands it will reach a point where it stops growing but ejects material toward the other star (b). This is known as a semidetached binary. If both stars fill such volumes, the system is called a contact binary (c). Contact binaries exchange mass, which drastically alters their evolution.*

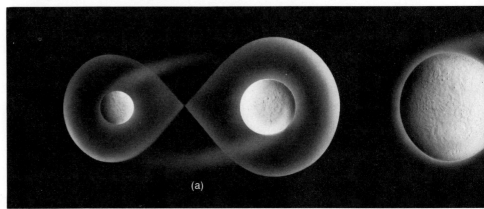

(a)

much of his life studying this system. It is thus sometimes called "Beta Struve" in jest.

An exchange of matter between close binaries may alter their evolution drastically. The more massive star will, as we have seen, evolve faster, and swell into a red giant first. But in swelling, it may reach a size large enough to transfer some of its material onto the other star. The less massive star may then become the more massive, and evolve faster, finally swelling to become a red giant as well. The mass-transfer process may reverse, perhaps several times. The detailed effects of such transfers of mass back and forth have not yet been worked out.

Sirius A and B are an interesting case where this may have happened. Sirius A is an A-type star, on the main sequence. Sirius B is a white dwarf, which, as we know, means it has gone through most of its life. Since the stars would have been formed at the same time, the white dwarf must have been the more massive originally. Today, however, it is less massive than Sirius A. Either it lost mass somewhere along the line, or it exchanged mass with Sirius A.

Mass transfer in close binary systems is thought to proceed relatively quietly in most cases. It is known, though, that several stars which became novae are members of binary systems. The question occurs whether all novae may be binaries. Astrophysicists are working on models to show how a large mass transfer from one star to another might trigger an explosive increase in the fusion rate of a star. Many more data need to be gathered before this question is settled. In support of it, however, is the observation that all or almost all stars of the class of eruptive variables called U Geminorum stars seem to be binaries, as are most stars which seem to have circumstellar clouds of gas.

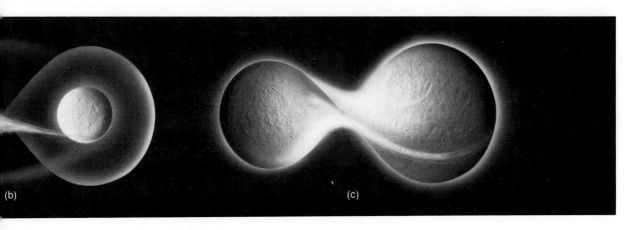

(b)

(c)

MULTIPLE SYSTEMS

Often more than two stars will be gravitationally bound to one another, forming a triple, quadruple, or higher order system. One famous example is the double-double Epsilon Lyrae. The unaided eye may see two stars, very close together. A small telescope clearly separates the pair of stars, and a larger instrument discloses that each of the pairs is itself double. Further, one star of each pair is a spectroscopic binary, so that this system is really six stars, all moving about one another.

The motion in such a system is quite complicated and no longer obeys Kepler's laws, which hold only for two objects. The motion of such systems must be calculated step by step, using computers, for there is no simple formula giving the stars' positions for any time in the future.

Another well-known triple system is Alpha Centauri. This is a system dominated by a star of spectral type G2, identical to our Sun, plus a companion of type K5. These two orbit relatively close to one another (although they are far from being a contact binary). Orbiting this pair is a faint M5 star, thought to be slightly closer to the Sun than the other two, making it the nearest star to our solar system. For this reason it is known as Proxima Centauri. Figure 7–9 shows this system and how the Sun would appear from there.

Observations have shown that many, perhaps a majority, of the stars have one or more companions. Of the 45 nearest star systems, 22 have, or are suspected to have, more than one star. In terms of single stars, there are 67 in the group, of which 41 are in multiple systems. Binaries are particularly prevalent for stars on the upper main sequence. At least 50 percent of the O and B stars are binaries. Many of these very luminous stars also rotate about their axes very fast. It seems that angular momentum of a star or star system is very important for its evolution, and for determining the possibility of companions.

IMPLICATIONS FOR EXISTENCE OF PLANETS

We know that the range of masses for binary stars is large, as is the range of the ratio of the mass of the smaller star to the total mass of the pair. Thus, we might extrapolate downward to the case of a stellar companion which was of too low a mass to start nuclear reactions, about $\frac{1}{100} M_\odot$ and even further to, say, $\frac{1}{1000} M_\odot$ comparable with the mass of Jupiter. Thus, we have bridged the gap from stars to planets.

Other evidence which supports the existence of planets is the matter of stellar rotation. It is observed that early-type stars rotate faster, while stars later than spectral type F5 rotate so slowly that we cannot detect the rotation (the Sun rotates with a speed at its

FIGURE 7–9   *The multiple star system Alpha Centauri, the nearest to the Sun. A G2 star (right) and a K5 star orbit one another, while a small red dwarf orbits them in the distance. This red dwarf, Proxima Centauri, is the nearest star to the Sun, which is seen in the upper left portion near the constellation Cassiopeia as seen from Alpha Centauri.*

equator of about 2 kilometers per second, much below what we could detect were we astronomers on a distant planet). Why there should be such an abrupt change in rotational speed at this spectral type has been a mystery. To explain it, the suggestion has been made that for stars below the mass corresponding to spectral type F5 (about 1.25 $M_\odot$), planets are formed instead of a companion star to take up the angular momentum. Something must slow down the rotation of the protostar; it could be strong magnetic activity due to flare-ups before the star reaches the main sequence. The theory checks out well with our ideas of stellar evolution, if we recall that the variable stars, known as T Tauri stars, show irregular eruptions and are believed to be stars not yet on the main sequence.

It is likely that there is a strong relationship between double stars and stars with planetary systems. Conceivably, there may be planets around multiple stars, but this is unlikely since their orbits would be unstable and the planets would eventually be thrown out of the system by interactions of the stars. It is likely that many, and perhaps most stars later than F5, have planetary systems. If so, there may be planets of the right size, composition, and distance from the parent star to be hospitable to life. We may conjecture that somewhere out there are other observers, wondering if there is life anywhere else.

REVIEW

1. Over one-half of all observed stars have one or more stellar companions.

2. The orbit of one object about another is an ellipse with the primary body (which is being orbited) at one focus of the ellipse (Kepler's first law).

3. The line from the primary object to the secondary object sweeps over equal areas in equal times (Kepler's second law). This means that the orbiting (secondary) body moves faster when closer to the primary body.

4. If T is the time for one revolution and A the average separation of the two objects, then $T^2 = kA^3$, where k is a constant whose value depends on the units of measurement used (Kepler's third law).

5. Occasionally we happen to be in the plane of the orbit of a double star so that the stars eclipse each other. Such eclipsing binaries yield data on the sizes and masses of stars.

6   Stars of spectral type later than F5 rotate very slowly whereas more massive stars have fast rotations.

QUESTIONS

1   We observe a binary star in which the radial velocity of the primary varies between +30 and −20 km/sec. The radial velocity of the secondary varies between +15 and −5 km/sec. What are (a) the radial velocity of the center of mass, (b) the orbital speeds of the two objects with respect to the center of mass, and (c) the ratio of the masses of the two stars?

2   Suppose we observe a pair of nearby stars in the sky, but can detect no evidence of motion of one about the other. How could you decide whether they are not associated with one another, or they are associated but have a period of revolution too long to produce a noticeable motion? Can we make such a determination if we cannot find their distances?

3   An observation of a double star consists of a measurement of the angular separation in seconds of arc between the two stars, and the position angle. What additional piece of information is needed to convert the angular separation into a knowledge of the true separation between the stars in a.u.?

4   Explain how observations of eclipsing binaries yield information about stellar properties.

READINGS

Aitken, R. C. *The Binary Stars.* New York: Dover Publications, Inc., 1964.

STAR CLUSTERS

CHAPTER 8   When clouds of interstellar matter contract to form stars, often
more than one star is formed. Occasionally a large group of stars
is formed from the same cloud. The stars are too far apart to be
classified as multiple star systems, but each star is gravitationally
bound to the others, at least temporarily. These groups we call
star clusters.

The two types of clusters are named according to their form:
globular clusters and open clusters. A third type, the association,
is a variety of open cluster. The two types are found in different
parts of the Galaxy, contain different types of stars, and move
around the Galaxy quite differently from each other. Since stars
in a cluster were presumably formed at about the same time,

The Pleiades star
cluster, often called
the "Seven Sisters."
Six stars are visible to
the average unaided
eye, but the cluster
contains several
hundred stars. (Lick
Observatory)

evolutionary effects which depend on the mass of a star should show up for cluster stars of different mass. Because of this excellent laboratory in which to test our theories of stellar evolution, star clusters have been studied extensively and play an important role in our attempts to describe the biography of a star.

GLOBULAR CLUSTERS

Globular clusters are so named because they appear as roughly spherical groups of stars. Sometimes they show a slight flattening in one direction, probably caused by rotation of the cluster, but in all cases they show a high degree of symmetry and compactness. Over a hundred globular clusters are known, ranging in diameter from about 60 light-years to about 300 light-years, and containing from 10 000 to several hundred thousand stars. They are the largest groups of stars smaller than galaxies.

Globular clusters are found in the nucleus and halo of a galaxy, that roughly spherical volume surrounding the plane of the galaxy. (See Chapter 9.) They were presumably formed from the primordial material from which the galaxies themselves condensed, and hence represent a sample of this unaltered material. The density of stars in a rich globular cluster is many times the average density in the galaxy. In a cubic light-year near the cluster's center there may be more than 30 stars. In the vicinity of the Sun, that many stars would be spread over a volume of several thousand cubic light-years.

A good example of a globular cluster, and one which is just barely visible to the unaided eye, is the globular cluster in Hercules, known as M13. The eye sees a faint point of hazy light, a small telescope reveals that it is composed of stars, and a long photograph with a large telescope reveals the shape shown in Figure 8–1. The stars are so numerous toward the center that they blur together in the photograph, although really they are far apart. Remember that we are looking through the entire cluster, and this makes the density seem higher than it really is. The individual stars in a globular cluster move along complicated orbits, each star perturbing the orbit of all the others. Thus we can no longer speak of simple, elliptical paths. When describing the motions within a cluster, we must use average motions for large numbers of stars. We find in this way that it is possible for a cluster to eject stars, usually the lower mass members. This would explain some of the occasional single stars observed in the halo of the galaxy.

Globular clusters do not participate in the general rotation of

FIGURE 8–1
One of the best known and most easily observed globular clusters, M13 in the constellation Hercules. On a very clear night this can be seen by the unaided eye as a faint, fuzzy "star." The cluster is about 20 000 light-years away and contains several hundred thousand stars. (Hale Observatories)

the rest of the galaxy. Each is following an orbit around the nucleus of the galaxy which is often highly elliptical. The clusters must pass through the plane of the galaxy twice each revolution. Even though the density of stars in a cluster is relatively high, and the density of stars near the galactic nucleus is also high, the chance of much interaction between the cluster and the rest of the galaxy is small.

Because most globular clusters are above or below the plane of the galaxy, our line of sight to them is unobscured by the interstellar matter which reddens and dims the objects we observe near the galactic plane. Thus, except for a few clusters near the nucleus of the Milky Way, we have probably discovered most of the globular clusters associated with our galaxy. We can observe globular clusters surrounding nearby galaxies, and they have the same distribution about those galaxies as the globular clusters in our galaxy do about the Milky Way. (At a great distance, a globular cluster would look like a single star of absolute magnitude −5 to −10, about the same brightness as the very brightest ordinary stars in a galaxy. We can tell that they are clusters, however, from their location and their spectra.)

Globular clusters are old. The stars in them were formed long

FIGURE 8–2
The globular cluster
Omega Centauri,
about 13 000 light-
years away. (Yerkes
Observatory)

ago, as the galaxy was condensing. Hence, all interstellar matter has been used up, and no new stars are being formed. Any gas that was thrown off by the stars has been swept out of the cluster during its passages through the plane of the galaxy. The clusters are so old that all stars except the very lowest mass stars have evolved off of the main sequence onto the red giant branch, or even onto the super-giant region. Thus, the brightest stars we see in globular clusters are red. All the former bright blue, high-mass stars have left the main sequence long ago.

Spectra of globular cluster stars show that the stars are deficient in metals. This is to be expected if the original material from which these stars formed was mostly hydrogen and helium. The metals (that is, anything besides hydrogen and helium) are thought to be produced mainly in supernova explosions. The metals enrich space and become the raw material, which then goes into forming more stars. Star formation in globular clusters stopped before the interstellar gas became enriched. Thus, globular clusters enable us not only to study the original material out of which the galaxy formed (which may be the primordial material of the universe), but also to study the evolution of stars having greatly different chemical composition from that of stars near the Sun.

FIGURE 8–3
Two of the best-
known open clusters.
(above) Praesepe, or
"Beehive" cluster,
located in the
constellation Cancer.
(below) Pleiades
cluster, in the
constellation Taurus.
The Pleiades cluster
is about 100 million
years old, and
Praesepe is about 1
billion years old.
(Yerkes Observatory)

OPEN CLUSTERS

Open clusters are loose congregations of stars, such as shown in Figure 8–3. They show little or no suggestion of overall shape, and they vary from a few dozen up to more than a thousand stars, in a volume usually less than about 30 light-years on a side. About a thousand open clusters are known so far, and there are probably hundreds of thousands scattered throughout the Milky Way.

Open clusters are also known as *galactic clusters* because we find them strongly concentrated in the plane of the galaxy. The ones so far discovered are those close enough to be distinguished from the background of distant stars. It is often a problem to tell whether a group of stars on a photographic plate is really a cluster, or just a chance alignment of distant stars and foreground stars. One way of distinguishing is to study the spectra of suspected cluster members and obtain radial velocities (of course, this can be done only for those stars bright enough to yield high-quality spectra). Stars in a cluster, which are bound gravitationally to one another and moving more or less together in space, will all have roughly the same radial velocity. A suspected cluster star which has a radial velocity significantly different from the average of the cluster stars is obviously

FIGURE 8–4
The Hyades star cluster. This famous group forms the "face" of Taurus the Bull. The bright star Aldebaran is not a member of the cluster, which is about 130 light-years away. (Warner-Swasey Observatory)

not a member. Open clusters are such loose groupings that it is often difficult to tell where the cluster stops.

Many open clusters are visible with the unaided eye, or with small telescopes. The Pleiades (Figure 8–3b) and Hyades (Figure 8–4), both in the constellation Taurus, are the most famous. Others are the Beehive in Cancer (Figure 8–3a) and the "Double Cluster" in Perseus (Figure 8–5).

Because of the strong concentration of these clusters in the plane of the Milky Way, they are usually associated with much interstellar matter, often imbedded in a nebula. For example, the Pleiades are associated with a reflection nebula, which shows a reflected spectrum of the stars in the cluster.

Open, or galactic, clusters are at least *second-generation stars,* formed from interstellar clouds enriched in metals by previous generations of stars. They are stars rather like the Sun, and, depending upon the age of the cluster, the brightest stars may be either red or blue. If the most massive stars have been around long enough to evolve off the main sequence, then the brightest stars will be red. If the massive stars are still on the main sequence, the brightest stars will be blue. Later in this chapter, where we discuss the H-R diagrams of clusters, we will see that this gives us a way of estimating the age of a cluster.

FIGURE 8–5
The Double Cluster,
h and χ Persei. These
clusters are only
about 5 million years
old. (Yerkes
Observatory)

FIGURE 8–6
A "poor" open
cluster, M11. Clusters
such as this one are
sometimes hard to
distinguish from the
general stellar back-
ground. (Yerkes
Observatory)

Some aggregations of stars are so loose that it is almost too much to call them clusters. These are mostly very luminous O and B stars. We choose to treat these groups as somewhat different objects, called associations.

FIGURE 8–7
The open cluster M2
in Aquarius. (Yerkes
Observatory)

ASSOCIATIONS

An *association* is a very loose group of stars which often seem to have a common origin but are only weakly bound gravitationally. Actually, these associations are usually flying apart, since the density of stars is too small to keep them together. Two kinds of associations are usually recognized. The most apparent are the OB associations, which contain very luminous O- and B-type stars. The other type is the T association, whose members are T Tauri stars, of rather low luminosity.

Associations contain from a dozen to a hundred stars and are always found with interstellar clouds of gas and dust (Figure 8–8).

FIGURE 8–8
The Orion association with the Orion nebula, M42. The young, hot stars excite the gas to glow. This is an extremely young group of stars, recently formed from the interstellar medium. Observations indicate that star formation is still going on in the cloud. (Lick Observatory)

They have a special place in our theories of stellar evolution, since they must be very young objects, recently formed from the interstellar clouds. O and B stars have short main-sequence lifetimes, and thus were formed within the past few million years, a very short period in astronomical terms. T Tauri stars are believed to be stars which have not yet reached the main sequence. They are variable in light in erratic ways and are also always found with gas and dust in space.

We can measure the motions of a few associations, and we find that the member stars are expanding—that is, the association is dissociating. The size of an association may range from 10 to 100 light-years, and probably a group of stars larger than this would not be recognizable as being connected or having a common origin. Like all very young objects in the Galaxy, they occur in and very close to the disk of the Galaxy, concentrated in the spiral arms. Because of the high luminosity of O and B stars, we can see them at great distances from the Sun. Somewhat less than a hundred are known so far. Many more undoubtedly exist, obscured from our view by interstellar dust. Their proximity to interstellar clouds means that we often see them imbedded in HII regions, clouds of

FIGURE 8–9
The spiral galaxy M74. Many of the starlike objects tracing out the spiral arms in this galaxy are young open clusters, associations, and HII regions.
(Lick Observatory)

gas ionized by the flood of ultraviolet radiation from the hot stars within.

The total light output from an association may be a million or more times the luminosity of the Sun. Thus, we can see associations in other galaxies as well, such as the spiral galaxy M74 seen in Figure 8–9. They have been used to estimate extragalactic distances and sizes. The stars in associations are the youngest objects in the galaxies in which they are seen.

CLUSTER DIAGRAMS

A star cluster presents an opportunity to study the evolution of stars. As before, we can do this with the aid of the H-R diagram. For clusters, we actually have a slightly easier problem, since all the stars in a cluster may be thought of as being at the same distance from us. Thus, the distance modulus, $m-M$, and the interstellar absorption and reddening are, at least to a first approximation, the same for all the stars. Therefore, we do not have to derive their actual absolute magnitudes, since any differences in apparent magnitude reflect differences in absolute magnitude. Further, it is quicker and easier to obtain color indices for the cluster stars than to get spectra, and so we usually use the color index as a substitute for detailed spectral classification. The graph we obtain is a plot of apparent magnitude versus color index, and is called a *color-magnitude diagram* rather than an H-R diagram. The two terms are practically interchangeable, so long as we remember what is being plotted. Figure 8–10 shows a typical color-magnitude diagram for an open cluster and a comparison of many clusters.

First, notice that most of the stars lie along a rather narrow band, which we call the zero-age main sequence, or ZAMS, since it is here that the stars appear which have not evolved far from their initial locations on the diagram. Also note that a few of the upper main-sequence stars are not on the ZAMS, but are slightly to the right of it. This can be attributed to the fact that these massive stars have just left the main sequence and are evolving toward the red giant branch. In some open clusters, a few stars have already reached the red giant stage. A few stars near the lower main sequence are above the main sequence. These are stars which have not reached the main sequence, stars still in the process of forming.

Notice in Figure 8–10b that each cluster has a different point at which its stars begin to turn off the main sequence. The *turn-off point* enables us to estimate the age of the cluster. Since all the stars were formed at once (astronomically speaking), the only evolutionary differences will be those due to different masses. We can see which stars are just beginning to turn off the main sequence, and, from our theories of stellar aging, we know how long a star of that spectral type will have been on the main sequence. This age must be the age of the entire cluster.

Recall Figure 6–5 on page 139 showing the post-main-sequence evolution of stars. The paths of the stars across the diagram would cause the main sequence to turn off, just as observed. The farther down the ZAMS the turn-off comes, the older the cluster is. Think of a candle wick burning; the farther down it has burned, the longer it has been burning.

The color-magnitude diagrams of open clusters differ quite a

FIGURE 8–10 *(a) If the magnitude and color index of each star in a cluster are measured, we can plot a color-magnitude diagram for the cluster. This graph shows a typical cluster. Note that some stars deviate from the zero-age main sequence, the ZAMS. Actually no cluster would show deviation at both ends of the main sequence. (b) A composite color magnitude diagram of many open clusters. Note that the bright stars in each cluster deviate from the main sequence. The point at which the stars turn off is an indication of the age of the cluster.*

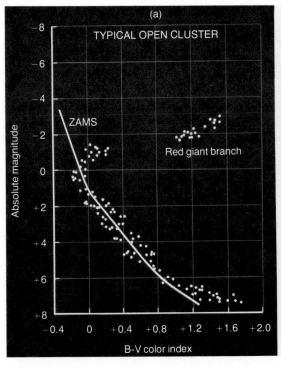
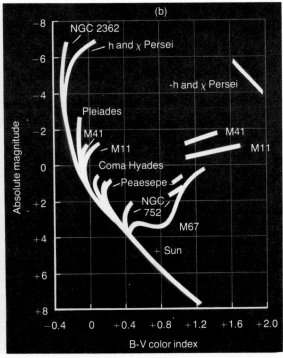

bit from one another depending on the age of the cluster. However, the diagrams of globular clusters are much the same as seen in Figure 8–11. The stars in a globular cluster are all old. The low-mass stars are still on the main sequence, but many have evolved off and are found in the red giant region, and even in other areas where stars are heading toward the white dwarf stage. The brightest stars on the main sequence are about absolute magnitude 3 or 4, that is, slightly brighter than the Sun. The red giants are somewhat brighter than the red giants found outside of globular clusters. Stretching downward and to the left of the red giant region is a region called the "horizontal branch." In the middle of this horizontal branch is a gap, called the *RR Lyrae gap*, since every star found here is an RR Lyrae variable star. This fits in well with our ideas of advanced stages of stellar aging.

The chemical composition of globular cluster stars is quite different from that of stars in open clusters. Stars of different com-

FIGURE 8–11 (a) A color-magnitude diagram of a typical globular cluster. Compare this with Figure 8–10(a). The differences are the result of differences in age and chemical composition. (b) A composite color-magnitude diagram of many globular clusters.

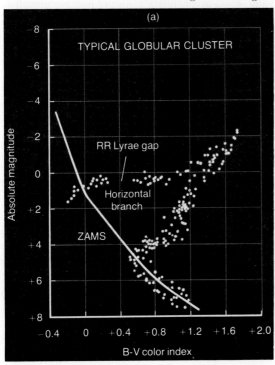

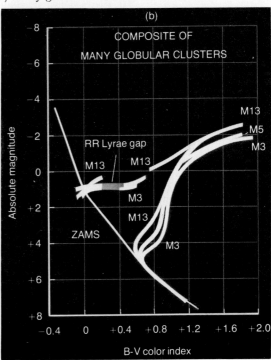

positions will evolve differently across the H-R diagram, so separate calculations and models must be made for various compositions and then compared with the observed color-magnitude diagram of a particular cluster. An approximate idea of the chemical composition can be obtained from the spectra of the cluster stars, but very slight differences in the amounts of metals which may change the evolutionary path may not be detectable in the spectra.

FINDING THE DISTANCE TO CLUSTERS

The best way to find the distance to a cluster is to use variable stars: type I Cepheids for open clusters, and type II Cepheids and RR Lyrae stars for globular clusters. Most globular clusters have one or more variables, but open clusters often have none at all. Thus, some other method must be found to determine their distance.

The fact that all the stars in a cluster are at the same distance from us provides the method. We can plot the color-magnitude diagram for a given cluster and compare it with the known relation between color index and absolute magnitude. For example, suppose a G5 main-sequence star in a cluster has an apparent magnitude of 10. From other observations, we know that a G5 star has an absolute magnitude of 5. This gives us the difference $m-M$, the distance modulus, of 5. The distance modulus, corrected for any interstellar absorption, gives us the distance by the formula on page 65. What we are really doing is sliding our cluster main sequence up and down vertically until it coincides with our theoretical or observational ZAMS. The amount we must slide it is just equal to $m-M$. The method of *main-sequence fitting* is the one most commonly used to get distances to open clusters.

Another method can be used only for very nearby clusters for which the proper motions of the stars are known. This is known as a *moving cluster distance*. The members of a cluster travel in space roughly parallel to each other. From our viewpoint in space, however, their paths will seem to converge owing to a perspective effect, rather like the way railroad tracks seem to converge in the distance. If we can determine the point of convergence, we can obtain the distance to the cluster. For an analogy, think of a flock of birds flying parallel to one another. The apparent paths of the birds, measured close to the observer, will seem to converge. The most important cluster for which this method is useful is the Hyades, shown in Figure 8–12.

STELLAR POPULATIONS

There are quite definite differences between stars found in open clusters and stars found in globular clusters. A list of some of these properties is given in Table 8–1.

Table 8–1 *Properties of open clusters and globular clusters*

	Open Clusters	Globular Clusters
Brightest stars	Blue or red	Red
Metal abundance	Rich in metals	Poor in metals
Age	Relatively young	Relatively old
Location	In spiral arms and disk	In halo and nucleus
Interstellar matter	Abundant	None or little

Because of these differences, it has been found useful to label such stars as to "type." The stars typical of open clusters are called *population I* and stars typical of globular clusters are known as *population II.* Of course, as in any classification, there are gradations in between. However, population II stars are older, redder, metal-poorer, and less concentrated in the plane of the galaxy than stars in population I. The concept of populations of stars has proved very useful as a general way of categorizing most stars.

One property not listed above is the star's motion, measured with respect to the Sun—that is, from our moving frame of reference. Since the Sun is a population I star, and participates in the general rotation of the Galaxy, other objects which have similar motions will seem to move rather slowly with respect to us. Thus, we may say that population I objects are low-velocity objects. On the other hand, population II objects, such as globular clusters and other stars in the halo of the Galaxy, do not revolve about the galactic center in the same way the spiral arms do. They seem to have a fast motion with respect to us. The orbits of stars of population I about the galactic center are roughly circular, whereas those of population II are highly elliptical. Figure 8–13 shows why they have high speeds inward or outward from the center. (A similar situation occurs in the solar system: the planets have relatively slow speeds with respect to one another, but the comets in highly elliptical orbits move quite rapidly with respect to the planets.)

FIGURE 8–12
The moving cluster method of determining the distance to an open cluster. The cluster must be close enough to show proper motions for its member stars. The stars are moving through space along parallel paths, which are the direction of motion of the cluster. If we could determine the angle between the motion of the cluster and the line of sight to the cluster we could, by simple geometry, find the distance to the cluster. This angle equals the angle between the line of sight to the cluster and the line of sight to the convergence point (main illustration). This, with a measurement of the radial velocity of the cluster, gives us the distance. The graph shows what is actually observed: an area of sky around the cluster, with small arrows indicating the proper motions of the individual stars. They seem to converge to a point. The angle from the cluster to this point is the angle needed in the calculation.

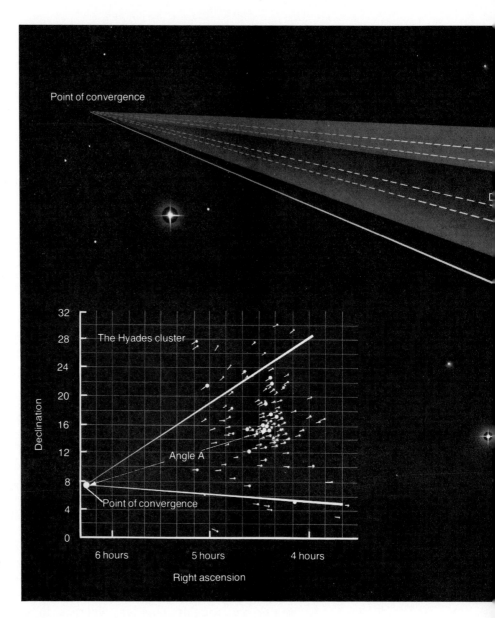

TESTING THEORIES OF EVOLUTION

Because of the common origin in time and space of the stars in a cluster, we can use them as comparisons with our computations of

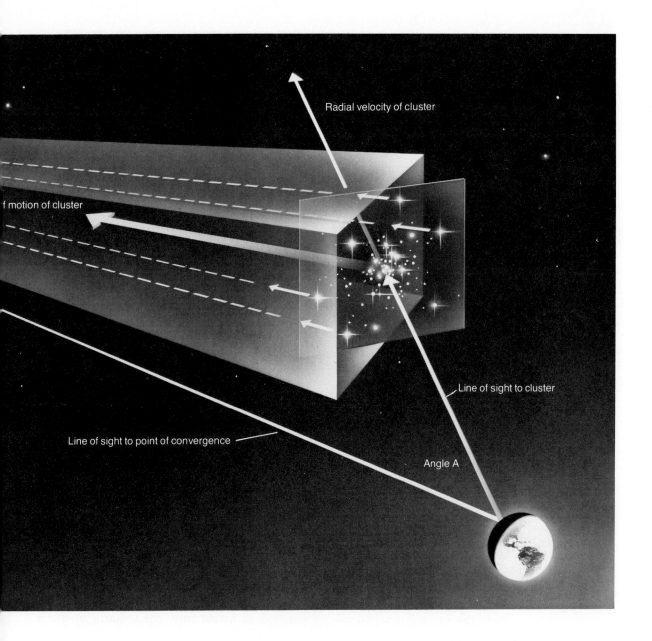

Radial velocity of cluster

f motion of cluster

Line of sight to cluster

Line of sight to point of convergence

Angle A

stellar evolution. We assume a given initial chemical composition as well as certain rules by which stars of different mass evolve. We can then compare the computed form of the color-magnitude diagram with actual observational results. In this way, we can ascertain the

FIGURE 8–13 *The orbits of high- and low-velocity objects about the galaxy. In this figure, the extent of the motions away from the plane of the galaxy are exaggerated. A population II star will have an orbit which is much more elliptical and carries the star further from the galactic plane than does the orbit of a star like the Sun. Thus, relative to the Sun, a star in an elliptical orbit will seem to have a high velocity.*

effects of different ages, different initial compositions, and even the effects of continuing stellar formation in the cluster. This is by far the most important use of clusters to astronomy, particularly open clusters.

Globular clusters provide us with important information on the original composition of the universe. The amounts of hydrogen and helium, and particularly the ratio of the two, yield information on the Big Bang which presumably created our universe 10 to 20 billion years ago. In this cosmic explosion, hydrogen, some helium, and also deuterium (heavy hydrogen) were formed. The temperature and density of the explosion determined the proportions of each that were produced. Also important is the question of how many, if any, elements heavier than helium were formed in the early stage of the universe. All these are questions to which we do not yet have complete answers, but it is fascinating to note that the study of globular clusters in our Milky Way enables us to look back to the very beginnings of time and space.

REVIEW

1 Globular clusters, roughly spherical in shape, are collections of tens to hundreds of thousands of stars. Globular clusters are found in the halo of the galaxy and in the nucleus, and are composed of old stars.

2 Open clusters are groups of tens to thousands of stars found within the Galaxy. They are generally composed of younger and more metal-rich stars than are globular clusters. Another name for open cluster is galactic cluster.

3 An association is a loosely bound group of stars recently formed from the interstellar medium.

4 When color-magnitude diagrams of open clusters are made, the spectral type at which the stars seem to "turn off" the main sequence is an indication of the age of the cluster. The later the spectral type of stars just leaving the main sequence, the older the cluster.

5 Most stars can be placed in one of two stellar "populations." Population I objects are younger, bluer, richer in metals, and more concentrated on the plane of the galaxy than stars of population II.

6 Since stars in a cluster were presumably formed at about the same time and from the same cloud of interstellar gas, differences in mass between the stars should appear as evolutionary differences. This is an important tool for checking theories of stellar evolution.

QUESTIONS

1 How might we determine the age of a stellar association whose member stars were dispersing?

2   What ways do we have of finding the distance to a globular cluster? To an open cluster?

3   Explain how we obtain a color-magnitude diagram for a cluster and how it is used to obtain the distance to the cluster.

4   Sometimes it is hard to determine just which stars on a photograph of a cluster actually belong to the cluster, and which are foreground or background objects. If you are plotting a color-magnitude diagram of a cluster and one star does not seem to fit with the others, how might you check to see if it really is a member of the cluster?

5   Explain how spectral analysis of globular clusters gives information about the origin of the Galaxy.

RECIPE FOR A SPIRAL GALAXY—THE MILKY WAY

CHAPTER 9

In the first three chapters of this book we discussed the scope and methods of astronomical investigation. In the next five chapters, we assembled the observations into facts about the major ingredients of the universe. Now we are in a position to combine these ingredients in the proper proportions and places, stir lightly, and produce what we believe is a typical example of the largest known form in the universe: the Milky Way Galaxy (Figure 9–1).

Other interesting astronomical objects yet to be discussed are the members of the solar system, which will comprise the last half of the book. On a galactic scale, however, planetary systems are quite negligible and are merely by-products of stellar evolution of some types of stars.

The Milky Way Galaxy as it might look to an observer several hundred thousand light-years away. The solar system is shown in its proper orientation and location in the Galaxy.

FIGURE 9–1
A composite photo-
graph of the Milky
Way as seen from
Earth. The coordinates
are galactic latitude
and longitude. The
direction to the center
of the Galaxy is near
the center of the
photograph. Note the
Magellanic clouds,
two small satellite
galaxies of the Milky
Way, in the lower
right. The band of
the galaxy is quite
narrow and streaked
with clouds of
absorbing material.
The Andromeda
galaxy, a nearby spiral
galaxy, is visible
below center near the
left side. Compare
this with Figure 4–8.

Our ingredients are stars, and groups of them, the open and globular clusters; the discrete and diffuse clouds of gas and dust, the molecules of the interstellar medium; and the electromagnetic radiation pervading all of space. The interactions of these parts are governed by the known (and perhaps some yet unknown) laws of physics, and described, for scientific purposes, by the language of mathematics. The result is the beautiful stellar display we can see on a dark night, and particularly the hazy band of light arcing across the sky, from horizon to horizon.

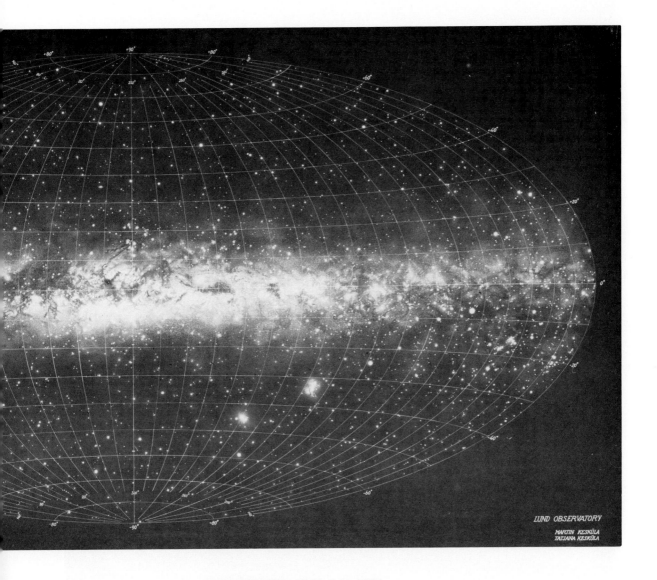

LUND OBSERVATORY
MARTIN KESKÜLA
TATJANA KESKÜLA

DISCOVERING THE GALAXY

The basic distribution of the material in the Galaxy is obvious to even an untrained eye on any clear night. The band of light we call the Milky Way was seen by the ancients just as we see it. The Greeks called it *galaxias kyklos*, the "milky circle." The entire Galaxy is now called after this ancient name for the belt of light across the sky. Most of the ancients thought of it as a road to heaven, or gave it some other such mythical significance. However, a few perceptive individ-

FIGURE 9–2
A close-up of a region of the Milky Way in the constellation Scorpius. This profusion of stars greeted the eye of Galileo when he first turned his telescope to the sky. The complex distribution of stars, gas, and dust is readily apparent. (Yerkes Observatory)

FIGURE 9–3   Thomas Wright, in the middle 1700s, thought that our galaxy was shaped like a grindstone, with the solar system near the center. This explains our perception of the band of the Milky Way as the result of our looking through a greater depth of stars along the plane of the grindstone.

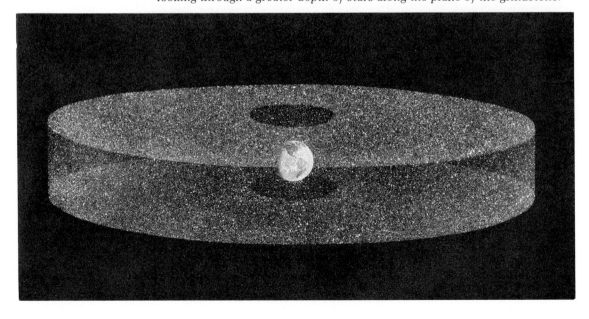

uals, such as Democritus in the fifth century B.C., thought it was composed of stars.

Proof that the band of light was indeed the combined effect of myriad stars was not obtained until the early seventeenth century when Galileo (1564–1642) turned his newly improved telescope to the sky. With this optical aid, he could see that there were many stars in the band, and it could be surmised that the patches of haze beyond the light-gathering power of his first crude instrument were also even fainter stars. The dark lanes, in which fewer stars were found, were thought to be "holes" in the sky through which one could see out into the emptiness of space (Figure 9–2).

Somewhat more than a century after Galileo, an English philosopher named Thomas Wright deduced from the shape of the visible Milky Way that our galaxy was shaped like a huge grindstone, with the Sun and solar system in the middle (Figure 9–3). We saw the band of light because we were looking through a thicker portion of stars, as opposed to looking perpendicular to the grindstone. Wright had very few detailed observations to back up his theory, but a book he published set another man to thinking. The man was William Herschel (1738–1822).

Star Gauging

In order to obtain the first crude map of the Milky Way, Herschel had to construct what were then the largest and finest telescopes in the world (Figure 9–4). Assuming that all stars were of the same intrinsic

FIGURE 9–4
William Herschel's large telescope enabled him to carry out his program of "star gauging" and to confirm the idea of Wright that the Galaxy is shaped like a flat disk. (Yerkes Observatory)

FIGURE 9–5
Star gauging. As we look to fainter and fainter stars, we are seeing to greater distances. If space is uniformly populated with stars, we can calculate how the number of observed stars should increase. Deviation of the observed number from the predicted number indicates different densities of stars in space. If our line of sight passes through a discrete cloud of absorbing material, we can tell how much it absorbs and its distance, but we must correct for the absorption when calculating the space densities of stars. By making such measurements in different directions we can obtain information about the structure of the Galaxy near the Sun.

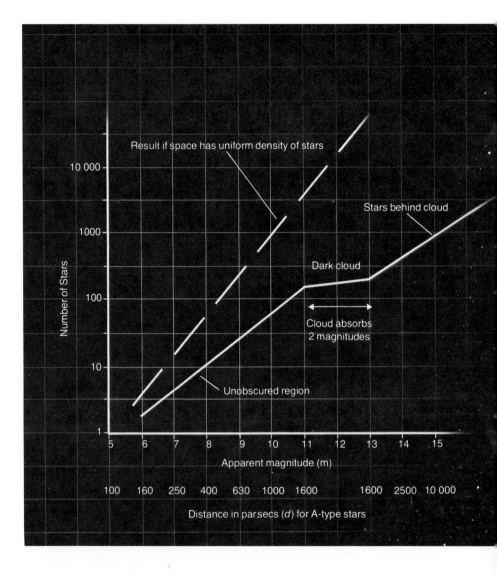

brightness (that is, absolute magnitude, although the term had not yet been invented), he reasoned that the fainter stars must be more distant. By this method, which he called "star gauging," he was able to map the distribution of the stars (Figure 9–5).

Star gauging works like this: if all stars have the same intrinsic brightness, the fainter ones must be more distant. Also, if stars are uniformly distributed (the same number of stars per cubic parsec, say, throughout the Galaxy), then as our telescopes look to fainter and fainter stars, for every increase in our viewing distance by a

factor of 2, the number of stars contained in the volume we are surveying should go up by a factor of 4. This is because the volume of a
spherical shell increases as the square of the radius. Thus, we must
simply count stars in different "shells" of distance, which means
observing the number of stars in different magnitude intervals.

We know how the number of stars should increase as they become fainter and fainter. If the actual number we count is greater
or less than expected, we know that the density of stars is higher
or lower than the average, respectively, and the magnitude at which

the star count starts to deviate from the expected number tells us the distance to this region of higher or lower density.

In this first detailed, careful, systematic study of the sky, Herschel showed that the Sun was indeed in the center of the known universe, for no matter in what direction he looked, the stars thinned out with distance. Along the band of the Milky Way, they thinned out at great distances from the Sun. But in directions perpendicular to the Milky Way, the stars thinned out much closer. Thus, Wright's "grindstone model" was shown to be essentially correct. It became accepted as the true picture of our galaxy (which at that time was The Galaxy) (Figure 9–6).

Minor refinements were made to the Herschel model over the following hundred years. During the nineteenth century modern astronomy came of age, with the development of photometry, spectroscopy, and photography, with the increase in size and quality of telescopes, and with the compilation of catalogs of large numbers of stars. The history of astronomy during this period is a fascinating study of the interdependence of the sciences.

The Spiral Nebulae

In the middle of the 1800s, an Irish nobleman, Lord Rosse of Parsonstown, had made the largest telescope in the world. When he turned his attention to the giant, wispy nebulae, he discovered that

FIGURE 9–6
By his efforts at star gauging, Herschel arrived at this picture of the Galaxy. The Sun is near the center and the number of stars thins out in all directions. When we look toward the thicker portions we see the band of the Milky Way. At this time, 1784, the Galaxy was the entire known universe. (Yerkes Observatory)

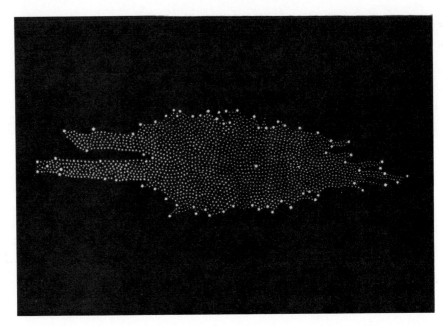

some seemed to be in the shape of pinwheels of gas. These became known as *spiral nebulae*, and just what they were was not known, although they engendered a lot of heated debate. At the time, not many people suspected they would have something to do with the structure of our galaxy (Figure 9–7).

FIGURE 9–7
In the mid-1800s, Lord Rosse built the largest telescope in the world. With it he was able to detect a suggestion of a pinwheel pattern in some "nebulae" such as M51 as shown in the drawing by Rosse below. (Yerkes Observatory)

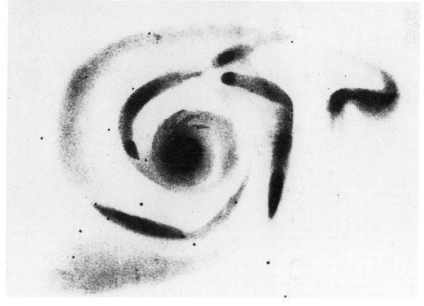

Kapteyn Universe

Using the wealth of data and technique acquired over the previous century, the Dutch astronomer J. C. Kapteyn (1851–1922), around the turn of the twentieth century, was able to improve on Herschel's work. He had by then a rough idea of the absolute magnitudes of stars of different types. But his result was essentially the same as Herschel's: the Sun was in the center of a highly flattened system of stars thinning out in all directions. The Kapteyn universe was larger than the Herschel universe, since Kapteyn could see fainter stars, but the shape was roughly the same, and we were in the center of it all (Figure 9–8).

The Shapley Galaxy

Upsetting this picture were the investigations of Harlow Shapley (1885–1973), an American astronomer. He studied the distribution of the globular clusters. It had been long known that almost all of them are seen above or below the band of the Milky Way. Applying the

FIGURE 9–8 *The Kapteyn universe. Note that it is essentially similar to Herschel's result. The Sun is near the center.*

recently discovered technique of using variable stars as distance indicators, he was able to discover the three-dimensional distribution of these clusters in and around the Galaxy, shown in Figure 9-9. He found that the center of the system of globular clusters was a point in the direction of the constellation of Sagittarius, and about 40 000 light-years away. Since clusters would be expected to be distributed around the center of mass of the Galaxy, he reasoned correctly that this and not the Sun must be the galactic center.

Hubble and M31

About this time the nature of the spiral "nebulae" became known. As mentioned before, Edwin Hubble had found Cepheid variables in one of them, M31, the "nebula" in Andromeda. He found that it was really far beyond the boundaries of our galaxy. It was, in fact, a galaxy itself, another "island universe" containing at least a hundred billion stars. The spiral nebulae are not gaseous but are really

FIGURE 9–9   *Harlow Shapley determined the distances to globular clusters using variable stars as metersticks. Combining distance and direction information, he could obtain a three-dimensional picture of their distribution in space. They seemed to be centered on a point lying in the direction of Sagittarius, tens of thousands of light-years distant. Shapley correctly concluded that this point must be the center of the Galaxy, not the Sun.*

galaxies. Almost every highly flattened galaxy we see is a spiral galaxy. It was thus concluded that the Milky Way is a spiral galaxy, and we are located somewhere out in one of the spiral arms. Thus, once again, we were removed from the center of the universe (Figure 9-10).

The spiral galaxies look like pinwheels, and we expect pinwheels to turn. Measuring the rotations of external galaxies is difficult. In trying to establish the nature of the galaxies, astronomers had looked for proper motions of stars in them. At galactic distances, these motions are too small to measure. However, today we can use the Doppler principle to get radial velocities of the approaching and receding edges of galaxies.

Oort and Rotation of the Galaxy

We had no proof of the rotation of our galaxy until the Dutch astronomer Jan Oort found evidence for differential rotation in the solar neighborhood, the region within a few hundred light-years of the Sun. *Differential rotation* means that, as in our solar system, objects at different distances from the center revolve about the center at different speeds. There is thus a shearing effect between different parts of the Galaxy. If we look at stars farther from the center than we are,

FIGURE 9–10
A cluster of galaxies. With the advent of large telescopes and photography, astronomers were able to obtain views of distant spiral galaxies. Just about every flat galaxy we see is a spiral, and since the Milky Way is flat, we concluded it was a spiral also. (Lick Observatory)

we should see them, on the average, going "backwards." What is really happening is that we are passing them, just as Earth passes Mars in the solar system. On the other hand, looking at stars closer to the center should show them to be passing us. These effects can be studied using either the proper motions or radial velocities of nearby stars. We will return to this important effect later.

Dust Clouds

About 1930, R. J. Trumpler (1886–1956) made an important discovery which explained why Herschel and Kapteyn had thought we were in the middle of the Galaxy. He found that the gas and dust in space are not only concentrated into definite clouds, but that there is also a thin distribution of this material throughout space. The dust dims the light of stars, by as much as a magnitude or more, for every thousand parsecs through which the starlight travels. Thus, when Herschel looked perpendicular to the Milky Way, he was seeing an actual decrease in the number of stars. But, when he looked along the plane of the Galaxy, the unknown dust dimmed the starlight, making him think the stars were farther away than they really were, and it dimmed stars in the direction of the galactic center beyond the light-gathering power of his instruments.

Star Populations

In the 1940s, Walter Baade (1893–1960) discovered the existence of two basically different groups of stars, the "populations" discussed in Chapter 8. This has proved to be a useful way of categorizing stars and other objects as to location in the Galaxy and velocity relative to the Sun. Population I objects are those which make up the flat part of the Galaxy, the disk, spiral arms, and the interstellar matter. Population II objects are distributed roughly spherically, or at least they are not very concentrated in the plane of the Milky Way.

Radio View

Throughout the 1930s and early 1940s, radio astronomy got its quiet start. An engineer with the Bell Telephone Company, Karl Jansky, built a large, rotatable antenna to try to discover the source of radio interference on long-distance telephone lines (Figure 9-11). He found, to his surprise, that some of the "interference" was coming from space, and further work showed it was coming from the direction of Sagittarius, from the center of the Galaxy. Since his results were published in engineering journals, not many astronomers saw them. A radio engineer and amateur astronomer named Grote Reber did see them, however, and he built what can be called the first radio tele-

FIGURE 9–11
In the early 1930s, Karl Jansky built this rotatable antenna to try to locate the source of radio interference in telephone communications. He serendipitously discovered radio emission from the Milky Way. (Bell Laboratories)

scope, seen in Figure 9-12. After several years of work, he had produced maps of the "brightness" of the Milky Way in radio waves. These maps showed that most of the radiation comes from the band of the Milky Way, just as does most of the optical radiation. After some difficulty in getting his work accepted, he finally published his discoveries.

During World War II, a few copies of this journal were smuggled into occupied Holland to astronomers there. One of them, H. C. van de Hulst, reasoned that there should also be radiation emitted in the radio part of the spectrum by the most abundant element, hydrogen. The radiation found by Jansky and Reber was continuum radiation. Van de Hulst predicted that hydrogen should emit a spectral line at a wavelength of 21 centimeters, or a frequency of 1420 megahertz (1 MHz = 1 million Hertz). After the war, which had produced giant strides in the technology of radio and radar, his prediction became known. Astronomers looked for, and found, the 21-centimeter hydrogen line. This discovery revolutionized the study of the Galaxy.

Spiral Structure

In the early 1950s, American astronomer W. W. Morgan and others gave us the first indications of the spiral structure of our galaxy.

They first studied other spiral galaxies, to find out what kinds of objects tended to trace out the spiral arms. These *spiral tracers* were young open clusters, OB associations, HII regions, and other young, highly luminous objects. Then they looked for similar objects in our galaxy. Because of the effects of interstellar absorption they could not see very far from the Sun, but far enough to establish indications of spiral arms in the Milky Way (Figure 9-13).

The study of the Milky Way has proceeded rapidly. This brief summary of the history of the discovery of the Galaxy serves as prologue to the modern discoveries of interstellar molecules, pulsars, black holes, and things yet to be found, each of which has fit, more or less well, into our total picture of the Galaxy.

FIGURE 9–12
Engineer and amateur astronomer Grote Reber followed Jansky's work and built the first true radio telescope. With it he made the first maps of the radio structure of the Galaxy. (National Radio Astronomy Observatory)

FIGURE 9–13
Comparing our galaxy
with others. By
studying what types
of objects outlined the
spiral arms in distant
galaxies, we were
able to get the first
indications of spiral
structure in our own
galaxy.

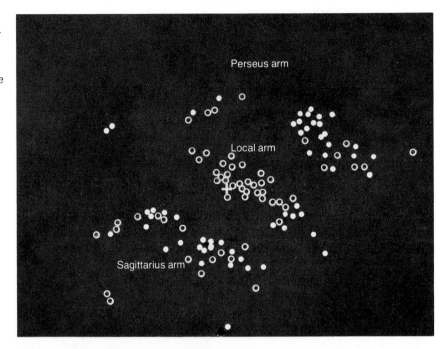

PROFILE OF THE MILKY WAY

The overall appearance of the Milky Way as seen by someone out-side it is that of a typical spiral galaxy, perhaps like M51, the beau-tiful "Whirlpool" galaxy seen in Figure 9–14. Seen from above, it is a vast pinwheel of stars, gas, and dust. From the side, it is a very thin disk, with a slight bulge in the middle, the *galactic nucleus*. The Milky Way is about 100 000 light-years, or 30 000 parsecs, across. But it is only about 1000 light-years thick. Relatively, it is thinner than a dime. The nucleus is spheroidal in shape and may be several thousands of light-years thick (Figure 9–15).

The thin plane of material is called the *galactic plane*, or *disk*. In this plane lie the *spiral arms*. Surrouding the Galaxy is the *halo*, or *corona*. Here are most of the globular clusters, high above and below the plane of the Galaxy, and a few stars with high velocities. The entire Galaxy is rotating, or, more correctly, each star, grain of dust, and molecule of gas is revolving about the galactic center at a speed which depends on its distance from the center.

Unlike the Sun in our solar system, the Galaxy is not a single point of mass, but a complex distribution of stars and interstellar

FIGURE 9–14
M51, the "Whirlpool" galaxy in the con-stellation Canes Venatici. Except for the small satellite galaxy hanging on one arm, this is how the Milky Way might look from 14 million light-years away. (Hale Obseravtories)

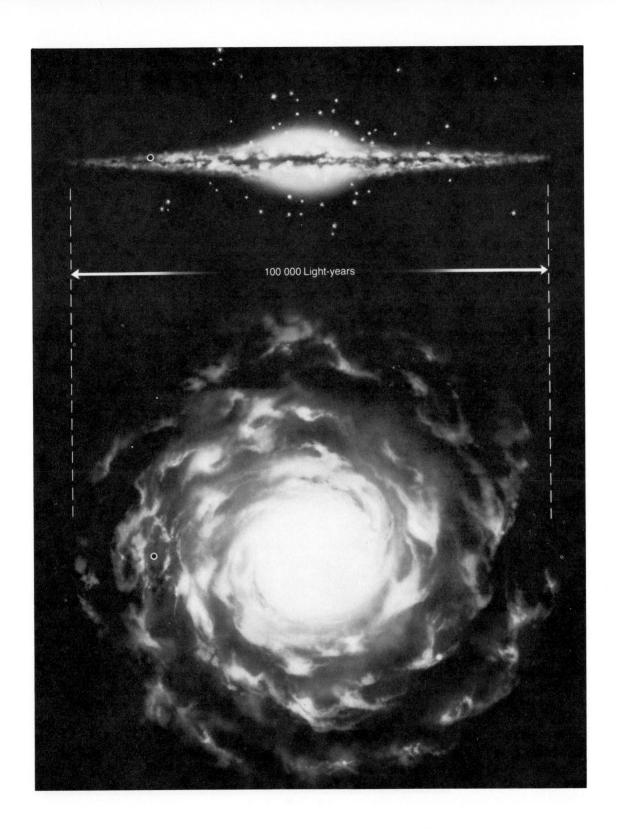

100 000 Light-years

matter. Thus Kepler's laws are only approximate, and the motion of a star or gas cloud around the center depends on the distribution of mass. In practice, we turn the problem around and study the motions of stars to try to derive the mass distribution of the Galaxy.

In the neighborhood of the Sun, this motion can be detected by studying the motion of nearby stars relative to the Sun, in the manner worked out by Oort. Refer to Figure 9–16. Consider the *proper motions* of stars, that is, the motions across the line of sight which change the star's position in the sky. If we look toward the galactic center, on the average we should see the stars moving in the direction of galactic rotation, since they are closer to the center and hence moving faster than we are. Looking away from the center, we should see, on the average again, stars going "backward," seemingly in the opposite direction from galactic rotation. This is because we are passing them. Looking "forward" and "backward"—that is, in the direction we are going and in the opposite direction—we should see no proper motion, since these stars should have the same speed as the Sun, neither passing us nor being passed.

We could, instead, observe the *radial velocities* of the nearby stars, that part of their velocity that is toward or away from us. Looking toward and away from the center, we would see no radial velocity, since the stars are moving at right angles to our line of sight. Observations "ahead" and "behind" us should also result in zero radial velocity, since stars in those directions have the same speed as the Sun. But if we look in directions 45° to these, we see a maximum radial velocity as either we catch up to these stars, or they catch up to us.

Although the principle of these studies is simple, the actual observations are often difficult, and many bits of data are needed. This is, nevertheless, a powerful tool for probing the Galaxy. The proper motions and radial velocities we measure, and how they change with distance, tell us not only how other stars are moving, but also how the Sun is moving, and where it is in the Galaxy.

Sun's Location

We can briefly state the results of these studies. The Sun is 10 kiloparsecs (kpc), or 32 000 light-years from the galactic center. (Actually, the technique is not exact enough to give the answer this accurately, but we know with good certainty that the distance is between 9 and 11 kpc, so we adopt 10 kpc as a convenient number subject to revision.) We are moving around the Galaxy at a speed of about 250 kilometers per second, in a direction toward the constellation of Cygnus. It will take us about 250 million years to make one trip around the Galaxy. Some people call this a "cosmic year."

*FIGURE 9–15
(opposite page)
From the side, our
galaxy is flat with a
slight bulge in the
center. The circle
marks the location of
the solar system.
Notice the globular
clusters in the halo of
the Galaxy. From the
"top," our galaxy is
a giant pinwheel.*

FIGURE 9–16   The motions of stars resulting from galactic rotation. In (a) we see the true motions of stars at different distances from the center of the Galaxy. Note that stars farther from the center move more slowly. In (b) we see the motions of the stars with respect to the Sun; that is, the Sun's motion has been subtracted from that of each of the other stars. Stars closer to the center than the Sun are passing us, and those farther from the center are being left behind. In (c) we see only the radial velocity part of these motions (black arrows). In (d) we look at only the transverse motions of the stars (black arrows).

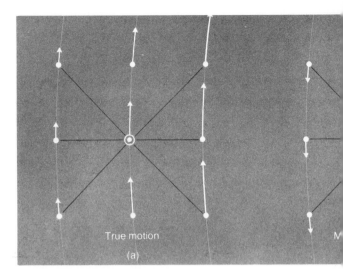

True motion

(a)

One could say that since the formation of the solar system about 20 cosmic years have elapsed, and we have about another 20 cosmic years to go before the Sun becomes a red giant.

Galactic Coordinates

In order to locate objects within the Galaxy, astronomers have adopted a special galactic coordinate system. *Galactic longitude* is measured around the plane of the Galaxy, with the galactic center taken as 0° and the direction of our motion, toward Cygnus, as 90°. *Galactic latitude* is measured above and below the galactic plane, which is called the *galactic equator*. One can find the galactic equator in the sky since it runs almost along the middle of the band of the Milky Way. It is seen in Figure 9–1.

All of the interstellar matter is strongly concentrated toward the galactic plane, with an average height above the plane of less than a few hundred parsecs. The material is spread almost evenly over the disk, with a slight concentration in the arms. We know, primarily from radio astronomy, that there are massive flows of gas in the disk, and outward from the galactic nucleus. Just why this is so is not yet known, but we suspect that some energetic phenomenon is going on in the nucleus. Because of the interstellar matter, a clear view in the plane of the Galaxy is limited to only a few kiloparsecs from the Sun. We cannot see the center optically. Radio waves are not greatly absorbed by the interstellar matter and enable us to probe the galactic nucleus, and even the far side of the Milky Way.

The stars in our galaxy also are spread almost evenly over the

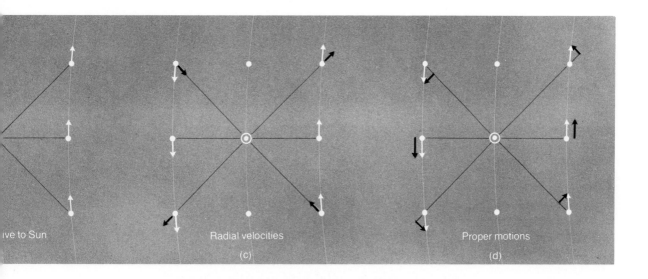

ve to Sun   Radial velocities   Proper motions

(c)   (d)

entire disk, much more evenly than we would think from looking at photographs of other spiral galaxies where we see striking spiral arms. The arms are, however, something of an optical illusion. In the arms we see only the young, hot, luminous stars, and the gas which happens to be near them, glowing in HII regions. These stars have recently formed from the interstellar clouds. Most of the stars are evenly distributed, but most of them are less massive, hence less luminous, and do not show up on our photographs. We see only the brightest stars, rather like the frosting on a cake, as one astronomer has called it.

Most of the stars in the disk are of intermediate age, between the very young objects tracing out the spiral arms and the old population II objects. They make up around 99 percent of the mass of all stars in the Galaxy, and the very luminous stars about 1 percent. Look back at the photograph of M51 in Figure 9–14. In our imaginations we might add to the picture another 99 billion stars, and then imagine ourselves on a planet revolving about one of these insignificant stars about two-thirds of the way from the center to the edge. We begin to see the problem men on Earth have when they try to discover the structure of the Milky Way from inside.

SPIRAL ARMS AND STAR FORMATION

We do not yet know why some galaxies have spiral arms. We can make some educated guesses, but much more observational data are needed before the question is settled. We do, however, have a workable hypothesis of why, once spiral arms have formed, they persist and do not fade away.

At first it was thought that spiral arms were composed of physically connected stars, gas, and dust, like the arms of a pinwheel. Large-scale magnetic fields in the Galaxy were thought to keep them together. There are two things wrong with this idea. First, we have found that the galactic magnetic fields are too weak to perform this task, and second, there is the problem of differential rotation. Consider a spiral feature: the inner parts would move faster than the outer parts, and gradually the "arm" would be stretched out, growing thinner and longer, winding up so tightly that eventually no arms could be distinguished. An example of this is provided by a cup of coffee: a few drops of cream stirred into black coffee will show up as spiral arms. Soon, however, they will wind up and be smeared out, and there will be no spiral features left. The question is, then, Why do spiral arms in a galaxy persist?

Spiral Density Wave

Basing his theory on preliminary work by Swedish astronomer Bertil Linblad, mathematician C. C. Lin has developed what is known as the "spiral density wave" hypothesis. In this scheme, the arms are not solid bodies, but merely wave patterns. Lin's theoretical studies have shown that once formed, spiral arms will remain if certain conditions are met. For instance, there must be an even number of arms, and the arms must trail. The arms are a pattern of slightly higher density of interstellar matter and stars, and the pattern is rotating at a speed different from the speed of the stars, gas, and dust in it. It is thought that the spiral pattern in the Milky Way is rotating at about two-thirds the speed of the Sun about the center. Thus, stars, gas, and dust catch up with the trailing inner edge of the pattern.

If the concept of a moving density wave seems strange, consider an analogy commonly encountered on the highway. If we were to observe a busy expressway from a helicopter we would see most cars moving along at about the speed limit. However, there is always the slow driver somewhere on the road holding things up. We would observe a group of cars behind the slow one, waiting to get by. Cars traveling at the speed limit rush up to this group, and must slow down because of the higher density of cars. After moving slowly forward through the group, they reach the forward edge and speed on ahead, leaving the group behind. This group of cars is a moving density wave. That is, it is a region of higher density of cars, and the region is moving. Moreover, the group is not always composed of the same cars. Cars catch up from behind, then leave after passing through.

The same thing happens in our galaxy. The moving density wave of a spiral arm is moving slower than the "speed limit" set by Kepler's laws. Stars catch up from behind and slow down slightly; thus, the density is slightly higher in the arm. They pass through and speed off again, leaving the arm. What causes the stars to slow down is the higher density of stars, which causes a slight alteration in the gravitational field, which causes stars to slow down, which causes a higher density, and so on. This feedback process maintains the spiral arms, which are thus merely patterns.

So far we have mentioned only stars, but the same thing happens to the gas and dust, with one big exception: whereas stars interact very little with each other and with the clouds, the clouds interact with each other quite a lot. And, whereas stars are rather small, clouds may be very large. Thus, when a cloud of gas and dust encounters the spiral arm, the front edge may slow down before the back edge, which compresses the cloud, and the cloud then slows down quite rapidly. Thus, the density of interstellar matter on the trailing inner edge of an arm is higher than usual. Recall from Chapter 5 that when interstellar matter is sufficiently compressed, star formation may begin. Thus, we would expect to find young stars on the inner edges of spiral arms, and associated with a lot of interstellar matter.

Among the stars formed when a large cloud condenses are stars of all spectral types. Most will be of later spectral types, but a few will be O or B stars, very luminous and capable of ionizing the interstellar matter remaining near them, producing HII regions. Thus, these types of stars, and HII regions, trace out spiral structure in a galaxy. There are more common stars there, too, but they do not show up. Further, a very luminous star will live a lifetime, both on the main sequence and even later as a red giant, which is much shorter than the time it takes it to go around the Galaxy— say, 1 percent of its cosmic year. Since typical stellar speeds with respect to nearby stars are a few tens of kilometers per second, in their lifetime these bright stars cannot have drifted far from the original pattern or from each other. So, we see these massive stars close to where they were formed. (In figuring stellar speeds on a galactic scale, a neat conversion factor is that 1 kilometer per second equals 1 parsec per megayear!)

The average stars formed in the clouds have much longer lives. They may go around the Galaxy many times (20 so far for the Sun) in their lives, and gradually get well "mixed" by interactions with all the other material in the Galaxy. And, while the clouds from which they formed were strongly concentrated to the galactic plane, these stars, after many trips around, will be spread a bit more widely from the plane. This explains the observed fact that the later-type stars are less concentrated to the plane of the Galaxy. Most of them will not, at any particular instant, be in a spiral arm. If they are, they will be just passing through.

In support of this idea, we have evidence that O and B stars and HII regions trace out spiral structure, and there seems to be higher stellar density on the inner edges of some spiral arms. Much more work needs to be done, however, to explain the details.

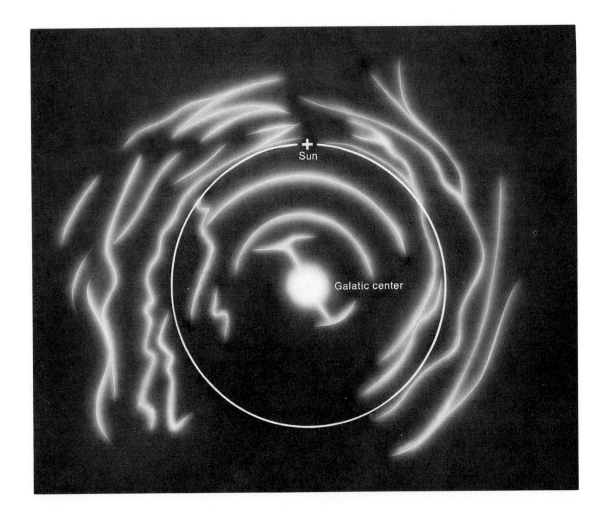

SPIRAL STRUCTURE IN THE MILKY WAY

We have been able to trace out, using both optical and radio techniques, the broad outlines of the spiral structure of our galaxy. This is illustrated in Figures 9–17 and 9–18. Optical studies have used the "spiral tracers" mentioned earlier. In addition to the young stars and HII regions, it appears that bright Cepheid variables, and even pulsars, lie nearly along the spiral arms. This is to be expected, since both these types of objects are thought to be the evolutionary result of massive stars.

Radio studies are much more extensive, since we can see farther from the Sun. The 21-centimeter line of hydrogen has been

FIGURE 9–17 (opposite) The spiral arms of the Milky Way as deduced from radio observatories. The bright areas indicate concentrations of neutral hydrogen gas. Compare this to Figures 9–13 and 9–18.

FIGURE 9–18 The spiral arms of the Milky Way. This map is obtained from a combination of radio and optical data. The arms are named for the constellations which appear the same direction.

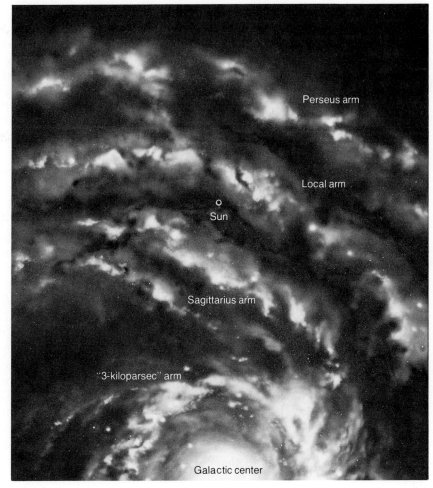

227

most important and deserves special consideration.

The 21-centimeter line results when the electron orbiting the hydrogen nucleus flips over. The probability of this happening for a particular hydrogen atom is about once every 11 million years. But, to quote Dutch radio astronomer Gart Westerhout, "Therefore, even though the probability is fantastically small, space fortunately is fantastically large, and thus we are indeed able to observe the line with rather high intensity."

When we look along a line of sight toward the inner parts of the Galaxy, somewhere along that line we will cross the point closest to the center, that is, the tangential point. Here any gas would be expected to be moving either directly toward or away from us, and should have the highest radial velocity of any gas cloud along the entire line of sight. The observational result of our radio telescope "view" is a graph, or line profile, in a given direction. Because of the Doppler effect, the line will be spread out into a broad feature, with different intensities at different wavelengths around the 21-centimeter value. By noticing how the maximum observed velocity and intensity vary in different directions, we can plot a map of the hydrogen gas distribution in the Galaxy. Sometimes this is done as a graph of velocity at different galactic longitudes, but such a graph has little meaning for the layman. It can be converted into a map of the density of hydrogen in space, which then gives a radio picture of the Galaxy.

There is far from complete agreement between the radio results and the optical results. Some of the reasons for this include the uncertainty in the assumptions we must necessarily make in converting the radio data into the "picture." Also, optical observations are restricted to a much smaller volume of the Galaxy. Obviously, much more work needs to be done. But a look back at the photograph of M51 will remind us of the problem we are facing. Someone has compared it to trying to draw a map of New York City by looking from the top of the Empire State Building on a foggy day!

Despite the problems, we have a general outline of the spiral structure. We know there are several spiral arms, and a few unexplained features. One debate centers on the question of whether or not the Sun is in a spiral arm. Observations indicate that we are, but theoretical studies say there shouldn't be a major spiral feature where the Sun is. We call this feature the Local arm, or Cygnus arm. Other spiral arms we recognize are mostly named after the constellation toward which we must look to see them: the Carina arm, the Sagittarius arm, the Norma-Scutum arm, and the Perseus arm. Beyond the Perseus arm is the so-called Outer arm, and there is a hint of an arm beyond that.

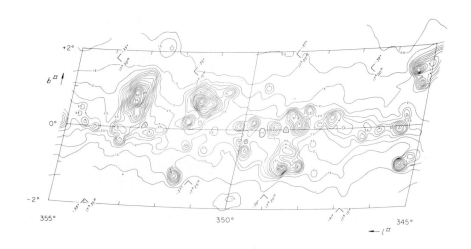

FIGURE 9–19   *A radio map of the center of the Milky Way. Many radio sources are seen by the contours of intensity on this computer-generated display. The galactic center is located at the center. Most of the other sources are either emission nebulae or supernova remnants. This map was made at a wavelength of 11 centimeters with the 140-foot radio telescope at the National Radio Astronomy Observatory. (National Radio Astronomy Observatory)*

Two particularly interesting items have been revealed by radio investigations. One is that the disk of interstellar matter gets thicker in the outer parts of the Galaxy. Another is that the galactic nucleus, within about 3 or 4 kiloparsecs of the center, is a very strange region. The density of gas there is very low, presumably used up in star formation in the very early stages of the Galaxy. The population II stars there outnumber the population I objects 1000 to 1. And what gas there is streams outward from the center, for reasons not yet understood. The subject of galactic structure will be an exciting and fruitful field of investigation in the years to come (Figure 9–19).

SUMMARY

We have by now a general idea of the structure of our Milky Way Galaxy, with its spiral arms. Just as we study Earth as the only planet we can examine close-up, and study the Sun as a typical example of ordinary stars, we study the Milky Way as an example of a spiral galaxy. We also turn the problem around, however, and look outside of our island universe, to others, in the anticipation that this cosmic view will tell us something about our own corner of the universe.

In the following chapter we will examine these distant galaxies and find out that spiral galaxies such as ours are not the rule, but the exception, in the universe of galaxies.

REVIEW

1 The Milky Way Galaxy is a vast pinwheel of stars, gas, and dust about 100 000 light-years across.

2 The Galaxy is in differential rotation; that is, objects at different distances from the center of the Galaxy move about the center at different speeds. The Sun moves around the Galaxy once every 250 million years at a velocity of about 250 km/sec.

3 The Sun is an average star located about two-thirds of the distance from the galactic center to the edge. The center of the Galaxy is in the direction of the constellation Sagittarius about 30 000 light-years away.

4 Very young objects outline the spiral arms of our galaxy. Examples of such "spiral tracers" are O and B stars, young clusters, HII regions, and some types of variable stars.

5 The spiral arms are not solid objects but are instead patterns of slightly higher density of stars and interstellar matter. The spiral patterns trail during galactic rotation.

QUESTIONS

1 Explain how star counting enables us to get a picture of our galaxy. Why did Herschel and Kapteyn conclude that the Sun was in the center of the Galaxy?

2 Why do we expect to find young stars on the inner edges of the spiral arms?

3 Suppose our galaxy encountered a star not attached to any galaxy moving on a path which would take it through the plane of the Milky Way. What would happen to the star as it passed through? What if it were a cloud of gas and dust instead of a star?

4 Why can we not use the usual technique to map the radio structure of our galaxy in the directions toward and away from the galactic center?

5 Explain the difference between the motion of the Sun around the Galaxy, and its motion with respect to the nearby stars.

6 How can we determine whether the Magellanic clouds are really satellite galaxies of the Milky Way?

READINGS

Arp, H. C. "The Evolution of Galaxies." *Scientific American*, January 1963.

Bok, B. J. "The Arms of the Galaxy." *Scientific American*, December 1959.

Bok, B. J., and P. F. Bok. *The Milky Way*. Cambridge, Mass.: Harvard University Press, 1974.

Jaki, S. *The Milky Way: An Elusive Road for Science*. New York: Science History Publications, 1972.

Page, T., and L. W. Page. *Star Clouds of the Milky Way*. New York: The Macmillan Company, 1968.

Sandage, A. R. "Exploding Galaxies." *Scientific American*, November 1964.

Sanders, R. H., and G. T. Wrixon. "The Center of the Galaxy." *Scientific American*, April 1974.

Whitney, C. A. *The Discovery of our Galaxy*. New York: Alfred A. Knopf, Inc., 1971.

GALAXIES
OF THE UNIVERSE

CHAPTER 10

In the last chapter, we saw how astronomers discovered our galaxy. During their investigations, they also discovered the nature of some of the hazy, pinwheel objects that had been observed for many years. In this way, two seemingly separate fields of investigation, spiral nebulae and the Milky Way, came together as astronomers learned that they are the same kind of object. To see how our present perception of galaxies came about, we should trace the detailed observations of the nebulae from the beginning of the seventeenth century.

The spiral galaxy M81 in the constellation Ursa Major. The galaxy is 6.5 million light-years away. (Hale Observatories)

CLASSIFYING THE GALAXIES

In 1611, soon after the invention of the telescope, the German astronomer Simon Marius (1570–1624) observed a fuzzy patch of light in the constellation Andromeda. He said it was "like a candle seen at night through a horn." Marius did not discover the formation; it appears on star charts constructed centuries earlier by Arabian astronomers. But with his telescope Marius could see the luminous cloud in more detail. It looked like several other wispy clouds that had been classified as nebulae, and so came to be called the Andromeda "nebula."

In the next several decades, many other nebulae were identified, and catalogs were prepared. These efforts culminated in the work of William Herschel (1738–1822) and his son John Herschel (1792–1871). With their large self-made telescopes they scanned both northern and southern skies. From their observations John Herschel compiled the *General Catalogue*, which contained some 5000 entries. It provided the basis for J.L.E. Dryer's *New General Catalogue* (NGC), which is still a standard reference.

Messier Objects

In nineteenth-century France, a sure claim to fame was to discover a comet. Observers anxiously scanned the skies and discovered luminous clouds that appeared to be comets. Their excitement was often quickly deflated when they found that the hazy patch of light they thought was a new comet turned out to be an object that was not new at all. To avoid such troubles, the comet hunter Charles Messier (1730–1817) compiled a list of some hundred-odd fuzzy objects, making it possible to distinguish a newly discovered comet from those known objects which look like comets. His list was a catalog of things to avoid. Today it is a list of objects sought by users of small telescopes, for many of the Messier objects are objects that are most easily visible. References to the catalog appear commonly. For example, the Andromeda "nebula" is known as Messier 31 (M31); it is also known as NGC 224 .

The Spiral Nebulae

With the advent of astronomical photography, many more "nebulae" were seen, and more details became discernible. We recall from Chapter 9 that earlier the Irish astronomer William Parsons, Third Earl of Rosse (1800–1867), had detected traces of a spiral pattern in some of the faint luminous clouds. They came to be called "spiral nebulae." Determination of their nature intrigued astronomers. In

1920, a debate was held before the National Academy of Sciences, dealing with the nature of the spiral nebulae and with the size of the Milky Way. Harlow Shapley (1885–1973) maintained that the spiral nebulae were gaseous formations within or on the edge of the Galaxy, while H. D. Curtis (1872–1942) believed they were "island universes" beyond our galaxy and similar to it.

Curtis was proved correct in 1924 when Edwin Hubble, the American lawyer-turned-astronomer, measured the distance to the Andromeda "nebula," a distance far beyond the boundaries of our galaxy. The formation was not simply a gaseous cloud, but a galaxy, a collection of billions of stars, in many ways like the Milky Way. The 1920s became not only a time of discovery of more island universes, but a time for interpreting this new knowledge. Just as an overabundance of data about stars had made it essential to develop a scheme of classification, so also it became necessary to classify the galaxies. At first the classification was based upon the shape of the ·formation.

Types of Galaxies

The pioneer in this field was Edwin Hubble. He found that there were basically three types of galaxies: *ellipsoidal* (also called ellipticals), shaped like footballs; *spirals*, some of which have bars through their centers, some of which do not; and *irregulars*, with no particular shape. Later, *lenticular*, or lens-shaped, galaxies were added to the scheme (Figure 10–1).

The ellipsoidal galaxies range from almost spherical, called *E0* ("E-zero"), to flattened galaxies about three times greater in length than in thickness, called *E7*. The spirals were given the symbols *S* for normal spirals, and *SB* for barred spirals, with a small letter *a, b, c*, or *d* after them to indicate how much the arms were spread out. *Sa* or *SBa* galaxies are tightly wound, *Sd* or *SBd* galaxies are very loosely wound. As sort of a transition type, the category *S0* was invented for the lenticular (lens-shaped) galaxies, similar to spirals, but with no arms. The irregular galaxies were abbreviated *Irr*.

Later, extensions and modifications of Hubble's original scheme were made. More categories were added, and types between the major shapes were allowed for. In particular, a symbol was added to indicate whether the spiral galaxy shows a ringlike structure or an S-shaped structure, or both. Despite all these pigeonholes into which galaxies may be classified, there are many which defy classification, or whose type must be amended with the symbol *p* for peculiar.

FIGURE 10–1
The Hubble "tuning fork" classification of galaxies. At the bottom are the elliptical galaxies, then the spirals, both barred and normal. The number after the E indicates the amount of flattening of elliptical galaxies. The lower-case letters after S and SB galaxies indicate the amount the arms are spread out.

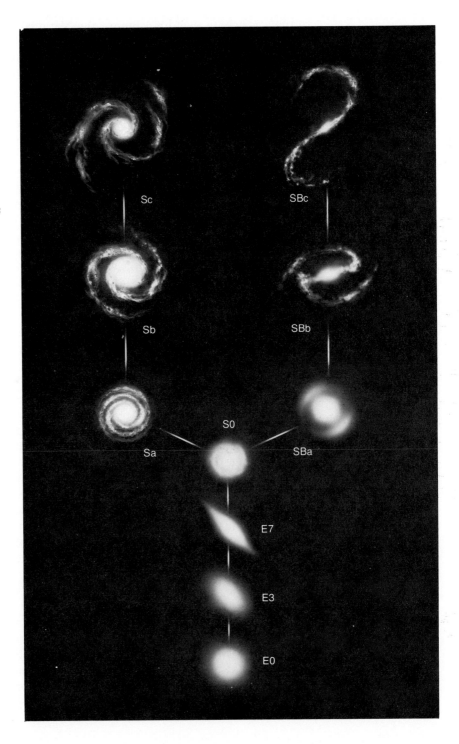

The Andromeda galaxy would be classified *SA(s)b*, indicating a normal spiral, with S-shaped arms, moderately spread out. The Milky Way might be the same, or there might be an *r* in the classifications to indicate a ring structure (a part of which is actually observed by radio telescopes about 4 kiloparsecs from the center).

THE CONTENTS OF GALAXIES

Galaxies are made up of stars, gas, and dust, in various amounts, and radiation. The proportions of these ingredients vary greatly from type to type, reflecting different initial conditions and evolutionary changes of the galaxies.

Spiral galaxies, such as the Milky Way, are composed of both population I and population II stars, while irregular galaxies are purely population I. Ellipsoidal galaxies are pure population II. We can discern individual stars only in the nearest galaxies, and use them to plot an H-R diagram as we do for stars in our galaxy. For the brighter of these stars we can obtain spectral types, and we find that the division into only two stellar populations is an oversimplification. We find some old stars rich in metals and some young stars poor in metals, in contradiction to our original definitions of the populations. Nevertheless, the general idea of stellar population types is useful for broad generalizations (Figures 10–2 and 10–3).

In spiral galaxies, the older objects are concentrated in the nucleus and halo, just as in the Milky Way, and the young objects in the disk and arms. Attempts to figure out the composition of the nuclei of galaxies lead to some surprising answers. For instance, compare the following calculations for M31:

Type of star	Contribution to total light	Proportion of stars
Orange giants (K0 III)	50%	1%
Yellow dwarfs (G0 V)	10%	2%
Red dwarfs (M1V)	40%	97%

Notice that very few, but very luminous, giants plus very many, but very faint, red dwarfs combine to give about 90 percent of the total light output of this nucleus of M31. Stars similar to the Sun contribute very little; they are not luminous enough, and not numerous enough to offset the deficiency in brightness. These conditions are typical of the nuclei of spirals, and of ellipticals in general. In the spiral arms, most of the light comes from O- and B-type stars, which again make up only about 1 percent of the total number of stars.

Gas and Dust

There is very little, or no, dust and gas in ellipsoidal galaxies. A few ellipsoidals show a faint bright-line spectrum indicating some gas, but the source of energy needed to excite the gas is unknown. The energy may come from a few bright, hot stars that are still left in the galaxy. In spirals, there is much gas and dust. The gas is detected by emission lines in the visible spectrum, and by radio waves given off by neutral hydrogen. The dust is readily apparent as dark patches or bands in photographs of spiral galaxies which are seen edge-on. The amount of gas and dust varies greatly over the galaxy, and in some unusual galaxies the gas seems concentrated to just one side. Why this should be so is still a mystery.

Rotation of Galaxies

All galaxies rotate. They have to, or else they would collapse as a result of the attraction of each mass for the others. Details of the

FIGURE 10–2 On the left, the photograph of nearby galaxy M33 was taken in red light to emphasize the nucleus and other population II objects. On the right, M33 was photographed in blue light to bring out the population I objects. (Copyright National Geographic Society–Palomar Observatory Sky Survey)

rotation, though, vary with the form of the galaxy. In the nearby galaxies, the rotation can be detected and plotted as a function of distance from the center, by the Doppler method (Figure 10–4). The spectral lines of the galaxy will be tilted due to the fact that one edge of the galaxy is approaching us, the other receding. The inner parts of most galaxies seem to rotate almost as a solid body. In the outer parts, the rotation is more like the Keplerian motions of the planets, with the more distant parts taking longer to make one circuit than those closer to the nucleus. The bar in barred spirals rotates as a solid body, but there seems to be much circulation of material within the bar itself.

The period of rotation of the central part of a galaxy is quite well correlated with its shape. The ellipsoidal and lenticular galaxies rotate fastest of all, on the order of once in 5 to 10 million years. As one proceeds from *Sa* through *Sb, Sc, Sd,* to *Sm* (which is almost irregular, so loosely are the arms arranged), the period of rotation increases; that is, the galaxies are rotating more slowly. A typical value for an *Sb* galaxy, such as ours, might be 20 million years. Recall, however, that this is for the inner parts; at the Sun's distance from the galactic center, the period is ten times as long.

It is from the rotation of galaxies that we obtain our best data

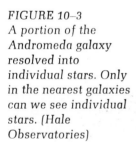

FIGURE 10–3
A portion of the
Andromeda galaxy
resolved into
individual stars. Only
in the nearest galaxies
can we see individual
stars. (Hale
Observatories)

FIGURE 10–4
The rotation of the galaxy M31 is observed by measuring the radial velocities of objects in that galaxy. Note that one edge is approaching us, and the other edge is receding. These recorded observations were made with 36-inch and 6-inch reflecting telescopes. (Yerkes Observatory)

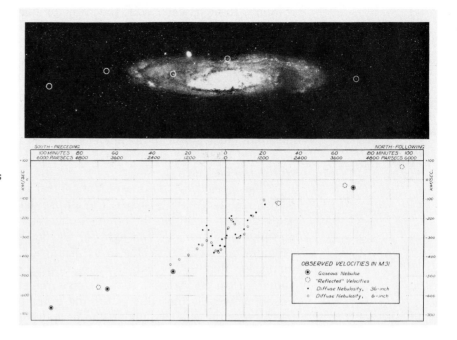

about the masses of galaxies. Just as the mass of the Sun and planets can be deduced from the details of motion within the solar system, we can obtain at least a lower limit for the masses of galaxies whose rotation can be measured in detail. Other methods, such as studies of clusters of galaxies, or the motion of one galaxy about another, give less precise results. There are many difficulties, but we do have a good estimate of the range of masses of galaxies of different types (in terms of the Sun's mass): irregulars, 10^8 to 3×10^{10}; ellipticals, 10^6 to 10^{13}; spirals, 10^9 to 2×10^{11}. We think the Milky Way is near the high end of the latter figure.

Giant and Dwarf Galaxies

Note that some ellipticals are much more massive than the largest spirals. These are the so-called *supergiant ellipticals*, the largest known galaxies. On the other end, there are small, low-mass ellipticals, called *dwarf ellipticals*, only slightly more massive than a large globular star cluster. Just as there are many, many more small, red dwarf stars in our galaxy than supergiants, so too there are many more dwarf elliptical galaxies than supergiant ellipticals.

Another way to classify galaxies is by their *integrated spectral type*, that is, the spectral type resulting from the blending of the light of all the stars in the galaxy. The irregulars, with large numbers of population I stars, look rather like A- or F-type stars. The ellipticals look like G- or K-type stars, and the spirals can look like almost any type from A to K.

All of the light, from all of the stars and luminous gas, is combined when we measure the *integrated magnitude* of a galaxy. In terms of absolute magnitude, a typical spiral would have $M = -19$. Some of the supergiant ellipticals would be as luminous as $M = -23$, about 40 times brighter. Some of the dwarf ellipticals, however, are scarcely brighter than the brightest single stars in the Milky Way. Because of this, they are impossible to see at great distances. We know they are the most common type; we just can't see them.

CLUSTERS OF GALAXIES

Galaxies are often members of multiple systems of galaxies, just as stars in the Milky Way may be members of multiple systems. Probably less than half of all galaxies are single. More often they are found with companions, or in clusters of galaxies, or both. Our

FIGURE 10–5
The Small Magellanic cloud, an irregular companion galaxy to the Milky Way. (Cerro Tololo Interamerican Observatory)

galaxy is typical. It has two companion galaxies, the Magellanic clouds, both of which are irregulars and are visible only from the Southern Hemisphere of Earth (Figures 10–5 and 10–6). It is believed that they are actually gravitationally bound to the Milky Way as satellites. The Andromeda galaxy, M31, has two companions also, known as M32 and NGC 205 (Figures 10–7 and 10–8). These are both elliptical. Another nearby spiral, M33, in the constellation Triangulum, several more dwarf ellipticals, plus the Milky Way, and the Magellanic clouds make up a cluster of galaxies known as the *Local Group*. There are about 19 members, including 3 spirals, 4 irregulars, 10 ellipticals—of which 6 are dwarf ellipticals—and 2 recently discovered galaxies, Maffei I and II, believed to be ellipticals. The diameter of our Local Group is about 1 megaparsec (slightly over 3 million light-years, abbreviated Mpc). There are other clusters of galaxies nearby, typically spaced about 3 Mpc apart. About 50

FIGURE 10–6
The Large Magellanic cloud, another irregular satellite galaxy of the Milky Way. (Cerro Tololo Interamerican Observatory)

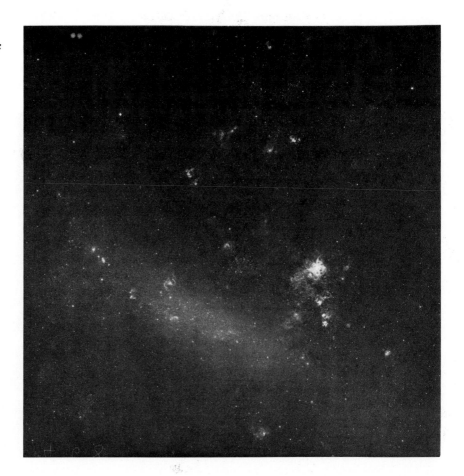

FIGURE 10–7
The Andromeda galaxy, NGC 224, and its satellite galaxies NGC 205 and NGC 221. This group, plus several other galaxies, and the Milky Way, make up the Local Group of galaxies. (Hale Observatories)

FIGURE 10–8
An elliptical galaxy, NCG 205, a satellite of the Andromeda galaxy. (Hale Observatories)

FIGURE 10–9
NGC 1300, a type
SBb(s) galaxy in the
constellation
Eridanus. (Hale
Observatories)

such clusters are within 15 Mpc of us, and when we plot a diagram of their distribution, we obtain a suggestion of a larger arrangement, a *local supercluster* with a center located in the direction of the constellation Virgo. The supercluster is flattened and is rotating. We are near the outer edge of the supercluster, which we think is about 40 Mpc across and 10 Mpc thick, rather like a galaxy of galaxies.

Superclusters
Other superclusters have been noticed, each about the same size as

FIGURE 10–10
NGC 1398, a type
SBb(r) galaxy. Notice
the ringlike structure.
(Hale Observatories)

FIGURE 10–11
The Sc galaxy M33, or
NGC 598, in the con-
stellation Triangulum.
(Hale Observatories)

FIGURE 10–11
The Sc galaxy M33, or
NGC 598, in the con-
stellation Triangulum.
(Hale Observatories)

ours. The question naturally comes up as to whether there is some larger-scale organization, a "local superdupercluster" or galaxy of galaxies of galaxies. We are not yet certain, but it seems that there is some such larger system. We are reaching very far from home, and the data are difficult to obtain. Is there a still larger group? If so, we may never know of it, since its boundaries would be beyond the light-gathering power of our telescopes. But it does appear that the universe is definitely arranged in a hierarchy of groups. We will see later that this idea has important implications for cosmology.

FIGURE 10–12
A cluster of galaxies
located in the
constellation Coma
Berenices. (Hale
Observatories)

THE EXPANDING UNIVERSE

Edwin Hubble was the first to notice that the more distant galaxies seem to be moving away from us faster than those nearby. He could find fairly accurate distance indicators for the relatively close galaxies, such as Cepheids, HII regions, and clusters. From these he found a definite correlation between red shift (which is interpreted as a Doppler shift) and distance. Except for a very few members of our Local Group, every galaxy is rushing away from us, and the more distant it is, the faster it is going. The whole universe is expanding! (Figure 10–13.)

FIGURE 10–13
The relation between red shift and the distance of galaxies. (Hale Observatories)

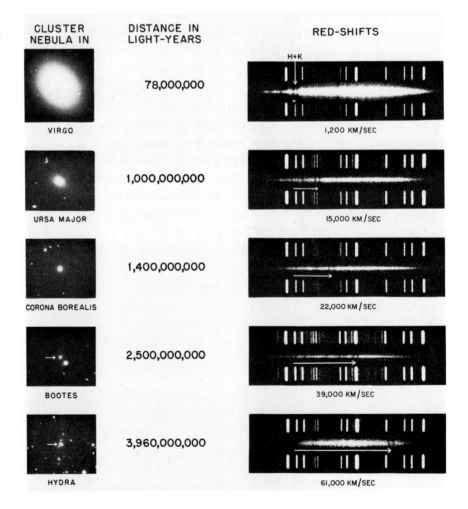

CLUSTER NEBULA IN	DISTANCE IN LIGHT-YEARS	RED-SHIFTS
VIRGO	78,000,000	1,200 KM/SEC
URSA MAJOR	1,000,000,000	15,000 KM/SEC
CORONA BOREALIS	1,400,000,000	22,000 KM/SEC
BOOTES	2,500,000,000	39,000 KM/SEC
HYDRA	3,960,000,000	61,000 KM/SEC

We saw in Chapter 1 how we need not assume that we are in the center of the universe, that this expansion is really the same as viewed from anywhere in the universe. The fastest galaxy we see is moving away at about half the speed of light. In the early 1960s, astronomers discovered small, almost point-sources of radiation which were found to be moving away at 15 percent of the speed of light, which seems to indicate that they are among the realm of the galaxies and subject to the general expansion of the universe. These objects are called *quasars*, for *quasi-stellar radio sources*, because they appear starlike and give off strong radio radiation. Since the first few were discovered in the early 1960s, many more have been identified. The speed record for the galaxies has been broken many times by the quasars. The fastest one we know of so far is moving at about 91 percent the speed of light. Figure 10–14 shows two quasars, BSO-1 and 3C-9.

Quasars are thought to be related to the many types of *peculiar galaxies* discovered by large telescopes. Many seem to be exploding. An example is M82, seen in Figure 10–15, a galaxy which not only looks as if it were tearing itself apart, but also has been found to be the source of strong radio radiation. Other exploding galaxies, such as M87, in Figure 10–16, show a highly luminous "jet" of material being expelled from the nucleus at thousands of kilometers per second.

FIGURE 10–14 The quasi-stellar objects BSO-1 (left) and 3C-9 (right). (Copyright National Geographic Society–Palomar Observatory Sky Survey)

Other sources of radio waves have been identified with otherwise ordinary looking galaxies, as well as with such strange objects as Centaurus A, seen in Figure 10–17. Other sources seem to be galaxies which are colliding, as in Figure 10–18. Often the regions of radio emission from a galaxy do not coincide with the optical emission. Typically there are two sources of radio waves, one on either side of the optical galaxy. Some emit only a small fraction of their energy in radio wavelengths, others emit more than the amount of visible energy.

Radio astronomers have recently discovered at a very great distance what appears to be the largest objects in the universe. It is an unusual galaxy with greatly extended regions of radio emission. The "object," called 3C236, is approximately 18 000 000 light-years across (compare the size of the Milky Way Galaxy which is 100 000 light-years across).

Seyfert Galaxies

Another connection between quasars and galaxies may be the galaxies with small, energetic nuclei first studied by the American astronomer Carl Seyfert and named *Seyfert galaxies* after him. They are all strong radio emitters, and may also emit large quantities of

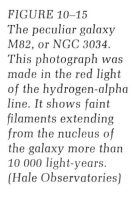

FIGURE 10–15
The peculiar galaxy M82, or NGC 3034. This photograph was made in the red light of the hydrogen-alpha line. It shows faint filaments extending from the nucleus of the galaxy more than 10 000 light-years. (Hale Observatories)

gamma rays. An example is NGC 4151, in Figure 10–19. Details of the relationship among all these types of peculiar galaxies and quasars are far from understood. We do know that if the quasars are really at great distances from us, as they seem to be, then they are emitting more energy than any known physical process can produce. Quasars must be very small, since they are seen to change brightness over a period of only a few days or months. Thus, they must be only a few light-days or light-months in size. If they were bigger, any brightness fluctuations would average out and we would

FIGURE 10–16   *A long exposure of the E0 peculiar galaxy M87 (center photograph). The short exposure (lower left) shows the jet of energetic gas erupting from the galactic nucleus. The upper-right graph shows the radio radiation received from M87. (Hale Observatories; graph by National Radio Astronomy Observatory)*

FIGURE 10–17
The peculiar galaxy
Centaurus A, NGC
5128. This galaxy is
also a strong source
of radio emission.
(Hale Observatories)

FIGURE 10–18
The twin peculiar
galaxies NGC 4038-9
in Corvus. These
galaxies also emit
strongly in the radio
region. (Hale
Observatories)

FIGURE 10-19
The Seyfert-type
galaxy NGC 4151.
Notice the small,
almost stellar nucleus.
(Hale Observatories)

*FIGURE 10-19
The Seyfert-type
galaxy NGC 4151.
Notice the small,
almost stellar nucleus.
(Hale Observatories)*

not see them. They are unresolved at their great distances from us (the nearest, 3C273, in Figure 10–20, is about half a billion parsecs away). Some have faint jets or wisps associated with them. For some of them, the region of radio emission is larger than the optical object, and is just barely resolved. And they are all quite blue in color, emitting much ultraviolet radiation.

A helpful piece of data would be proof of the existence of both quasars and ordinary galaxies in a cluster of galaxies. There is a problem as to whether the quasar is actually a member of the cluster, or is merely seen in the same direction accidentally. Recent research has produced conflicting results, and much more data are needed. So far, quasars, and their presumed relatives, are a mystery.

EVOLUTION OF GALAXIES

When Hubble first drew his "tuning fork" diagram, some astronomers proposed that the life of a galaxy proceeded from elliptical to spiral. Others suggested it was the other way around. Today it seems likely that there is no such drastic evolution from one type to another, with one possible exception. If a spiral galaxy were to collide with another galaxy, the gas could be swept out of the spiral, halting star formation. Thus, after a while, the stellar population would come to resemble that of an *S0* galaxy. (In such a collision,

FIGURE 10–20
Four quasars: 3C-48,
3C-147, 3C-196, and
3C-273 (3C stands for
the third Cambridge
catalog of radio
sources.) (Hale
Observatories)

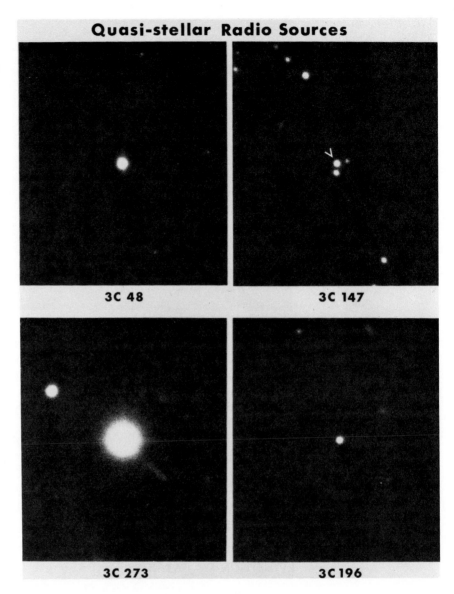

the stars would still be far apart and not interact directly with one another. Only the gas and dust would be greatly affected.)

To get an idea of how galaxies might have originated, try to imagine a very large cloud of matter, very far back in time, which fragmented to form smaller clouds, which in turn split up into smaller fragments, and so on, until finally galaxy-sized clouds began to collapse. As they contracted, the form of galaxy they would become was determined by the amount of rotation of the cloud. In some, star formation began suddenly and stopped soon thereafter, and an elliptical galaxy resulted. In the case of the Milky Way, at some point the cloud became dense enough to form some stars, and these became the globular clusters (still smaller fragments of the cloud) and other old stars. The contraction stopped when the forces of contraction and repulsion became equal, and since that time, the younger stars have formed. The first very massive stars became supernovae, formed heavy elements, and enriched the interstellar medium from which later generations of stars formed in the disk of the Galaxy.

This rough outline of evolution is thought to be substantially correct, but we shouldn't be too surprised if many unsuspected things have been left out. Our knowledge of galactic evolution is still very uncertain.

THE DISTANCE SCALE OF THE UNIVERSE

We have reached, with our telescopes and in our discussion, the depths of the universe. Questions of the origin (if any), evolution, and interrelationships of the component parts of the universe hinge on our ability to measure their position and velocities. Without stating it explicitly, we have built ourselves a ladder out from Earth, upon which we base our distance measurements. Figure 10–21 shows the rungs of this ladder.

The first step is surveying Earth, from which are obtained its size and shape. In order to extend the distance scale out into the solar system, we combine a knowledge of Kepler's laws with actual distance measurements. The laws of celestial mechanics give us the scale of the solar system, that is, the relative distances between the planets at different times. This scale is not in absolute units, such as kilometers, but is relative to some choice of unit distance. This unit is the average Earth-Sun distance, which is called the *astronomical unit* (a.u.). We do not know how many kilometers are in an a.u. until we measure the true distance to some planet whose relative distance we know. This can be done by bouncing radar off Venus or one of the asteroids. The time from sending to receiving the radar signal is twice the time it takes the radio wave to travel to the planet. We know the speed of radio waves, and can thus figure out the distance to the planet. Radar distances, combined with knowledge of the position of the planet in its orbit, allow us to determine that the a.u. is equal to 150 million kilometers.

FIGURE 10–21
Finding the distance scale of the universe. (1) Measuring distances to planets with radar. (2) Using distances and the laws of celestial mechanics to find the size of the astronomical unit. (3) Trigonometric parallax to nearby stars. (4) Distances to nearby clusters establishes the ZAMS. (5) Color-magnitude diagrams of clusters (6) fit to the ZAMS allow us to measure brightnesses to Cepheid variables (7) and other variables. Variables in nearby galaxies (8) allow us to calibrate the Hubble law (9) and extend this (10) to the depths of the universe (11). See page 253 for a more detailed explanation.

Because of the motion of Earth around the Sun, nearby stars will seem to shift slightly in position, an effect called *parallax*. The parallax of a star is the angular size of the radius of Earth's orbit (1 a.u.) as seen from the star. The angle is very small. Even for the closest star, Alpha Centauri, it is only 0.76 second of arc. For convenience we adopt a new unit of measurement, the *parsec*, or parallax-second, which is the distance at which a star would have a parallax of 1 second of arc. We can determine that 1 parsec equals 3.26 light-years, or 31 trillion kilometers. This method of determining distance has no built-in assumptions except the laws of trigonometry. For this reason it is sometimes called *trigonometric parallax*.

We use the method of trigonometric parallax to find distances to nearby stars of different spectral types. We also measure the apparent magnitudes of the stars. Then using the equation $m - M = 5 \log d - 5$ we obtain the absolute magnitudes of the stars. Most of these nearby stars will be lower main-sequence stars, since they are more common than the more luminous stars.

We next look for open clusters in which are located other stars of the same spectral types observed closer to the Sun. We now know both the apparent magnitude and absolute magnitude and so we can calculate the distance to the cluster. Then we calculate the absolute magnitude of more luminous stars in the cluster. In this way we build up a knowledge of the location of the main sequence on the H-R diagram.

The method of finding distances to nearby clusters by the moving cluster method (see page 196) is also used to improve the data on the main sequence. If there are Cepheid variables in any of the clusters whose distance we can determine, we can calibrate the period-luminosity relation for the Cepheids, as discussed on page 153.

Knowledge of the absolute magnitudes of stars of each spectral type (or color index) allows us to obtain the distance to any cluster or association by observing spectral types (or color indices) for its member stars. Thus we can extend our data to include the most luminous stars.

Using all these data, we can observe similar objects in nearby galaxies and obtain their distances. For somewhat more distant galaxies, we also use the brightnesses of HII regions of entire globular clusters.

For each of these nearby galaxies we measure the speed of recession due to the expansion of the universe, and our knowledge of their distances allows us to calibrate the Hubble law and find the value of the Hubble constant, H. The currently accepted value of H

is about 55 kilometers per second per megaparsec. In distant galaxies we cannot observe individual stars, or even clusters, but we can measure their speed of recession by measuring their red shift. Using the Hubble law, $V = H \times d$, where V is the velocity and d is the distance in megaparsecs, we can determine the distances to the galaxies. From these data we obtain information on the large-scale distribution of galaxies, and thus on the density of the matter in the universe, which determines the eventual fate of the universe. And it all started, with a surveyor measuring Earth.

That question of density is important, for it determines the future of the universe. We can calculate the density of the local supercluster to be about 10^{-28} grams per cubic centimeter. But that small amount is still much higher than the 10^{-30} g/cm^3 believed to be the value for much larger regions. The crucial point is this: if the density of space overall is high enough, then the gravitational attraction of all the matter will be great enough so that eventually the universe will stop expanding and begin to shrink. If it is not high enough, the universe will keep on expanding. The dividing line is near 10^{-30} g/cm^3. A factor of 10 on either side will make the difference. So far the answer is not known.

REVIEW

1   Galaxies may be classified by form as ellipsoidal, spiral, lenticular, or irregular. There are variations in shape among these classes.

2   Spiral galaxies are composed of population I objects in the arms and population II objects in the halo and nucleus. Ellipsoidal galaxies are composed of population II objects.

3   The form of a galaxy seems to be determined by its amount of angular momentum (rotation).

4   Galaxies occur in clusters, and in clusters of clusters. Whether there are larger groupings is not known.

5   The universe is expanding.

6   The velocities of recession of the galaxies are proportional to their distances. This is the Hubble law, and the constant of proportionality between velocity and distance is known as the Hubble constant.

QUESTIONS

1 How might you determine if a galaxy which appears to be lenticular is really an inclined spiral?

2 Would you expect to see O-type stars or nebulae in ellipsoidal galaxies? Why?

3 How could Messier detect that some of the fuzzy objects he discovered were not comets, but objects much more distant?

4 What color sensitivity of emulsion and color of filter would you use to photograph (a) the nucleus of a spiral galaxy, (b) the spiral arms, (c) an ellipsoidal galaxy? Why?

5 What is the effect on the size and age of the universe of reducing the Hubble constant by a factor of two?

6 Explain the effect on our measurement of the distance to a distant galaxy if we overestimate the size of the astronomical unit.

READINGS

Hoyle, F. Galaxies, Nuclei, and Quasars. New York: Harper & Row, Publishers, 1965.

Hubble, E. The Realm of the Nebulae. New York: Dover Publications, Inc., 1958.

Page, T., and L. W. Page. Beyond the Milky Way. New York: The Macmillan Company, 1969.

Rees, M., and J. Silk. "The Origin of Galaxies." Scientific American, June 1970.

Sandage, A. R. The Hubble Atlas of Galaxies. Washington: Carnegie Institution of Washington, 1961.

THE SUN

The Sun is the star nearest to Earth, the supplier of energy for our planet and for all parts of the solar system. The total amount of energy it produces is stupendous. Only one two-billionth is received by Earth, yet this is enough to fulfill our total energy needs many times over. Compared with other stars, the Sun is not impressive; there are hotter stars and cooler, brighter stars and dimmer, newer stars and older. The Sun is an average star. Nevertheless, as far as we are concerned it is the most important of all stars, for without it life would disappear from Earth, and the planet would become a barren, utterly cold world. From the standpoint of the astronomer, the Sun is of prime importance also because it is close enough to be observed as a disk. All

The Sun produces visible light, as well as radiation we cannot see. Here the "white light" Sun is shown together with (from the left) the Sun in X radiation, ultraviolet light, hydrogen light, and radio waves.

261

other stars are so far away they can be seen only as points of light, and the amount of information obtainable from these points is limited. Study of the solar disk enables astronomers to learn a great deal about the Sun, and thus indirectly to understand the more distant stars.

DISTANCE TO THE SUN

As we have seen, distances in deep outer space are measured using the light-year and the parsec (3.26 light-years) as metersticks. Within the solar system, the astronomical unit (a.u.) is the meterstick. This is the average distance from Earth to the Sun—149 598 000 kilometers (or, in the old English system, 92 957 000 miles). Through the years, various techniques have been employed to measure the a.u., and the results have varied considerably.

Aristarchus
Aristarchus, a Greek astronomer who lived in the third century B.C., found the distance to be 1.15 million kilometers, or about twenty times farther than the Moon was believed to be. His procedure involved careful observation of the changing appearance of the Moon, as seen in Figure 11–1. When the Moon is at position 1, he reasoned, we see it as a perfect quarter Moon, exactly half illuminated. When it is at a right angle to the line connecting Earth and Sun, we see it as a *gibbous* Moon—slightly more than quarter. The

FIGURE 11–1 *The method used by Aristarchus to measure distance to the Sun involved changes in the appearance of the Moon.*

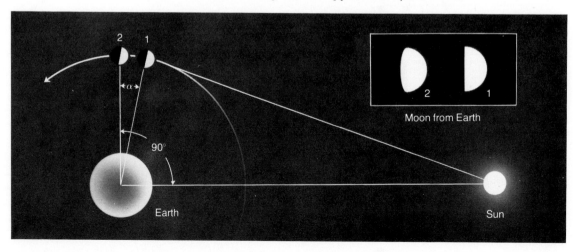

farther the Sun is from Earth, the smaller will be angle a. Aristarchus estimated that between these two positions the angle was 3°. His reasoning was right but his estimate was wrong, for the angle turns out to be only 9′, about twenty times smaller. Actually, angle a is unmeasurable, for it is virtually impossible to determine the moment when the angle between Sun and Moon is precisely 90°.

The Moon, according to Aristarchus, would take four hours to move from one position to the other; actually it takes only 12 minutes. The reasoning Aristarchus followed gave some feeling for the relative distances of the Sun and Moon, although they were actually much greater.

Hipparchus

Hipparchus, who lived in the second century B.C., ranks among the great astronomers of all time. He made several contributions; for example, he used lunar eclipse observations to determine the distance to the Sun. By watching the Moon enter and leave Earth's shadow, Hipparchus obtained the width of the shadow at the Moon's distance.

Scholars of that time knew roughly the sizes of the Earth and Moon and the distance between them. Therefore Hipparchus could construct a scale drawing of Earth, Moon, and Earth's shadow during a lunar eclipse. Lines marking the edges of the shadow were extended out into space. Hipparchus also knew the apparent diameter of the Sun, angle B in Figure 11–2. He extended the sides of the

FIGURE 11–2 *Hipparchus observed passage of the Moon through Earth's shadow during a lunar eclipse to provide data for measuring distance to the Sun.*

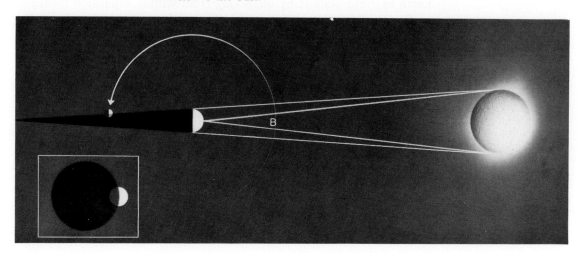

angle until they intercepted the lines drawn earlier. This would define the position of the Sun. From the information about sizes and distances available to him, Hipparchus could compute the size of the Sun and distance to it. His conclusions were incorrect, largely because of the inability to make accurate observations.

Other astronomers devised other procedures for solving the problem. One was to observe the Sun from two different locations on the Earth. Suppose two observers are on the equator but at different sides of the Earth. They observe the same point on the Sun. The angle of viewing can be measured, the distance separating the observers can be measured, and so the distance to the "point" can be computed. The procedure has merit; however, it is quite impossible to select a "point" on the Sun. In practice, the center of the Sun has been used.

A more precise "point" is available on those rare occasions when Venus transits (passes in front of) the Sun. While the procedure is good, there are few chances to check figures because transits are rare—the most recent occurred in 1874 and 1882, and the next transits will be on June 8, 2004, and June 6, 2012.

Johannes Kepler

Johannes Kepler (1571–1630), using the data compiled by Tycho Brahe (1546–1601), discovered that the squares of the periods of the planets are proportional to the cubes of their average distances from the Sun. The relationship enables one to construct a scale map of the solar system in which relative distances of the planets can be given correctly. A knowledge of the number of kilometers in an a.u. is needed, however. This is obtained by measuring distances from Earth to other planets by observing from widely separated locations on Earth. Using the planets works better than trying to find the distance to the Sun directly, because planets are more nearly points of light than is the Sun. In actual practice, the asteroid Eros serves as the "planet." It is a point in the sky, and it comes within 22 400 000 km of Earth; the closest a planet (Venus) gets to us is 33 600 000 km, and so computations based upon Eros are more accurate.

Doppler Shift

Measuring the speed of Earth's motion around the Sun is another way of finding the size of the a.u. In Figure 11–3 Earth is shown in three positions as it goes around the Sun. Measurements of the Doppler shifts of the background stars along the ecliptic made from position 1 indicate a shift toward the blue, equivalent to a speed of 29.6 km/sec. Either the stars are moving toward us, or

Earth is moving toward them. Six months later, similar measurements reveal a shift toward the red which also reflects a velocity of 29.6 km/sec. Stars cannot move first one way, then another. The only interpretation is that we are observing effects of Earth's motion. Since the distance Earth moves in one second is known, we can find the distance covered in one year—the circumference of the orbit.

FIGURE 11–3 As Earth goes around the Sun, light from distant stars along the ecliptic appears to shift toward blue at 1, toward red at 3; and there is no shift at 2 and 4. By knowing the amount of shift, distance to the Sun can be computed.

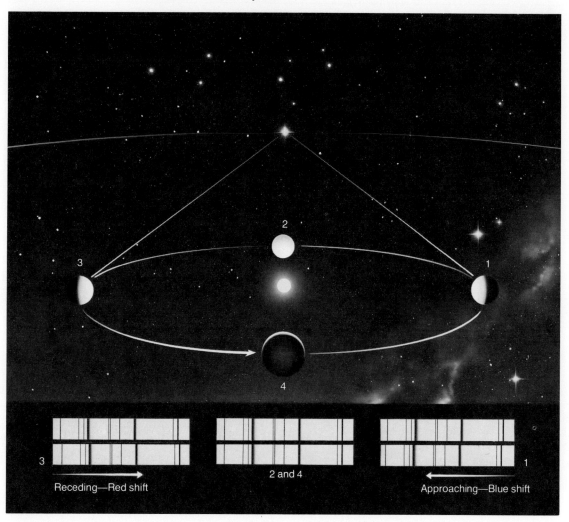

Knowing the circumference, we can find the diameter of Earth's orbit. The a.u. is the radius of the circle, one-half the diameter. The figure obtained by this method (148 607 370 km) is approximate because Earth's orbit is not circular, and its velocity is not always 29.6 km/sec. A more precise meterstick is needed.

In the Space Age, astronomers have other tools available to them such as radar, Earth satellites, and Sun-circling probes. Values of the a.u. obtained at different times by different methods vary by more than two-tenths of an astronomical unit. The a.u. as adopted by the United States Naval Observatory, which is the center in this country for such information, is 149 600 000 km.

MASS OF THE SUN

The Sun is an average star, but compared to the Earth its size and mass are awesome. In order to determine the mass of the Sun we must first know the value of the constant of gravitation, G, as well as the size of the astronomical unit, a. How G is found is explained in Chapter 12 on page 327.

We can measure the speed of Earth in orbit around the Sun, which we will call v. There must be a force exerted on Earth to keep it in orbit, and this is the force of the Sun's gravitational attraction, F. By Newton's law, this is

$$F = G \frac{S \times E}{a^2}$$

where E is the mass of Earth. But this must equal the force trying to keep Earth moving in a straight line, which is $(E \times v^2)/a$. Equating these we get

$$G \frac{S \times E}{a^2} = \frac{E \times v^2}{a}$$

Notice that Earth's mass, E, and one of the a's cancel out. Rearranging things gives

$$S = \frac{v^2 \times a}{G}$$

But v is just the length of the orbit, $2\pi a$, divided by the time for one orbit, T, so

$$v^2 = \frac{4\pi^2 \times a^2}{T^2}$$

Putting this in, we get

$$S = \frac{4\pi^2 \times a^3}{T^2}$$

We know all the items in the equation, so we can solve for the mass of the Sun, S. The Sun's mass turns out to be 2×10^{33} grams or 2×10^{30} kilograms, some 332 958 times the mass of our planet Earth.

SOLAR DENSITY

The average density of the Sun can be computed, since we know its size and mass. The volume of the Sun is 1.4×10^{33} cubic centimeters (3.37×10^{17} cubic miles), meaning that the density must be about 1.41, considerably less than that of Earth, which is 5.5. The density of the Sun increases greatly with depth, probably reaching a value of 100, and some believe as high as 400, in the central region. In the outer, visible region of the Sun the density is much less than our air at sea level.

THE SOLAR SPECTRUM

Spectroscopic studies of the Sun reveal that some 67 different elements occur there. The only parts of the Sun available to us for spectroscopic analysis are the *photosphere*, the *chromosphere*, and the *corona*. Additional elements may exist in regions inaccessible to us, or occur in such small amounts they do not appear in the spectrum.

The Sun produces a dark-line spectrum. The hot interior region, where gases are compressed, produces a continuous spectrum. The less dense outer gases produce bright lines. Those lines, in a sense filtered out by the atoms in the less dense layer, appear as dark lines.

The German physicist Joseph von Fraunhofer (1787–1826) identified some 574 dark lines, although he was unable to explain their presence. Today tens of thousands of them are identified, and most are associated with particular elements. The unidentified lines represent atomic changes not fully studied in laboratories.

In Figure 11–4 we show a small section of the dark-line spectrum of the Sun, and the bright lines of iron. Notice that the bright lines of iron (produced in a laboratory) match exactly certain of the dark lines of the Sun, informing the observer that iron exists in the Sun. Similar procedures reveal that the Sun contains 67 elements.

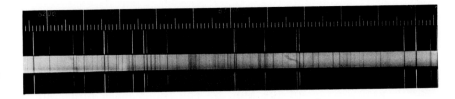

FIGURE 11–4 A small section of the spectrum of the Sun (dark lines in the central band) and the spectrum of iron (bright lines above and below the solar spectrum). (Hale Observatories)

For every million atoms of hydrogen in the Sun, there are some 100 000 atoms of helium, 500 of oxygen, 400 of nitrogen, 200 of carbon. The relative number of atoms then drops off to 33 for magnesium, the next most abundant material, 20 for silicon, and so on.

Because metals, which are defined as anything other than hydrogen and helium, occur in the Sun, the Sun is called a metal-rich star. This sets it apart from those stars that are almost purely hydrogen and helium.

The visible portion of solar radiation ranges from long waves at the red end of the spectrum to short waves at the violet end. By using instruments sensitive to infrared and ultraviolet wavelengths, the area of radiation open to study is broadened considerably. Because of the filtering action of Earth's atmosphere—which is an everlasting barrier to investigation—astronomers were most anxious to get instruments aboard space probes. This has been done and continues to be done routinely, and so we now have photographs of the Sun not only in ultraviolet light but also in X radiation, which is even shorter wavelength (see title page).

The Sun has been observed for as long as man has been on Earth. Except for the occasional appearance of sunspots (see page 273), the Sun was considered to be a perfect object—smooth and uniform. Galileo's telescope first revealed that the solar surface was not smooth. Today's instruments reveal it to be dynamic, ever-changing, roughly textured. And it is probably built up in distinct layers.

REGIONS OF THE SUN

We cannot see the greater part of the Sun, nor can we explore it with instruments. Therefore, much of the knowledge of its structure has been obtained by calculation and not by observation.

The Sun is made of gases. In the interior the gas is completely ionized, but near the surface and above there are both neutral and ionized atoms, and a small number of molecules, those hardy enough to survive high solar temperatures. A few additional molecules occur in sunspots where temperatures are slightly lower.

Density, concentration of matter, increases with depth. In the outer region of the Sun, the region called the *atmosphere*, density is very low—about 10^{-7} g/cm^3. This is 0.0000001 g/cm^3, or about 0.0001 the density of Earth's atmosphere at the surface. Inside the Sun, at the central core, the density according to some investigators is about 100 g/cm^3—and some calculate it is as much as 400.

Photosphere

The part of the Sun visible to us is the *photosphere*, the intensely bright region. Beneath the photosphere is the opaque, impenetrable mass which we cannot explore except by calculation. Above the photosphere is the *chromosphere*, a region that we can observe well only at rare moments just before and after solar eclipse totality. The photosphere appears to be a smooth surface. However, it is a layer, one that is a few hundred kilometers deep, and it is mottled with cells of hot gases that rise to the surface, cool after a minute or so, and then fall back to be replaced by additional hot gases that rise from underneath. The solar surface is granulated; the cells of hot gases are often called *rice grains* because they appear as grains in a dark surface—the cooler regions. The surface is dynamic, changing continually. It is a seething, boiling surface often interrupted by unusual activity (Figure 11–5).

At the center of the visible solar disk the photosphere appears brightest, dropping in brilliance toward the *limb*—the outer edge. Light from the center passes through a minimum depth of filtering gases in the chromosphere which lies above the photosphere. As the limb is approached, the depth of filtering material increases, and we cannot see as deeply, so we see layers at lower temperature and thus lower brightness; there is *limb darkening*.

Time-lapse photography reveals that streamers of bright hydrogen and calcium explode from the Sun for thousands—even millions—of kilometers, cool, and are then pulled back by the Sun's gravitation. Localized regions flare into brilliance. Areas of the Sun hundreds of times larger than Earth cool, become dark, and contain strong magnetic fields.

These phenomena, most of which originate in the photosphere, extend through and above the chromosphere which lies above the photosphere. In the solar interior, at the core where nuclear reactions generate the energy of the Sun, temperature is at least 15 000 000°K. The temperature of the photosphere is on the order of 6000°K. But the temperature of the upper region of the chromosphere and bending into the *corona* above is much greater—a million degrees, and more. (In this case temperature indicates the energy state of the

FIGURE 11–5   *Section through the solar surface showing dark, cooler sunspot, prominence, coronal streamer, granular surface.*

atoms, or the amount of ionization that has occurred.) In the corona the atoms are extremely energetic—so much so that radiation from them is in the ultraviolet and X-ray region of the spectrum invisible to the unaided eye, and fortunately filtered out by Earth's atmosphere.

Chromosphere

Ordinarily we cannot see the chromosphere. It is overpowered by radiation from the photosphere. However, during total eclipses, it can be seen as a narrow pink band just outside the disk of the Moon. The total phase of an eclipse is fleeting, usually lasting only a few minutes, and so the chromosphere has been observed very little

FIGURE 11–6
Photograph of the
flash spectrum of the
Sun (produced by the
chomosphere) taken
at Middletown, Con-
necticut, just before
and after totality
during the eclipse of
January 24, 1925.
(Hale Observatories)

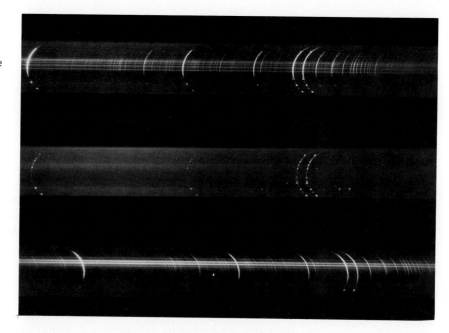

FIGURE 11–6
Photograph of the
flash spectrum of the
Sun (produced by the
chomosphere) taken
at Middletown, Con-
necticut, just before
and after totality
during the eclipse of
January 24, 1925.
(Hale Observatories)

against a dark sky. Using filters that block out the photosphere, the chromosphere can be seen at times other than eclipses, even though the sky is bright (Figure 11–6).

The lower part of the chromosphere, extending 1600 kilometers or so, is called the *reversing layer*. It is the place where the dark lines of the solar spectrum originate. This was first proved in a photograph of the chromosphere made during the eclipse of 1870. Bright lines showed up in the spectrum, and the lines coincided with the dark lines that had been analyzed initially in the early years of the nineteenth century by Fraunhofer.

Prominences

The chromosphere does not have definite boundaries. Rather, the height varies at different locations and at different times up to 12 000 kilometers, and has been observed much farther during some eclipses. It appears to be undulating much as a violent, rolling sea. Occasionally gaseous masses appear to float through the boundary of the chromosphere. At other times they burst out explosively. When a mass extends out very far, it is called a *prominence*. Hydrogen is the principal substance in a prominence, and ionized helium and calcium are quite apparent. Metals such as iron, magnesium, titanium are often identified, though only under the most favorable conditions.

We would expect all prominences to appear first at the surface, move upward, and then be pulled back to the surface. Many of them do, but very often a prominence is seen moving toward the Sun when there has been no apparent movement away from it. Astronomers do not completely understand the origin of the material. Probably it comes from the surface, but how does it get far beyond the surface without being observed? It may be carried upward in an invisible form, rather like the water vapor in our atmosphere which is invisible until it forms into clouds. Perhaps while moving upward the atoms (ions) are nonluminous and become bright as ionization is altered. The gases themselves move through the prominences, so that a given prominence is not always composed of the same material. The outer portion of the chromosphere blends imperceptibly into the corona, the outermost layer of the solar atmosphere.

Corona

During total solar eclipses the corona appears as a spectacular, usually irregular-shaped glow extending thousands of kilometers beyond the solar disk, and with occasional streamers that appear to extend for a million and more kilometers. Coronal material reaches to Earth, and even beyond We are enveloped in the solar corona (Figure 11–7).

FIGURE 11–7
Photograph of the Sun in total eclipse taken aboard S. S. Canberra, June 30, 1973. (M. R. Chartrand)

The spectrum of the corona was an enigma that challenged astronomers for decades since the lines could not be related to any known elements. They were broad and fuzzy, not sharp and clear as were those of the chromosphere. In 1942 Bengt Edlen, a Swedish astrophysicist, determined that the lines were being produced by highly ionized atoms—ions that could be produced only if the temperature were in the order of 700 000°. Here was an apparent paradox: an extremely hot layer overlying a much cooler (6000°) solar surface.

The broad lines seen in the spectrum of the corona are the so-called forbidden lines similar to the ones which appear in luminous nebulae. They can occur only in gases at very low density.

In the corona, hydrogen and helium do not show up. These elements are there, no doubt, but they have lost all their electrons and so do not produce any spectral features. Most of the ions of other atoms are so energetic that much of the radiation they produce is ultraviolet and so not readily observable. The lines which do appear provide clues that enable temperature readings, because temperatures needed to produce certain lines are known. For example, in a limited region of the corona a line produced by the calcium atom that has lost 14 electrons appears. The amount of energy needed to produce this particular atom indicates a temperature of 2.5 million degrees. The high temperatures are believed to be produced by shock waves caused by flares. This also drives the corona outward. The exact mechanism is far from completely understood.

SUNSPOTS AND SOLAR FLARES

Early observers of *sunspots* believed they were seeing opaque objects located between Earth and the Sun. After observing them with his 32-power telescope, Galileo concluded they were located in the solar surface. (He also went blind eventually from looking at the Sun through a telescope.) He noticed that the dark areas drifted across the Sun in about 13 days, and explained the observation by saying the spots were carried along by a rotating Sun.

Time has shown that Galileo was correct. Sunspots close to the equator go around the Sun in about 25 days, while those occurring at higher latitudes have longer periods. In fact, observations of the movements of sunspots serve to give measurements of solar rotation. The application is limited generally to the region between 5° and 30° north and south latitudes, the belt where sunspots are most common (Figure 11–8).

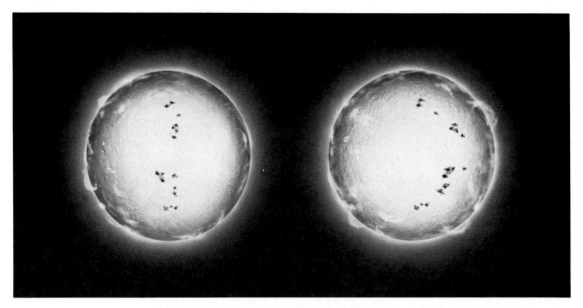

FIGURE 11–8 *Sunspots generally occur in the middle latitudes. They can be used to determine the rotation period of the Sun. At left a stylized arrangement, and at right the positions after a few days—the equatorial region rotates faster than the polar regions.*

Early reliable information about sunspots was obtained quite accidentally by Heinrich Samuel Schwabe (1789–1875), an astronomy buff and apothecary who was thwarted in his hobby by his assignment to night duty. Daylight sky watchers could not observe the stars. nor could they seek comets. However, around that time there was considerable excitement about the possibility of a planet existing in the region between the Sun and Mercury. If there were such a planet (it was called tentatively Vulcan) it would have to move in front of the Sun sometime during its orbit. The possibility of discovering a planet was enough to keep Schwabe observing the Sun at every opportunity. For 20 years he kept meticulous records of his observations of all black splotches seen in the Sun. None turned out to be a planet. However, his drawings of sunspots and their movements enabled him to discern the 11-year cycle of sunspot activity. Every 11 years the number of sunspots reaches a maximum.

Research on sunspots lagged for some fifty years after Schwabe's discovery. It was revived by George Ellery Hale (1868–1938), who invented the *spectroheliograph*. This is an instrument that enables photographs of the Sun with radiation from a single

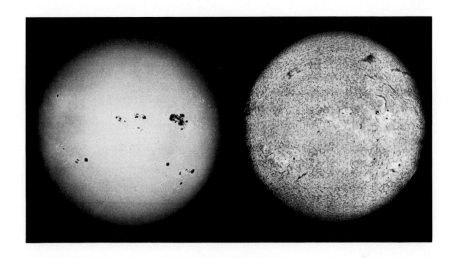

spectral line—a hydrogen alpha Sun for example, or a calcium line. Spectroheliograms revealed that sunspots cause spectral lines to be split. A Dutch astronomer, Peter Zeeman (1865–1943), established that the reason for the splitting of spectral lines is the presence of a magnetic field. The dark vertical line in the photograph in Figure 11–10 is the region of the sunspot scanned by the spectrograph. The dark vertical line on the left is an absorption line—a dark line of

FIGURE 11–10
The dark vertical line
at the left is a solar
absorption line.
Notice that when the
line passes through a
sunspot (indicated at
the right), the line is
split into three parts
—the Zeeman effect.
(Hale Observatories)

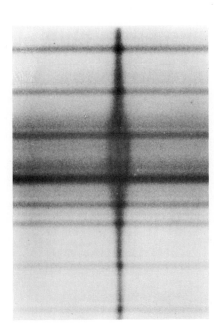

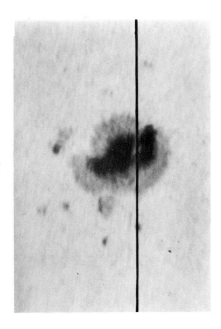

the solar spectrum. A careful look reveals that in the region of the sunspot, the line is split into three parts; this is positive proof of the presence of a magnetic field.

Origin of Sunspots

Magnetism and the appearance of sunspots in pairs proved to be valuable pegs upon which astrophysicists could fasten theories to explain the occurrence of sunspots. George Ellery Hale suggested that there are U-shaped whirls of gases within the Sun, which bend in toward the center. Occasionally, the ends of the whirls break through the surface, producing sunspots. This would explain the pairing, and also the alternation of clockwise and counterclockwise movements.

Wilhelm Bjerknes (1862–1951), a Swedish astrophysicist, agreed with Hale that swirls of gases were involved. However, Bjerknes suggested that the swirls encircle the Sun at right angles to the axis. Occasionally, the doughnut shapes break into horseshoe shapes, the ends of which break through the surface. Spots appear at the breaks and persist until equilibrium is again established—by a process presently unknown. The theory supposes there are four systems, two in each hemisphere. The systems move up and down in latitude while the gases in them are spinning in opposite directions. According to Bjerknes, this would account for the changing polarity of the solar magnetic field.

C. Walén and Hugo Alfvén, two Swedish investigators, proposed a different explanation. They assumed that the core of the Sun is magnetic. If an explosion were to occur at the edge of the core, a

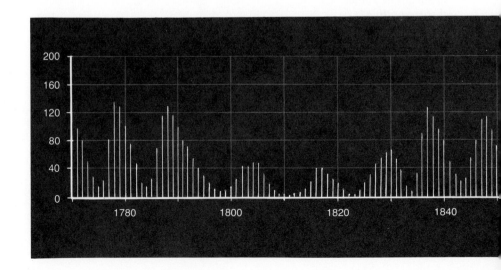

wave or pulse would be created. It would spread out and move toward the surface. The wave itself would carry a magnetic field that would persist as long as the pulse existed. Perhaps the pulses are rings that rise to the surface. Where they do, a sunspot is produced.

Although the causes of sunspots have not been determined, we do know that sunspots are cool regions—some two or three thousand degrees cooler than the solar surface. Still, they are very hot, and appear dark only because of contrast with the hotter material that surrounds them. The areas surrounding sunspots are often slightly brighter than the rest of the surface. Such bright areas are called *plages*. If the Sun were blotted out suddenly except for a single moderate sunspot, Earth would still be illuminated by a light equal to about one hundred full Moons. Sunspots are not whirlpools of gas, but more like heat pumps cooling off areas of the Sun's surface.

Schwabe was right when he reported that the number of sunspots varies in an 11-year cycle. We still do not know why. While large numbers of spots do appear in an 11.2-year cycle, the period is not regular. Maximums have recurred as little as 7½ years apart and as much as 16 years apart. The charts show sunspot activity over the past 200 years (Figure 11–11).

The cycle begins with spots appearing in high latitudes in both hemispheres, and as the years go by spots appear in successively lower latitudes. In one cycle a large spot, called the leader, will be magnetically positive. It will be followed by another large area

FIGURE 11–11 Sunspot frequency over a period of some 200 years. Note that frequency rises steeply, then drops off gradually.

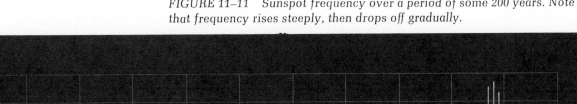

which is negative. During a given cycle, all leaders in the Northern Hemisphere will be positive and the followers will be negative; just the opposite is true in the Southern Hemisphere. During the next cycle the polarity situation will be reversed. These changes in polarity of sunspots actually mean that a sunspot cycle requires 22 years, and not 11, to be completed.

Solar Flares

When sunspots are at minimum, the Sun is quiet; that is, there are few prominences, solar flares, or other disturbances. When there are many sunspots, so also are there other activities on the Sun; prominences are more frequent and of greater dimension, so are flares. A flare begins as a brilliant fleck of light, usually near a sunspot. Rapidly the flare spreads over a surface of millions of square kilometers. Temperature goes up very fast as the flare moves with explosive speed—some believe we are observing movement of a wave of brightness and not matter at all. In half an hour or less the flare has subsided and there is no indication on the Sun of its presence.

A flare may be an indication of major upheavals occurring beneath the solar surface. For reasons not known, the area of a flare becomes greatly excited, so much so that light is given off, and also other forms of energy that affect Earth. For example, after flare activity there is a marked increase in ultraviolet radiation. The radiation that reaches Earth causes disturbances in radio transmission. Ordinarily, the ion concentration in the ionosphere of Earth is such that radio signals are reflected from it much as light is reflected by a mirror. When the ion concentration increases rapidly, as it does after flare activity, radio signals tend to be absorbed and so transmission is seriously interrupted.

AURORAS

Twenty or thirty hours after there has been flare activity, energetic auroras occur in Earth's upper atmosphere. In the Northern Hemisphere the display is known as *aurora borealis*, and as *aurora australis* in the Southern Hemisphere. Generally auroras appear every day along the outer rims of the arctic and antarctic regions. They are more numerous and more intense a day or so after a solar flare. As one moves toward the magnetic poles or toward the equator, intensity drops off rapidly, although on rare occasions displays are seen in lower latitudes—around Mexico City and in the Mediterranean region of Europe.

When light from an incandescent source passes through a prism, a spectrum is produced. Each chemical element produces characteristic lines in the spectrum which enable us to identify it. (Painting by H. K. Wimmer)

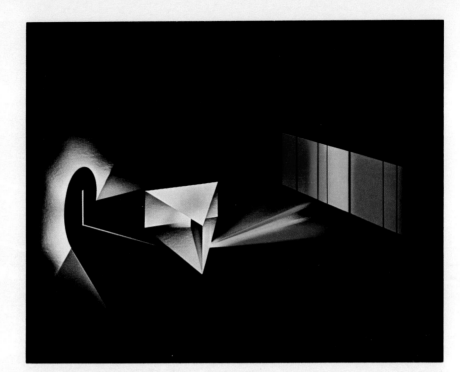

A "flash" spectrum of the Sun taken an instant before the total phase of the solar eclipse of July 20, 1963. In this way we can photograph the spectrum of the chromosphere of the Sun without the interfering light from the photosphere. (Tiara Observatory)

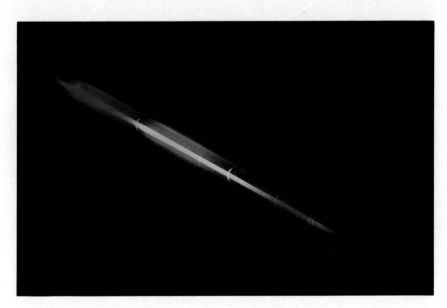

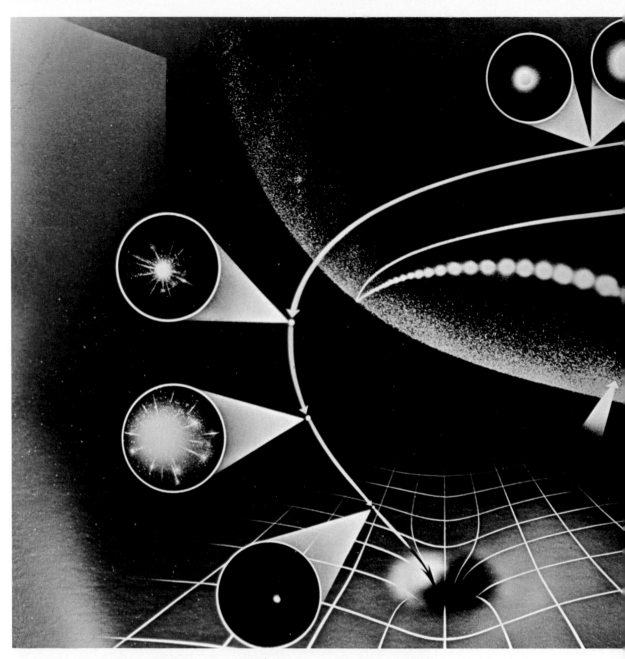

The life of a massive star, from dark cloud to black hole. In the lower right the cool cloud of interstellar matter contrasts and heats up. The protostar moves to the main sequence where it spends most of its life. After hydrogen is used up in the core, the star evolves to become a red giant, then a supergiant. Finally, having exhausted all nuclear fuels, it moves to the left on the H-R diagram. At some stage, it becomes a variable star. Later it may become a supernova, leaving behind it a neutron star, or a black hole.

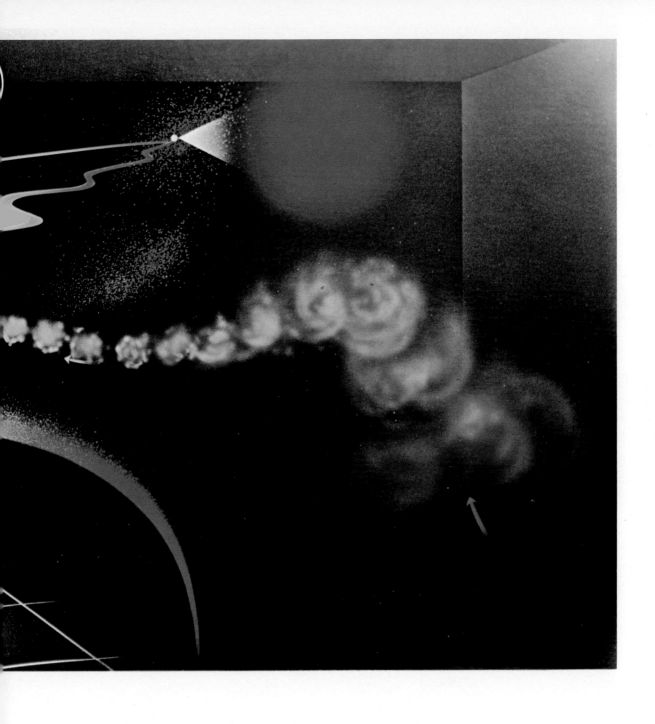

The Egyptians in their conception of the universe thought the Sun was a golden disk rowed across the sky in a boat. At night, the boat traveled around the horizon to bring the Sun back to the East in time for sunrise. In this view, the stars were lamps suspended from the dome of heaven by threads.

(Opposite page top)   Saturn, photographed by the 200-inch Hale Telescope at Mount Palomar. (Hale Observatories)

(Opposite page, left)   Earth as seen from space, photographed by Apollo astronauts near the Moon. (NASA)

(Opposite page, right)   The red planet Mars, photographed by Dr. Robert Murphy with the 24-inch telescope at Mauna Kea, Hawaii. (University of Hawaii)

(Opposite page, bottom)   Jupiter, photographed by Pioneer 10 on December 2, 1973, from a distance of 2½ million kilometers. In the upper right is Io, innermost of the four Galilean satellites. The dark spot on Jupiter's clouds is Io's shadow. Io is slightly orange due to surface ices. (NASA)

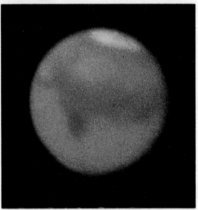

The Eta Carinae nebula. This cloud of gas and dust was almost as bright as the brightest star a hundred years ago. It has since faded to about eighth magnitude. (Cerro Tololo Interamerican Observatory)

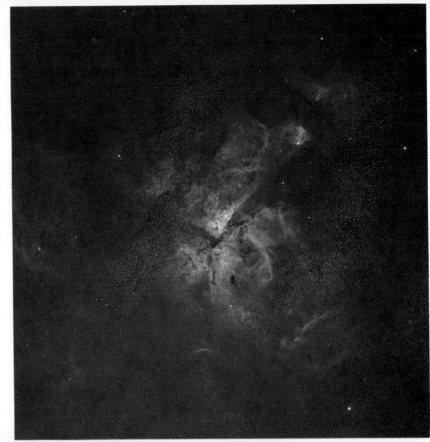

The nearby spiral galaxy M31 in Andromeda. In the upper left is a small companion elliptical galaxy. (Kitt Peak National Observatory)

The Crab nebula, the remains of a super-nova seen in A.D. 1054. The remaining star is a pulsar. (Hale Observatories)

The Pleiades star cluster, showing the hot, blue stars as well as the surrounding nebulosity which reflects some of the star light. (Hale Observatories)

The Great nebula in
Orion, photographed
through the new
4-meter Mayall Tele-
scope at Kitt Peak
National Observatory.

The dominant color in an aurora is a soft yellow-green produced when certain oxygen ions are stimulated by subatomic particles—fast-moving protons and electrons ejected from the Sun. The particles are captured in Earth's magnetic field, the region of the Van Allen radiation belts (see page 280), where they are accelerated to very high velocities.

Various theories have been advanced to explain the phenomenon of auroras. Frederick Carl Störmer (1874–1957), a Norwegian physicist, showed that low-energy electrons are guided to Earth's magnetic poles while those of higher energy spread out in a region about 22° from the poles. Sydney Chapman, an English astronomer, and Donald Menzel, an American, built theories around the idea that Earth is surrounded by streams of particles, some of which leak through to Earth's upper atmosphere. More recent knowledge ties the auroras to a system involving the solar wind, the magnetosphere, and the radiation belts.

Every day the Sun loses tons of material that move outward and away at high velocities. These subatomic particles comprise the *solar wind*. Earth's magnetic field, which extends some 80 000 kilometers, deflects these charged particles and captures a good many of them. Once captured, the particles move back and forth along magnetic lines of force from north magnetic polar area to south. They do not appear around the poles themselves; the region above the poles is almost devoid of radiation. The regions holding radiation particles (the Van Allen belts) are shaped somewhat like huge malformed doughnuts surrounding Earth, compressed on the Sun side of Earth and elongated on the night side (Figure 11–12). It has been suggested that particles "spill over" and out of the restraining magnetosphere (the doughnut). At altitudes of about 100 to 125 kilometers, and sometimes as high as 1000 kilometers, they react with ions of various gases to produce the auroras.

SOURCE OF SOLAR ENERGY

The Sun is a star and, as such, it is a nuclear furnace, a vast reactor in which matter is converted into energy. For at least 5 billion years the Sun has been giving off 400 trillion times more energy than the output of all the power plants of Earth put together. For millenniums the Sun was regarded as a beneficent god, and so there was no mystery concerning the source of so much energy; all things were possible so long as the gods were involved. When people sought more realistic explanations, it was understandable that they believed the

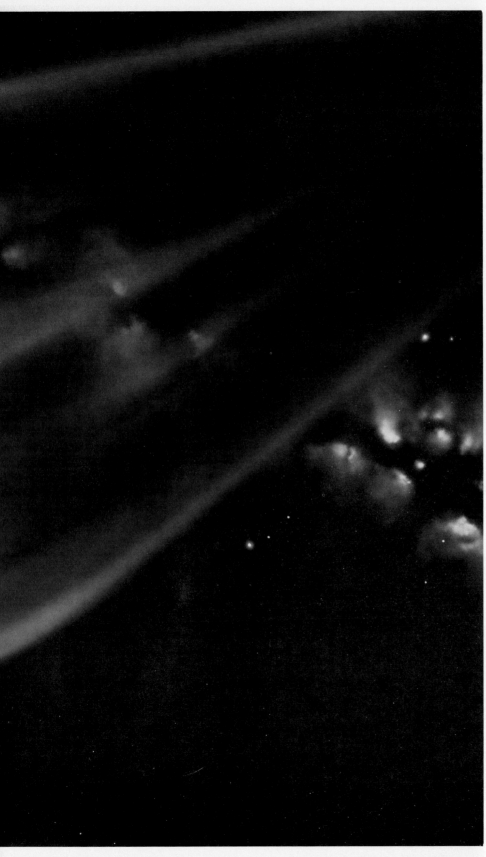

FIGURE 11–12
The solar wind, composed of subatomic particles, is shown blowing from the left. Earth's magnetic field traps particles forming radiation rings— the Van Allen belts. A shock front forms out beyond the belts. Particles deflected by the Earth trail the planet, producing a tapering "tail" extending thousands of miles into space. High-energy cosmic particles penetrate to Earth's surface.

Sun was a great celestial fire. This was actually most unrealistic, for if the Sun were pure carbon burning in pure oxygen, it would last only a few thousand years, burning at its present brightness. When thinking became more sophisticated and something was known about the structure of stars, investigators attempted to explain the energy of the Sun by saying it was produced by contraction. About a hundred years ago Hermann von Helmholtz (1821–1894), a German physicist, suggested that as the Sun collapsed, material packed together, and potential energy was converted to heat and radiation. Computations indicated that if the Sun were to shrink only a few hundred kilometers a year, enough heat would be generated to account for the measured energy. While such slight changes in the size of the Sun were not observable, they must be occurring. How else, it was reasoned, could the Sun provide so much energy? Also, the process would continue for several hundred million years, as long as the Sun and Earth had existed, or so it was believed. At the time of Helmholtz geologists did not know the age of Earth, nor of the Sun, and so the contraction theory was quite acceptable.

At the start of this century Albert Einstein opened the door that was to reveal the correct explanation. He theorized, and later was proven eminently correct, that matter and energy were related, and interchangeable. Under certain conditions energy could be converted into matter; also, small amounts of matter could be made into stupendous amounts of energy. Indeed, if 1 kilogram of matter were converted into energy it would produce 25 000 000 000 kilowatt-hours of electricity—enough to supply electricity for the entire United States for two months. Contrast this with 8.5 kilowatt-hours, the amount of power obtained from the burning of 1 kilogram of coal.

In the 1930s Hans Bethe (1906–), physics professor at Cornell University, determined how the Einstein relationship ($E = mc^2$) could be applied to the problem of explaining the source of solar radiation. Long before the process of nuclear fusion was demonstrated, Bethe reasoned that hydrogen atoms were reshuffled inside the Sun, resulting in new nuclei, and during the many steps in the process which took some 6 million years to complete, energy was liberated.

The nuclear process devised by Bethe is the *carbon cycle*, a fusion reaction which is triggered when a proton fuses with a carbon nucleus. The process is represented schematically in the diagram on page 110. Notice that four protons, hydrogen nuclei, go out of existence, and one helium nucleus is created—the carbon that initiated the reactions is unchanged. The carbon cycle is important for stars hotter than the Sun.

In the Sun, and in cooler stars, the dominant reaction is one involving the fusion of protons (hydrogen nuclei) and is called the *proton-proton reaction*. It is shown schematically on page 112. Notice that it is an ongoing process, and one that also involves converting four protons into a single helium nucleus. This process takes some 3 billion years to be completed, much longer than the carbon cycle discussed earlier.

In both processes, the mass of helium at the end of the cycle is less than the mass of the hydrogen involved. Calculations indicate that in these processes 564 000 000 tons of hydrogen go out of existence every second. The greater part, 560 000 000 tons, is converted into helium; some 4 000 000 tons are converted into energy. That is an astounding loss, yet the Sun is so massive that it will continue to radiate energy at this rate for another 5 billion years, as it has for the past 5 billion years.

SOLAR MOTIONS AND ECLIPSES

The Sun, Moon, and Earth, and all the other planets and satellites of the solar system, share in the motion of the Galaxy. At their location, some 30 000 light-years from the center, the galactic rotation is some 250 kilometers per second. The Sun itself is also moving some 19.4 km/sec toward the region of the constellations Hercules and Lyra. This motion is with respect to the nearby stars. The direction of motion is called the *solar apex*. The entire solar system is carried along by the Sun. As the system moves through space, each of the planets goes around the Sun, and each of the satellites in turn continues its motion around its mother planet.

Right now we are concerned with only those motions that place the Moon in just the right spot for its shadow to fall upon Earth, producing the phenomenon of a solar eclipse.

Total solar eclipses are among the most spectacular of all celestial displays, and eclipse expeditions are exciting occasions for both astronomers and laymen. In ancient days those who could predict the darkening of the Sun were considered godlike, and no doubt were held in as much awe as the event itself. There is a school of investigators which believes that the rings of stones at Stonehenge on the plains of Salisbury, England, were computerlike arrangements that enabled the prediction of eclipses. The idea has considerable merit, and may be correct. Those who could "read" the stones would have been considered to have magical powers. Almost universally, ancient people worshiped the Sun. It was a god that must be kept content at all times, for without its light and warmth life would be quite impossible. Ignorant as early man was, he knew this to be true.

The chances are that most people will someday see a total solar eclipse, if they have not seen one already. The experience makes it easy to comprehend why early man thought the Sun was being devoured by an evil monster, and why men shouted and yelled and beat on drums to drive away the dragon or at least to frighten it enough so it would disgorge the Sun and so return Earth to daylight. And it always worked.

A total solar eclipse occurs when the Moon moves in front of the Sun, cutting off its light. It can do this even though the Moon is much smaller than the Sun. The diameter of the Moon is 3456 kilometers, 400 times smaller than the diameter of the Sun. The distance to the Moon is about 380 000 km (it varies monthly from about 354 400 km to 404 000 km) and the distance to the Sun is about 148.8 million km—400 times greater.

The Sun and Moon are each very nearly 108 times their own diameters away from Earth. The edges of both the Sun and Moon just touch a cone that might be drawn from a location on Earth to the edge of the Sun, making it possible for the Moon to obscure the Sun without covering the solar corona.

For an eclipse to occur, the Sun, Moon, and Earth must be on the *line of nodes*, or nearly so. This is a line drawn through the points where the plane of Earth's orbit around the Sun intersects the plane of the Moon's orbit around Earth. Even then a total solar eclipse will not occur unless the Moon is at or near *perigee* (closest approach to Earth) so the shadow of the Moon reaches Earth. The mean distance of the Moon from Earth is 382 170 km, while the mean length of the Moon's shadow is 371 360 km. Most of the time the shadow of the Moon falls short of Earth, as well as falling above or below Earth's plane (Figure 11–13).

The plane of the Moon's orbit is tilted 5° to the plane of Earth's orbit. A total solar eclipse can occur only at new Moon (the Moon between Earth and Sun) when the Moon is moving downward, or upward, in its passage through Earth's plane. We can shorten this by saying the Moon must be at one of the nodes.

Even with these limitations, a total solar eclipse occurs somewhere on Earth once in an 18-month period. In 100 years, the average number of solar eclipses is 238—84 partial, 66 total, 77 annular (a ring of the Sun remains visible around the Moon), and 11 that are part total, part annular. However, the width of Moon's shadow on Earth averages only 96 to 112 kilometers, so it is rare to see an eclipse from any single region of Earth. On the average, about 350

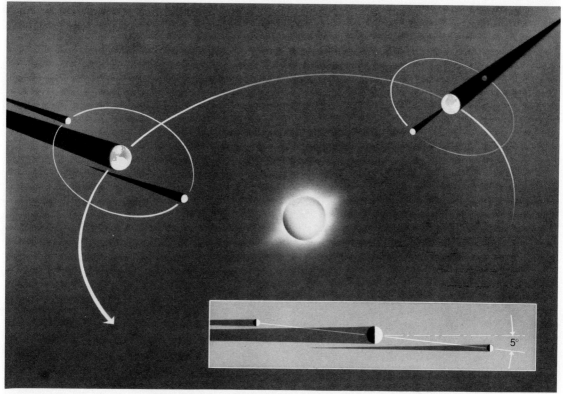

FIGURE 11–13   Eclipses occur when the Sun, Moon, and Earth are all on, or close to, the lines of nodes, the intersections of the planes of Earth and the Moon.

years elapse between total eclipses visible from any given location. However, there are exceptions; the last total eclipse in New York City was on January 24, 1925, and the next one will occur on April 8, 2024—only about a hundred years later. There was an eclipse in San Francisco on June 8, 1918, and the next will be on August 12, 2045.

The smallest number of eclipses possible during a given year is two, both of which will be solar, though not necessarily total. The largest number is seven, four or five of which will be solar. In 1805 and 1935 there were seven eclipses—five solar and two lunar—and the phenomenon will occur again in 2160. Only rarely are there so many eclipses in a 12-month period. The more usual number is four.

When an eclipse does occur, it is short-lived. The longest duration of totality (when the Sun is completely hidden) is about 7½ minutes, but usually the period is considerably shorter, only 2 or 3

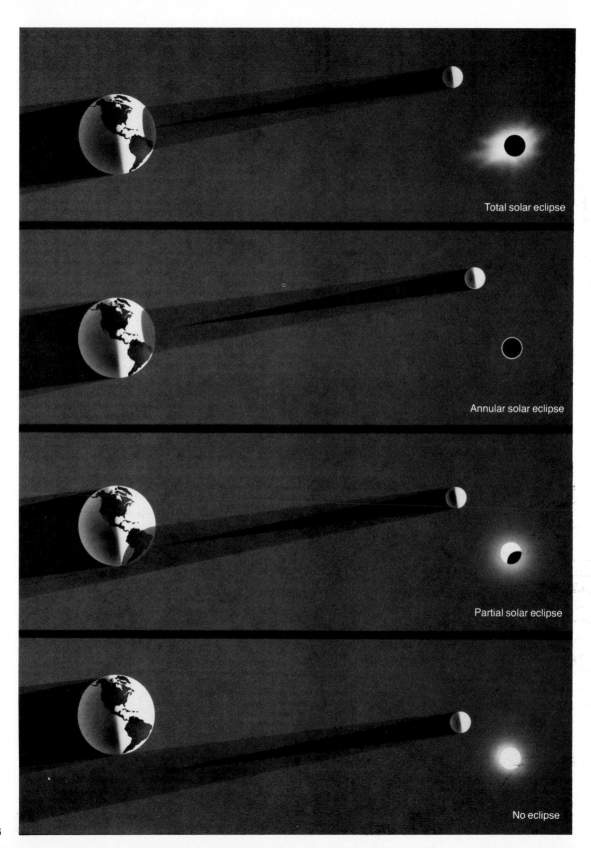

Total solar eclipse

Annular solar eclipse

Partial solar eclipse

No eclipse

FIGURE 11–14
(opposite page)
From top to bottom
the configurations
produce a total solar
eclipse, an annular
eclipse, a partial solar
eclipse, and no
eclipse.

minutes. The Moon's shadow moves rapidly past an observer—a bit over 1600 kilometers an hour. If Earth were not rotating, the shadow would move about 3360 kilometers an hour, the velocity of the Moon in its orbit around Earth. But Earth rotates in the same direction that the shadow moves, and so the actual speed of the shadow is the difference between the Moon's motion and Earth's. This amounts to about 1696 kilometers an hour at the equator (the slowest) and becomes faster with increasing latitude.

Table 11–1   *Total and annular solar eclipses 1976–1980*

1976	April	29	A	S. E. Asia, Africa, Asia Minor
"	October	23	T	Australia, S. Africa
1977	April	18	A	S. Africa
"	October	12	T	Pacific Ocean
1979	February	26	T	N. America
"	August	22	A	Antarctic
1980	February	16	T	S. E. Asia, Africa, India
"	August	10	A	Pacific, S. America

Solar eclipses may be partial, total, or annular. A partial eclipse is not very exciting; in fact, many people are not aware of anything unusual. An annular (from *annulus*, meaning "ring") eclipse is seen when the observer is beyond the cone of the Moon's shadow. It is observed as a bright ring around the disk of the Moon (Figure 11–14).

Baily's Beads

Total eclipses are spectacular displays. The lunar disk encroaches upon the Sun and, an instant before totality, a ring of bright lights appears around it. These are Baily's beads, named after the British astronomer Francis Baily (1774–1844), who first explained them. They are sunlight shining through the valleys between the mountains of the Moon. The last bright flash of light is called the "diamond-ring."

During totality light is reduced, for the only illumination received is that produced by the solar corona. When the Sun is active, streamers extend from the corona a solar diameter, 1.5 million kilometers and often farther. This is the only time the corona can be observed against a dark background. At the moment just before totality, the flash spectrum is revealed. This is a bright-line spectrum produced by the chromosphere but usually obscured by the brighter photosphere.

Light and Gravitation

In 1919, by making precise measurements of star positions during totality and comparing results with those obtained at other times, astronomers were able to provide evidence proving that Einstein was right when he said that light is bent by gravitational force of the Sun. Actually, in the language of the theory of relativity, the space around the Sun is warped by the Sun's mass, and the light rays from a distant star are "bent" by the distortion of space. At every eclipse since then additional measurements have been made, and refinement of results continues to be a challenge to eclipse observers.

Vulcan

In 1859, Dr. Lescarbault, a French amateur astronomer, reported that he had seen a black spot move rapidly across the face of the Sun. Immediately there was conjecture that the black spot was a planet, one that was closer to the Sun than Mercury; this "planet" was called Vulcan. The spot has never been observed by anyone other than Lescarbault; however, there are still believers in Vulcan. They photograph the region around the Sun during total eclipses, hoping sometime to see a tiny dot—sunlight reflecting from this mysterious world.

Total solar eclipses often are visible only over water or remote places, and cloud banks frequently obscure the Sun. Jet travel has solved pretty well the problem of remoteness—almost any place on Earth is now accessible within a day or so. Clouds are still a problem for ground-based observing. However, rockets carry remotely controlled equipment above the clouds, and eclipses can also be observed from space probes and from fast-moving jet planes.

Evolution of the Sun

The Sun is a population I star, one which is primarily hydrogen and helium but which also contains elements beyond helium. It is therefore a metallic star, as explained earlier.

For billions of years nuclear fusion reactions in the core have been producing energy and combining hydrogen nuclei to produce nuclei of helium. After billions of additional years, the hydrogen content at the core of the Sun will diminish sufficiently for the reaction to run down. At that time, hydrogen will "burn" in a shell surrounding the core. The Sun will become considerably brighter, it will expand, and the color will become redder. As the burning continues, the redness will increase, and so will the size of the Sun. Perhaps it will become large enough to reach the Earth, enveloping us in the hot solar atmosphere.

If the mass is great enough, and the loss of neutrinos is not too large (see page 140), the Sun may evolve in somewhat the fashion shown in Figure 11–15. There may be explosive periods, stages of

FIGURE 11–15   *The life cycle of the Sun. (1) The Sun originates from a vast cloud of plasma that condenses to form planets as well as a star. Mass of the original plasma places the star at about the middle of the main sequence. (2) Nuclear fusion reactions—hydrogen to helium— generate energy for perhaps 8 billion years. (3) Because of tremendous internal pressure, the Sun expands (4) to become a red giant star; the surface cools. (5) For thousands of years the Sun pulsates in luminosity; gradually it cools, or it may explode. If so, it becomes extremely brilliant (6), and "burns" itself out. (7) The remaining material collapses, and the star becomes a dwarf—perhaps no larger than Earth is now.*

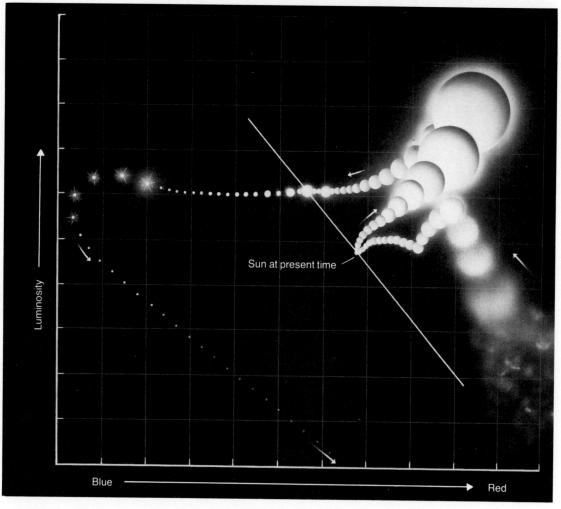

helium burning, and ultimate cooling. Judging by our present knowledge of neutrino loss, the Sun may follow a calmer existence, though one that will lead relentlessly to its demise.

Eventually the Sun will run out of all nuclear sources of energy. Gravitation will cause the star to collapse, resulting in the production of heat. The heat will cause expansion, and so the star will pulsate—changing in brightness, in color, and in volume. The remaining material, helium and the lighter metals (elements just beyond helium), will pack more densely together. The once-radiant Sun will become a white dwarf, a small, dull star not even discernible by our nearest space neighbors.

REVIEW

1   The average distance from Earth to the Sun is called the astronomical unit. It is equal to 1.496×10^8 km.

2   The Sun is entirely gaseous, with a surface temperature of 6000°K and an interior temperature of over 10 000 000°K.

3   The average density of the Sun is 1.41. The visible "surface," the photosphere, has a density less than that of air at sea level. The central density of the Sun is over 100 times that of water.

4   Sixty-seven elements have been identified in the spectrum of the Sun.

5   The solar spectrum is produced in the chromosphere, the layer of the Sun's atmosphere overlying the photosphere.

6   The corona, or outer atmosphere of the Sun, extends millions of kilometers into space. Although the density is very low, it is extremely hot.

7   The main source of energy for the Sun is the proton-proton process in which hydrogen is converted into helium with a release of energy.

8   Sunspots are most numerous every 11 years. Their origin is as yet unknown.

9   The Sun ejects high-energy particles into space, the solar wind. The interaction of strong bursts of particles from the Sun produces the auroras in Earth's atmosphere.

10   Solar eclipses occur when the Moon is near new phase and lies on or nearly on the line of nodes.

11   The age of the Sun is about 5 billion years, and it is ex-

pected to remain on the main sequence another 5 billion years.

QUESTIONS

1 What are some evidences of Sun worship by early man?

2 What is the astronomical unit? How was it computed by the Greeks? by the Doppler shift?

3 How does Kepler's work enable one to construct a model of the solar system?

4 How can the mass and density of the Sun be computed?

5 How does the reversing layer provide evidence of solar composition?

6 How are astronomers able to conclude that the Sun is built up in layers?

7 The temperature of the solar corona is much greater than that of the photosphere that lies beneath. How can this be?

8 What are some of the theories that have been offered to explain the origin of sunspots?

9 Why do we sometimes say that the sunspot cycle is 22 years and not 11, as commonly believed?

10 What is the nature of energy generation in the Sun and in other stars?

11 What circumstances are essential in order for there to be a total solar eclipse?

12 How might an investigator go about seeking the presence of Vulcan?

13 What changes in the Sun will occur as it goes through its life cycle?

14 It has been suggested that engineers must look to the Sun to provide energy. Does this seem reasonable? How might engineers go about utilizing solar energy efficiently?

15 Are there relationships between tree rings, stock-market cycles, birth rate, and the sunspot cycle? (You might include the work of Charles G. Abbott of the Smithsonian Institution in your research.)

READINGS

Ellison, M. A. *The Sun and its Influence.* New York: American Elsevier Publishing Co., Inc., 1968.

Gamow, George. *Birth and Death of the Sun.* New York: The New American Library Inc., 1952.

Hawkins, Gerald S. "Celestial Clues to Egyptian Riddles." *Natural History,* April 1974.

Kuiper, Gerard, ed. *The Sun*. Chicago: University of Chicago Press, 1954.

Menzel, Donald. *Our Sun*. Cambridge, Mass.: Harvard University Press, 1959.

Mitchell, Samuel. *Eclipses of the Sun*. 5th Ed. New York: Columbia University Press, 1951.

Pasachoff, Jay M. "The Solar Corona." *Scientific American*, October 1973.

CLOSE-UP OF A PLANET-EARTH

CHAPTER 12

Among all the planets, Earth is the only one we have been able to study firsthand. Probes sent to Venus have provided us with considerable information, and radar bounced from the planet gives us some idea of the contour of the surface. The surface of Mars and Mercury and the cloud cover of Jupiter and Venus have been photographed in considerable detail by space-borne equipment, and old questions about the Moon have been answered by the Apollo program at the same time new questions were raised. Three hundred years of telescopic observations and decades of investigations by advanced techniques have amassed considerable knowledge of our neighbor worlds, yet there are great gaps in our information. While each world appears unique,

Apollo 17 view of Earth taken while in transit to the Moon. Note heavy clouds over Southern Hemisphere. Almost the entire African continent can be seen, as well as the Arabian peninsula. (NASA)

it is reasonable to suppose that we can learn a great deal about the planets by learning about Earth.

Man is a curious creature. He has an everlasting discontent about the whys and hows of his existence, and certainly about his world: where did Earth come from, how did it come about, and when? No one has the answers to such questions. It was hoped that firsthand investigations of the Moon would provide some clues, and they did, but the answers led to new questions that still persist.

AGE OF EARTH

In the seventeenth century, the question concerning the age of Earth was answered by John Lightfoot, a scholar at Cambridge University. In 1642, he announced that the moment of creation was at 9 A.M. on September 17 in the year 3928 B.C.

Some years later, in 1658, the Primate of Ireland, Archbishop Ussher, wrote:

> In the beginning God created heaven and earth, Gen.I.V.I. which beginning of time, according to our chronologie, fell upon the entrance of night preceding the twenty third day of Octob, in the year of the Julian Calendar, 710.

FIGURE 12–1
In photographs of certain nebulae (NGC 6523 shown here) we may be observing early stages in the formation of stars and perhaps planetary systems associated with newly formed stars. (Lick Observatory)

Later on, Bishop Lloyd of England inserted the year 4064 B.C. in a commentary on Archbishop Ussher's statement. For more than a hundred years the Church accepted and taught that Earth began at that time—October 4064 B.C.

In the eighteenth century, geologists began to study rocks and knew from their observations that Earth was much older. Whatever processes were involved in its formation, they must have occurred not thousands of years earlier, but millions—even billions.

More recently, radioactive dating techniques—measuring ratios of breakdown products and the original elements (lead and uranium, strontium and rubidium, for example)—indicate that the age of Earth is in the order of 4.6 billion years. Moon rocks turn out to be the same age or a bit older, and so do meteorites. Theoretical considerations indicate that the age of the Sun is comparable.

The rather close agreement of the ages of worlds so far studied leads one to believe that all of them had a similar origin; whatever materials and conditions brought about the Sun also produced Earth.

EARTH ORIGINS

The Milky Way Galaxy is mostly composed of stars; only about 5 percent of it is unconsolidated materials, interstellar gases including nebulae which are formed of more dense gases. Studies of various nebulae such as NGC 6523, shown on page 296, reveal variations in density. These darker regions are probably concentrations of materials—plasma packing together more and more tightly. The masses of these darker regions appear to be about equal to that of the Sun, and distances separating them are about equal to those between the Sun and its neighboring stars. It is likely we are observing protostars, stars in a very early stage in their evolution. Eventually the gases will reach critical mass, at which point gravitation of the system will cause the formation to collapse. Temperature will rise, rapidly reaching the critical level where nuclear reactions will be initiated. Outward pressure will balance inward collapse—a star is born. (See also Figure 4–7 on page 76).

If some such process accounts for the formation of the Sun, it is logical to consider extensions of the theory in attempts to account for the manner in which Earth originated. Here are steps that may have occurred. The planets came out of the same mass that gave birth to the Sun—a (probably) spherical mass having a diameter of 15 billion kilometers, which contracted to a small fraction of the original volume. Collapse continued and, as the system rotated, it elongated and flattened, becoming a disk with a diameter of some 9.5 billion kilometers.

Most of the material became the Sun (today it comprises 99.86 percent of the mass of the solar system). The leftover dust, gases, and plasma churned about. Eddies formed here and there, many of them breaking up but some persisting, gathering in additional material and becoming more and more dense. Such a system became stable enough to hold together; it had become a protoplanet. Additional materials were attracted to it, and gravitation caused collapse; density increased steadily, and slowly the orbit changed from a flat ellipse to one much closer to circular.

The composition of protoplanets was mostly hydrogen and helium, and small amounts of heavier elements. Gradually compounds of hydrogen—ammonia, methane, water—would have formed. Since protoplanets were cold, these compounds would have condensed into ice crystals. Denser materials would have been pulled toward the central core. Ices and silicates of various kinds would have formed a cover for the denser materials, and surrounding the system there would have been an extensive mantle of gases, gases that had not become consolidated into the newly forming crust. At some stage there may have been violent, short-lived flare-ups of the emerging Sun, and newly formed heavier elements may have been ejected from it into space. The flare-ups may have been energetic enough to dissipate the gaseous envelopes of the early planets. Many of the materials in the crust may have been evaporated also. The permanent atmosphere of Earth and its oceans may have emerged later, formed by combinations of substances that were released from the crust, after the final solar flare-up, through volcanic action.

AIR (ATMOSPHERE)

Of all the known planets, Earth is the only one that contains an atmosphere composed primarily of nitrogen and oxygen. These two gases add up to more than 99 percent of the total. Argon and carbon dioxide comprise the main part of the remaining 1 percent. The breakdown of that fraction is shown in Figure 12–2.

Although we are concerned here with very small fractions, the total number of atoms is considerable, for the total weight of the atmosphere is about 4.5×10^{15} metric tons.

The chart shows those parts of the air mixture which remain essentially unchanged at all times and at all locations. In addition, the air contains numerous substances which are highly variable.

Water droplets and dust particles are obvious. Not so apparent are many other materials, some of which are of considerable concern because of their effects upon the human organism: ammonia, hydrogen sulfide, sulfur dioxide, sulfur trioxide, carbon monoxide, and methane.

The air envelope is separated into layers, many characteristics of which have been revealed by satellite studies. While the layers do exist, they are not sharply separated; they blend into one another.

The lowest layer, the *troposphere*, which contains about 75 percent of all the atmosphere, extends some 9 to 11 kilometers.

FIGURE 12–2   *The components of the atmosphere that remain essentially constant. In addition, the atmosphere contains varying amounts of water vapor, dust, pollen, spores, and the multitude of pollutants from industry and vehicles.*

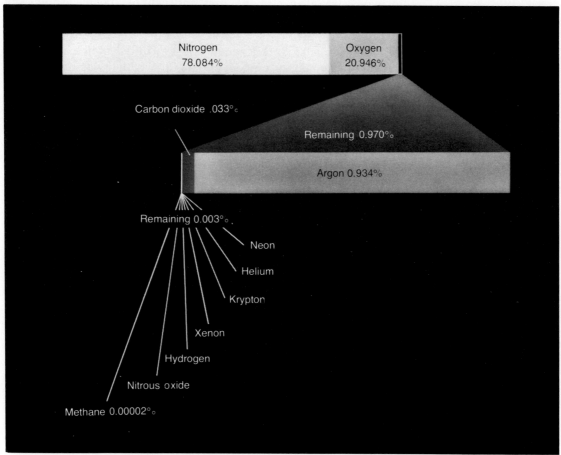

Weather occurs in this layer. Above it, and extending some 48 kilometers, is the *stratosphere*. Above this, and extending about 30 kilometers farther, is the *chemosphere*. This is the region where some new chemical compounds are found. For example, oxygen ordinarily occurs as O_2; it is *diatomic*, made of two atoms. In the chemosphere, we find elemental oxygen (O), or *monatomic* oxygen, occurring. Water molecules separate into free hydrogen atoms and the incomplete hydroxyl radical (OH) which recombines with oxygen to form water.

There is some ionization (the stripping of electrons) in the chemosphere. However, the process is more pronounced in the *ionosphere*, the region that extends some 400 kilometers beyond Earth's surface.

In 1901, when Marconi planned to send a radio message across the Atlantic, he was advised that the distance was too great since radio waves travel in straight lines and cannot "bend" around the horizon. Somehow, however, the waves did reach around the horizon.

Two physicists, A. E. Kennelly in the United States and Oliver Heaviside in England, offered an explanation which was tested some 25 years later when a signal was bounced off the electrically active ions of the ionosphere from station to station. It was established that the lower part of the ionosphere was 100 kilometers up. Later investigations revealed there are several different layers of the ionosphere, each reflecting radio waves of different frequencies. The ionosphere also plays a significant role in the generation of auroras, which were discussed in the previous chapter. The various layers of the atmosphere from ground level to interplanetary gases are shown in Figure 12–3.

Notice also a temperature line indicating initially a drop with increased altitude, but later on a steady rise as altitude gets above 120 kilometers. In the lower air, temperature drops because the primary source of heat is the warm Earth. In the stratosphere, temperature holds at about $-55°C$, then it rises sharply, reaching $10°C$ above zero at a height of 50 kilometers where a layer of *ozone* (O_3) catches solar radiation. Above this level, temperature falls off once more, reaching $-85°C$. Above this altitude temperature rises steadily, probably because of absorption of ultraviolet radiation by ions. Six hundred kilometers out, the temperature has soared to $1400°C$. But one wouldn't feel it, because temperature is a measure of molecular motion, and it gives no information about the number of molecules. Heat is sensible to us because of molecular motion as well as the number of molecules involved. At the surface, billions of molecules collide with us every second; in outer space the number drops sharply. A temperature reading for outer space of 1400°C means that the molecules (few as they are) are moving at a velocity they would possess if heated to 1400°C here on Earth.

From the standpoint of the astronomer, observation would be much more extensive and informative if there were no atmosphere, for it is a serious obstacle to observing. The atmosphere refracts starlight, making it necessary to correct stellar position—stars appear higher in the sky than they are actually. The atmosphere causes stars to twinkle, a phenomenon produced by interference as light passes through. The atmosphere glows, and it produces aurora displays; it scatters starlight, reducing the blue radiation. Without an atmosphere, observations could be made continuously, for the sky would be everlastingly dark. However, the atmosphere is an essential buffer that shields Earth from lethal radiation. As the diagram shows, most solar radiation is in the visible-light and infrared regions. However, gamma and X rays are also present. As the radiation strikes particles in the upper layers of the atmosphere, gamma rays, X rays, and most of the ultraviolet radiation are filtered out. When Earth is cloud-covered, radiation received drops considerably, but even on clear days a large part of the radiation is filtered out by the various layers of the atmosphere.

WATER (HYDROSPHERE)

Earth is unique among the plants in having an atmosphere of nitrogen and oxygen. Also, it is set apart because of the presence of large amounts of water. Seven-tenths of the surface of the planet is water-covered. Indeed, it appears from space to be a planet of water with islands here and there.

Previous to this century, little was known about the ocean bottom; in fact, few aspects of oceanography had been explored. Now we know the ocean is immensely deep, in some places a mile deeper than the highest mountains. The ocean is so deep and so extensive that all the land masses could be fitted into it seven or eight times over.

Extensive soundings of sea bottoms reveal that there are wide variations from ocean to ocean, more varied in contour and topography than are the dry-land areas. However, each of the oceans has areas in common: the continental slopes which lie beyond the shelf and often extend 150 kilometers or more, and the deep sea floor where there is an abrupt drop-off to utter cold and darkness.

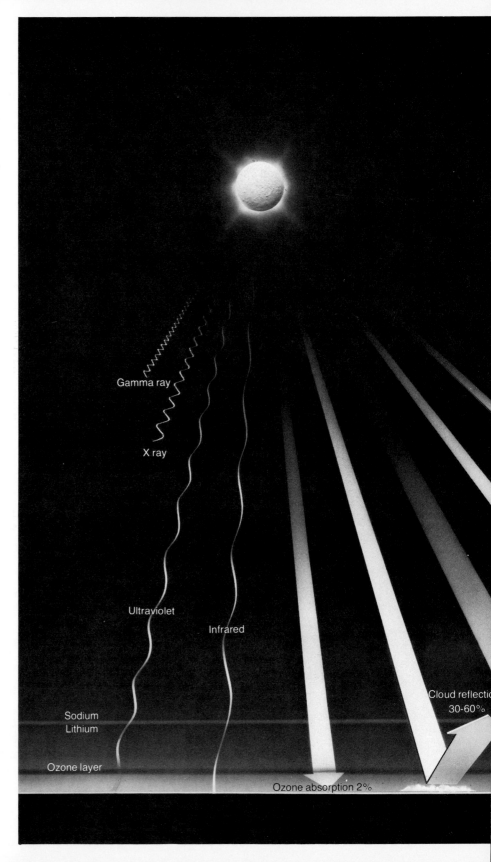

FIGURE 12–3
We live at the bottom of an ocean of air that is divided into layers having rather distinct properties. The atmosphere filters out most of the ultraviolet radiation of the Sun. Clouds which always cover large segments of the total area of the planet remove a large percentage of the total solar radiation.

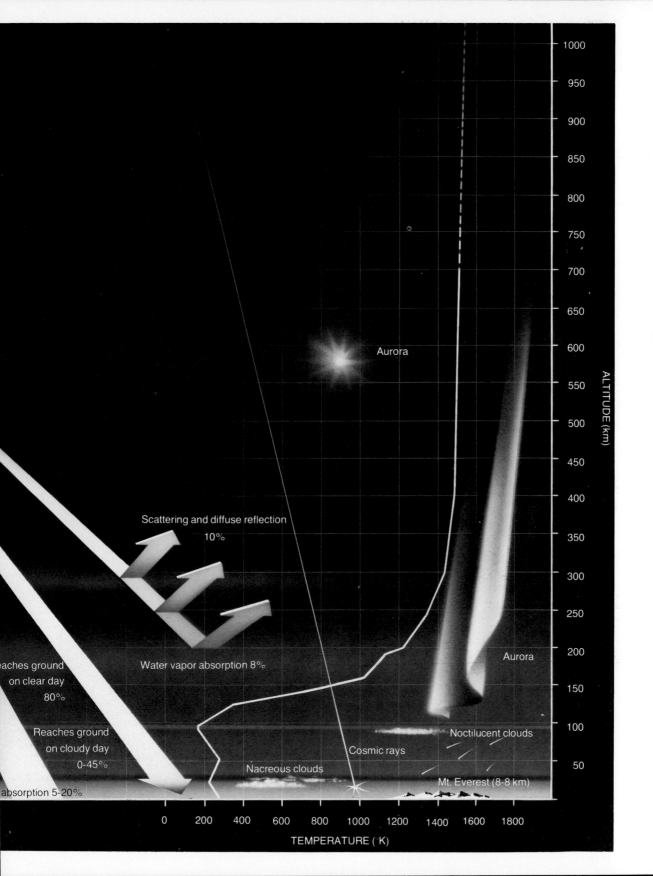

Scattering and diffuse reflection

10%

Aurora

Water vapor absorption 8%

Reaches ground
on clear day
80%

Reaches ground
on cloudy day
0-45%

absorption 5-20%

Aurora

Noctilucent clouds

Cosmic rays

Nacreous clouds

Mt. Everest (8-8 km)

ALTITUDE (km)

TEMPERATURE (°K)

The continental shelf area of a sea has been alternately dry and covered with water during the evolution of Earth. In some places, off the West Coast of the United States, for example, it is only about 16 kilometers wide. At other locations, off southern Asia, the shelf slopes gradually for some 1200 kilometers. Plant and animal life abound in the shelf area. This is the part of the sea that supplies fish and lobster, clams, and seafood of all varieties.

The shelf ends abruptly and slopes rapidly down several thousand meters. No plains or plateaus interrupt the steady fall-off, although deep fissures and canyons are cut into the slope. Occasionally, broad channels extend for kilometers into the sea. There is one

FIGURE 12–4   *Earthquake centers and volcanoes trace out boundaries of huge sections of the crust—the plates. Movements apart have produced great rifts, and the digging of one under the other has produced the mountain chains.*

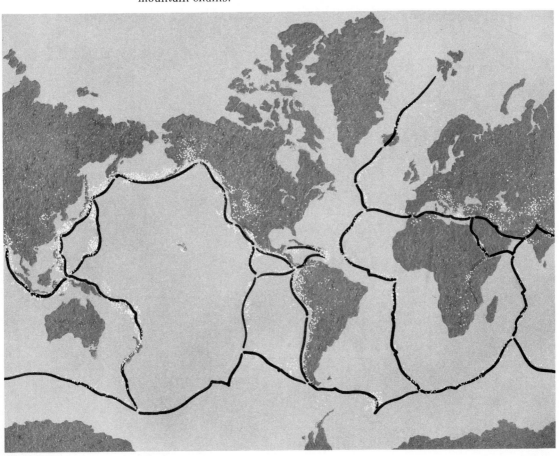

associated with the Hudson River, another with the Congo. Perhaps they are deep river beds cut out during the ages when these parts of Earth's surface were above sea level. Or they may have grown under the sea. Mud and silt may have slipped down high cliffs to lower levels and been carried away by persistent strong-flowing ocean current. The explanations are suppositions which require further investigation.

The continental slopes fall off to the deep ocean floor. "Floor" is not a good word, for the region is not smooth. On the contrary, the deeps are cut by ridges and canyons, by island chains and mountain systems. Sediments laid down for millions of years are about a kilometer deep, much deeper in some places. The sediments are the remains of plants and animals of the sea, also material eroded from Earth and carried into the sea by rivers and streams. Some researchers compute that each year 3 billion tons of solid earth are carried to the sea by the rivers of the world.

One of the great surprises of modern ocean research was the discovery of deep rifts and mountain systems under the sea. The mid-Atlantic ridge, a mountain system covering 19 000 kilometers and extending north to south of the Atlantic Ocean, was a key discovery. The ridge and deep gorge associated with it gave strong support to the *plate theory* of *continental "drift."* The shape of the ridge conforms to the shape of the west coast of Africa and Europe; also, the bulge of South America fits well into the indent of Africa. Great plates move 2 or 3 centimeters a year; they creep over the semiliquid rock lying 80 to 100 kilometers below the surface. It is believed that the South America plate broke free of the Africa plate some 150 million years ago. A motion of 2.5 centimeters would amount to about 4800 kilometers in 150 million years. Volcanic and earthquake activity is abundant along the ridge, perhaps because of movement of the underlying rock into the deep rift. If so, the rocks here should be younger than those along the coasts of Africa and South America. Investigations reveal this to be the case (Figure 12–4).

It is probably true, as many maintain, that chapters in the story of Earth's history will be revealed to us as exploration of the bottom of the sea continues.

LAND (GEOSPHERE)

The ball that is Earth is contained within a rather rigid, solid sheet of crustal granite that has an average thickness of only 15 to 30

FIGURE 12–5
Analysis of earthquake waves. Their velocity as well as areas where they are received and where they are shadowed out provide clues to the interior structure of Earth. The inner core appears to be solid, surrounded by a semimolten layer.

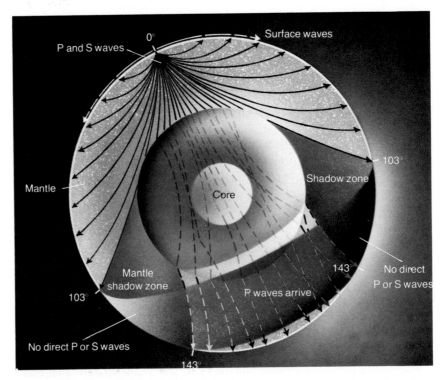

kilometers, thinner in relation to the whole Earth than the skin of an apple is to the apple itself.

The basic rocks of Earth, those that formed the original mass, are either *sialic* (those that contain large percentages of silica and aluminum) or *mafic* (rocks rich in magnesium and iron—Fe). *Granite* found in Earth's crust is sialic, while *basalt*, found below the surface, is mafic.

The interior of Earth cannot be explored firsthand. However, seismic waves (earthquake phenomena) and volcanic study provide clues to the region. Earthquake waves travel in all directions through Earth and pass through both solid and liquid rock. Primary (P) waves are compression-expansion waves, like those of sound; secondary (S) waves vibrate up and down, much as light waves do. Primary waves can move through both solid and liquid regions; secondary waves move through only the solid portion. Careful analysis of the waves therefore can reveal basic structure.

If Earth were solid, both types of waves could travel in all directions, and they could be received at locations opposite to their origin. Observations reveal there is a shadow region, a region where no S waves are received, directly opposite an earthquake. Since S waves cannot pass through liquids, the central region of Earth must be liquid (or semiliquid). By careful reckoning, the core of Earth has been computed to have a radius of 3500 kilometers, twice that of the Moon. The outer portion is liquid iron at high temperature and under pressure 2 million times greater than that of the atmosphere. The very center of the core, some 1200 kilometers in radius, appears to be solid (Figure 12–5).

Temperatures in Earth's interior must be extremely high, though not as high as might be expected from measurements made in the upper crust. In deep mines and borings for oil wells, temperature rises about 0.5°C for every 15 meters of depth, or about 50°C per kilometer. It is about 6400 kilometers to the center of Earth, so if we follow this reasoning, temperature there should be some 320 000°C. Such a figure is not reasonable. Perhaps crustal temperatures are high because of radioactive heating, which may not be a factor at lower depths. Interior temperature is perhaps 6400°C (some researchers believe it is closer to 12 000°). The pressure and temperature conditions that may exist in Earth's interior are indicated in Figure 12–6.

FIGURE 12–6   Inside Earth, pressure and temperature increase with depth. Temperature rises most rapidly in the upper crust, perhaps because radioactivity as a heating factor exists there but not in the deep interior.

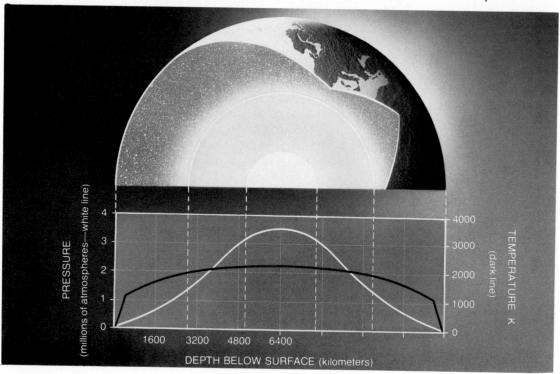

During 4.6 billion years this planet of ours has been evolving. The air, land, and water have changed drastically and no doubt will continue to change in the future. While Earth's structure has been changing, the planet has been moving everlastingly, and in a multitude of ways.

MOVEMENTS OF EARTH

Looking into the daytime sky, the casual sky watcher as well as the skilled observer sees the Sun rise in the east, move across the sky, and set in the west. In the night sky the Moon and the stars behave in the same fashion. And so do the planets, although not always in the smooth, unchanged manner of the Sun, Moon, and stars.

To ancient man it was quite obvious that Earth was motionless, and that the celestial objects moved around Earth. The objects could not move unattached, it was reasoned; they had to move upon something. In the fourth century B.C., two Greek astronomers—Eudoxus of Cnides (408–355 B.C.) and Callippus of Cyzicua (370–300 B.C.?)—worked out a complicated system of spheres within spheres. Each celestial object was attached to its own sphere which spun at its own speed; the stars were contained on the outermost sphere.

Later on—the exact dates are unknown, though surely in the second century A.D.—Claudius Ptolemaeus (Ptolemy), who was living in Alexandria, dispensed with the "spheres of Eudoxus" and substituted moving circles. He taught that the planets move evenly in circles, called epicycles. Centers of these epicycles move uniformly on other circles which are called deferents. This double system of circles moves around a point, the eccentric, which does not coincide with the position of Earth.

Ptolemy and Hipparchus before him rejected the ideas of Eudoxus because these ideas could not explain their own painstaking observations of the motions of the planets. They were well aware of the retrograde motions of the planets—the apparent slowing down and reversal of motion—and found it quite impossible to explain these motions with spheres. Ptolemy's moving circles enabled explanations for the observed motions. For more than 1400 years, the teachings contained in the 13 volumes of Ptolemy's *Almagest*, as it was called by the Arabs around A.D. 800, dominated man's conception of the universe in which he lived.

In 1473, Nicholaus Copernicus was born in what is now Torun, Poland. At the end of his life (1543), Copernicus was to publish *Concerning the Revolutions of the Heavenly Spheres*, a book that was to set Earth in motion. He placed the Sun at the center of the solar system (the universe, as it was then known) and put Earth, and the other planets, in motion around the Sun. Copernicus retained most of the epicycles, deferents, and eccentrics that Ptolemy had used to explain retrograde motions. But he argued convincingly that the Sun was at the center of things—so convincingly that Giordano Bruno, some 50 years later, lost his life in support of the idea, and Galileo was excommunicated for the same reason.

Copernicus did not originate the Sun-centered idea, nor did he ever contend that he did so. (However, in a handwritten copy of his *Revolutions*, Copernicus struck out reference to Aristarchus, the Greek astronomer who had proposed the idea some 2000 years earlier.) Copernicus made selections from ideas, supporting strongly those that were reasonable and provable by interpretations of observations.

The Foucault Pendulum

Copernicus, for example, explained the rising and setting of the Sun in terms of Earth's rotation. However, it was not until much later, in 1851, that rotation could be proved by the classic pendulum experiment of the French physicist Jean Bernard Leon Foucault (1819–1868). Foucault hung a bronze ball weighing 25.4 kilograms (about 56 pounds) from the ceiling of the Pantheon, a high-domed building in Paris. The ball, about one-third of a meter (or one foot) in diameter, was at the end of a fine wire 60 meters (about 200 feet) long. A needle extending a few centimeters below the ball was just long enough to pass through a layer of sand atop a table.

To start the experiment, the weight was pulled to one side and tied in place. After all vibrations of the system had quieted, the string holding the ball was burned. The wire was fastened to the dome with a "frictionless" bearing, free to move through 360 degrees. The weight swung back and forth for several hours, always in a plane that passed through the center of Earth.

As each swing of the pendulum was completed, the needle marked a new path through the sand, always a bit to the right of the previous one. The ceiling and floor of the Pantheon were turning because they were fastened to Earth. But the pendulum was not; it continued to move in the plane in which it had started. Careful timing of the drift toward the right indicated that the needle would cut through the original furrow after 32 hours.

If the experiment were conducted at the North Pole, the needle would cut through the original furrow in 24 hours. If the experiment were conducted at the equator, the needle would always cut through the same furrow. The time required for the sand table to return to the original position increases from the poles (24 hours) to the equator (infinity). This is because the support of the pendulum is shifted eastward continually during each swing (except at the poles), and because the north side of the swing covers less distance on Earth (the sand box) than the south side of the swing (except at the equator).

The amount the pendulum shifts at different latitudes from the original plane in degrees per hour can be found by multiplying the sine of the latitude by 15°, as shown below.

Location	Latitude	Sine of latitude	Deviation (degrees per hour)
Florida Keys	25°	.423	6.43
Los Angeles	34°	.559	8.38
Boston	42°	.669	10.03
Seattle	48°	.743	11.14

The day-night cycle is an obvious effect of Earth's rotation. Not quite so apparent is the effect upon our weather which results largely from the prevailing motions of large masses of air and water. Rotation causes winds and sea currents to move toward the right in the Northern Hemisphere and toward the left in the Southern Hemisphere. We might compare the motions to those of a bullet fired from a hypothetical rifle.

The Coriolis Effect

Suppose a bullet were fired due north from the equator. The rifle bullet is moving eastward 1600 kilometers per hour (about 1000 mph), the speed of rotation at the equator. The bullet is fired from a region moving 1600 kilometers per hour toward an area that is moving more slowly. The bullet has motion northward, and also eastward. Therefore, it lands east, to the right, of the north-south line.

If the rifle were north of the equator and fired southward, the bullet would also land to the right of a north-south line. The rifle and bullet are now moving eastward more slowly than the target area. During flight of the bullet, the target area moves faster than the location of the rifle, so, in effect, the bullet lags and veers to the right.

The effect of Earth's rotation on the prevailing winds, called the *coriolis effect*, is shown in the illustration. We orient ourselves by imagining that we are standing at the tail of the arrow and are facing toward the head. (Figure 12–7).

Day and Night

Because Earth is spherical, or nearly so, one-half of the planet is always illuminated by the Sun, the spinning of Earth constantly bringing different regions into and out of the illuminated half. As the year goes by, the location of the *terminator* (the line separating daylight from darkness) swings about through 47° because of the tilt of Earth's axis. Earth's orbit determines the boundaries of a flat surface, or plane. Earth's axis is tilted 23½° from a vertical to that plane. Because the tilt remains unchanged throughout the year, a location on Earth is sometimes tilted toward the Sun and sometimes

FIGURE 12–7
Winds and sea currents are deflected to the right in the Northern Hemisphere, to the left in the Southern Hemisphere. (Here the paths are idealized.)

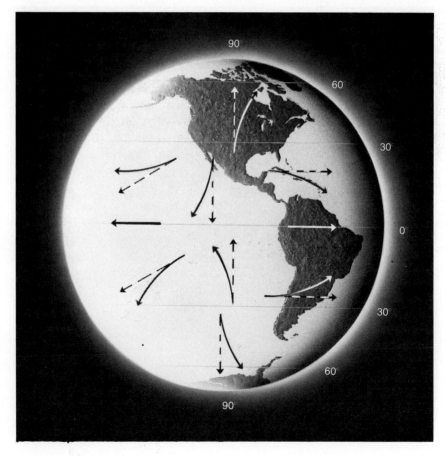

away. The total amount of daylight the place receives changes as the year goes by. In tropical regions, the change is minimal, but, as one moves toward the poles, the changes become greater. In the middle latitudes, for example, summer daylight may last 16 hours, whereas the length of winter days may be 8 hours or less (Figures 12–8 and 12–9).

If Earth's axis were vertical to the plane of its orbit, day and night would be the same length throughout the year. If the axis were tilted at a much greater angle, there would be several months of daylight followed by a long period of darkness. That is the situation with Uranus, which is tilted 8° to the plane of its orbit, 98° from the vertical.

FIGURE 12–8　The axis of Earth is tilted 23½° from a perpendicular to the plane of the ecliptic.

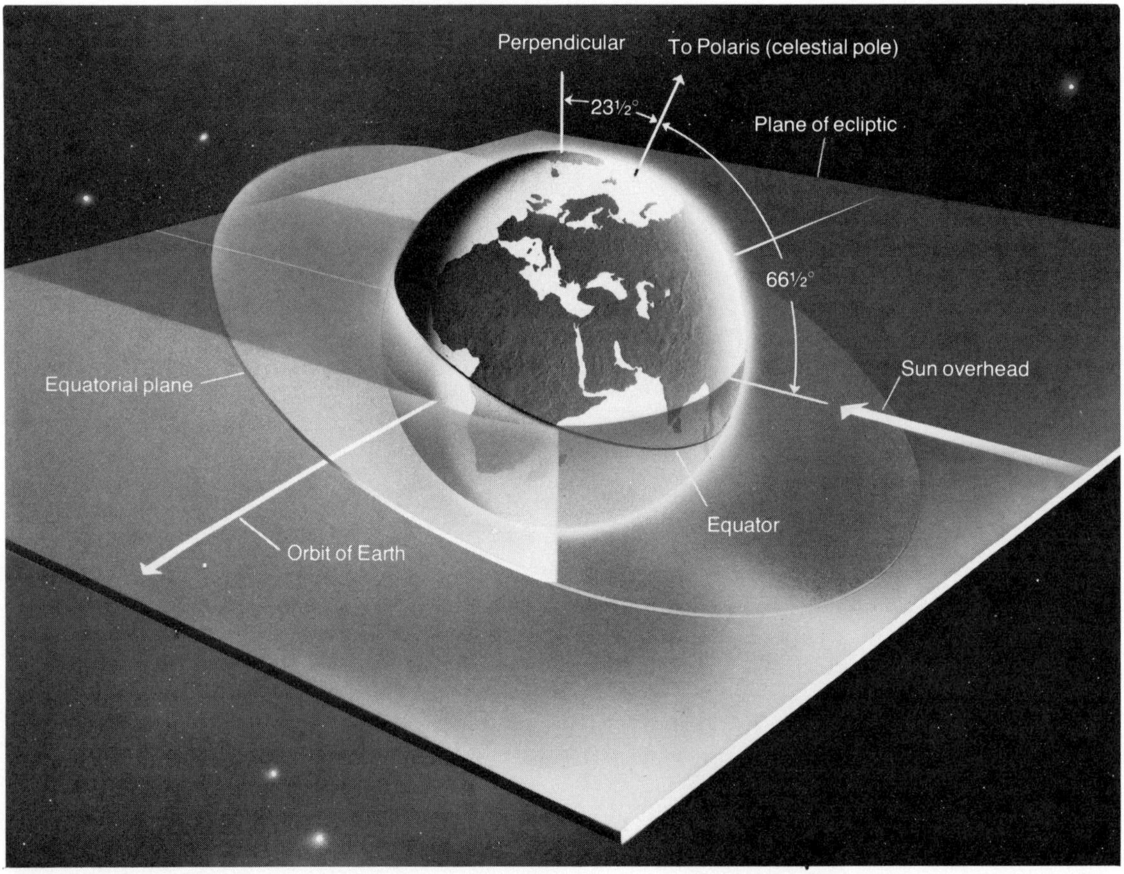

FIGURE 12–9
The number of hours of daylight and darkness that a region experiences is due to rotation of Earth and the latitude of the place. Three regions shown are from top to bottom: equatorial, middle latitudes, upper latitudes.

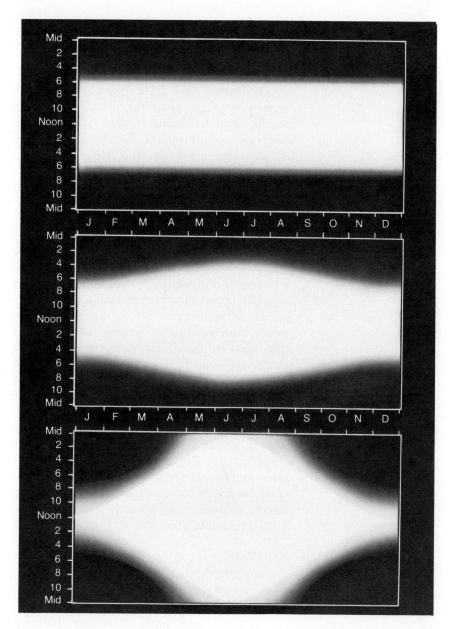

Revolution

As the seasons go by, the most casual observer is aware that the position of the Sun changes. In winter in the Northern Hemisphere, the noontime Sun is low in the sky. As spring approaches, the Sun appears higher, reaching its greatest elevation at the time of the summer solstice. (To find the greatest elevation of the Sun, we add $23\frac{1}{2}°$ to our *co-latitude*—the difference between our latitude and 90°.)

Today we have little difficulty in explaining these changes because we are aware of the tilt of Earth's axis and of the planet's motion around the Sun. In his *Almagest*, Ptolemy discussed many observations, and one of them concerned the changing elevation of the Sun. Ptolemy believed Earth was stationary in space. The Sun went around Earth in a spiral having 182 turns—one-half of 365. When the Sun was at the top of the spiral (greatest elevation), there was summer, and winter arrived when the Sun had reached the bottom of the spiral.

Whenever the Sun reached an end of the spiral, it stopped, reversed itself, and moved in the opposite direction. The idea, though wrong, served quite well to explain the changing locations of the rising and setting Sun, and its changing noontime position.

The spiral idea was not able to explain another observation— the yearly movement of the Sun eastward among the stars. One cannot observe the Sun among the stars directly; the glare of sunlight lights up the atmosphere, making it opaque. However, one can see the seasonal changes in the stars that rise in the east as the Sun disappears below the western horizon. In winter, for example, the Orion group appears, changing to Leo in spring, Scorpius during the summer, and Pegasus in the fall.

Annual changes such as these can be explained by supposing Earth is standing still and the Sun and stars are moving around it in complex ways. However, as instruments for observing were developed, precise observations and measurements were possible. These could be explained in only one way—by supposing Earth was revolving around a "stationary" Sun. One of these observations was the aberration of starlight.

Aberration

The word *aberration* comes from the Latin, meaning "to stray," and it is appropriate in astronomy, for stars do not always appear where one would expect them to be: they appear to stray.

In 1725, the English astronomer James Bradley (1693–1762) together with his friend Samuel Molyneux constructed a telescope in Molyneux's house. They pointed their telescope at the zenith—the point directly overhead. They observed for several months and noted that the star nearest the zenith seemed to shift about forty seconds (40″)—a very small amount to a layman but a large amount to an astronomer. Continued observations confirmed the original conclusion.

Neither Bradley nor Molyneux could explain the observation until Bradley, a sailing enthusiast, noticed that the wind vane on the mast shifted relative to the boat as the boat changed direction, because the wind blew steadily from the same direction. Starlight, he thought, was steady just as the wind was, so any displacements must result from Earth's motion. He was right, as later observations confirmed.

Think of starlight as a stream of particles. If Earth is moving, the particles will seem to come from a location other than their actual source. The analogy usually used is that of a man walking through a rain storm, the drops falling from directly overhead. If the man stands still, an umbrella held straight overhead protects him. If he walks, the umbrella must be tilted in the direction in which he moves. If he walks faster, the umbrella must be tilted more. The rain seems to be coming from a direction other than overhead.

So it is with starlight. One way of lining up a star is to point a tube at it so starlight goes in at the top of the tube and out the bottom. As shown in Figure 12–10, a single "particle" of light enters the top of the tube. While the "particle" is traveling through the tube,

FIGURE 12–10   *A definite proof of Earth's revolution is the aberration of starlight.*

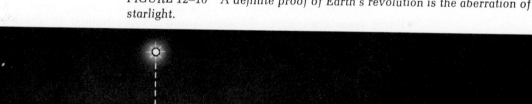

the tube is carried forward by Earth's motion. In order for the light and the tube to arrive at the same place simultaneously, the tube must be slanted.

Suppose the tube were 298 000 kilometers long; light would then take 1 second to travel through it. During a second, Earth would travel 29.6 kilometers through space. The tube would have to be slanted so the bottom trailed the top by 29.6 kilometers. In astronomy, telescopes must be slanted so the bottom trails the top. The amount of tilt is 20".37 (twenty and thirty-seven one-hundredths seconds)—the constant of aberration—an effect that can be explained only by supposing Earth is moving in an orbit.

Parallax of Stars

Parallactic displacement (the apparent shifting) of stars is another observation that can be explained only by assuming that Earth is revolving.

Aristarchus in the third century B.C. contended that Earth moved through space. His contemporary, Aristotle, maintained that if Earth did move, then distance to the stars should change and the configuration of the constellations should therefore vary seasonally. Since it did not, Earth must be stationary.

Aristotle was right about shifts of the stars. Because of the great distances involved, the shifts are slight, however, requiring precise instruments which he lacked to detect them.

Suppose, as in Figure 12–11, a nearby star is being observed. If Earth were stationary, the star would always appear in the same location among the background stars. If Earth moved, then the star would appear at A when Earth is in position 1, at B when in position 2, and so on. This is precisely what happens—nearby stars appear to scribe out ellipses in space, small counterparts of Earth's orbit around the Sun.

Additional observations resulting from Earth's motion around the Sun are regular changes in the radial velocities of stars. *Radial velocity* is the rate at which distance between Earth and the star changes. If the distance decreases, the star has negative radial velocity; if it increases, the star has positive radial velocity. Stars do not move alternately toward us and away from us, therefore; if seasonally a star shows alternate negative and positive radial velocities, the cause must be Earth's movement away from the star and toward it.

From the drawing it is apparent that at a given time of the year Earth will be moving toward a star on the plane of Earth's orbit. Six months later, Earth will be moving away from the star. As an example, in January Earth moves toward the stars in Virgo and away from those in Aries. In July, the conditions are reversed.

All stars have motions of their own. However, all stars in a given region would not move in the same manner, so when they appear to do so it must result from motion of the observer. In January, the radial motion of all stars in the Virgo region is about 30 kilometers per second greater and in the Aries region 30 kilometers per second less than during April and October when Earth is moving at right angles to the stars. Observations such as these are further proof that Earth is revolving around the Sun in a nearly circular path at a velocity of about 30 kilometers per second.

The Shape of the Orbit

In the study of astronomy one comes across the name Hipparchus frequently. He was probably the outstanding Greek scholar-astronomer of ancient days. Among his many observations was one concerning the length of the seasons. He noticed that 186 days were required for the Sun to move from the vernal equinox to the autumnal equinox, while the Sun returned to the vernal equinox in only 179 days. In ancient days celestial objects were considered to move uniformly, in perfect circles, around Earth. If Earth were at the center of the circle, then the Sun would take the same time to move through a given area, but it did not. Hipparchus explained the disparity by saying Earth was not at the center; it was off center and the Sun's orbit was eccentric.

While the explanation served quite well to explain observations, it was not correct. Today we know that the disparity in the two periods results from Earth's moving in an elliptical orbit, causing its velocity to increase and decrease as distance from the central Sun changes.

Since the distance between Earth and Sun changes as a year passes, the apparent diameter of the Sun changes also; the closer we are to anything, the larger it appears to be. Careful measurements of the diameter of the Sun throughout the year enable us to construct a model of Earth's orbit.

A point is selected to represent the Sun, and a line is drawn from that point to the vernal equinox. Other lines are drawn from the same point, the angle of each line with the line to the vernal equinox being the longitude of the Sun-angle east of the vernal equinox. If a line is drawn the first of each month, let us say, we have the direction of Earth as seen from the Sun at each of those times. To make an orbit, we need to know the relative distances between the Sun and Earth on each of those days. To do this, careful

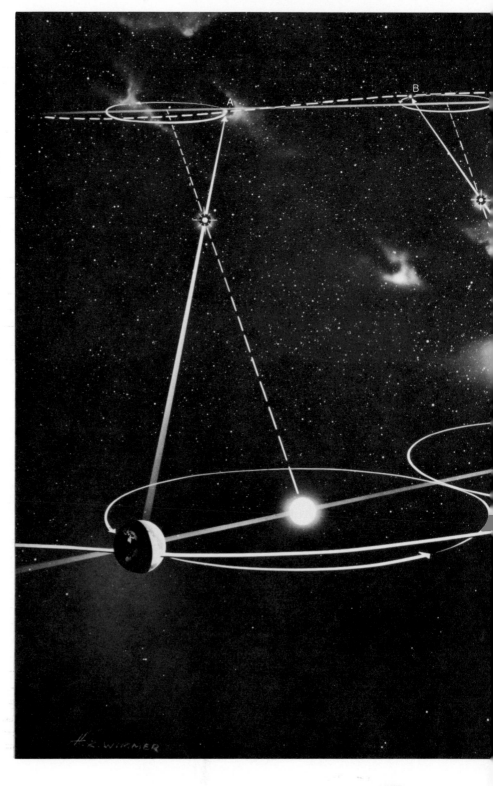

FIGURE 12–11
Because Earth revolves, nearby stars show parallactic displacements. Seen from a star, the Sun moves along a straight line and Earth appears to move from side to side. Seen from the Sun, a nearby star travels along a straight line (solid line at top). From Earth, the star appears to move along the dashed line.

measurements are made of the diameter of the Sun on the first of each month. Typical results are as follows:

Month	Solar radius (seconds of arc)	Length values
October	960	10.41
January	978	10.20
April	962	10.38
July	945	10.48

If the seconds measurement is divided into 10 000, a number chosen to give manageable results, we obtain relative values for the length of each line. Any convenient unit (centimeters, inches, or whatever) can be used. When the ends of the lines are connected, a model of Earth's orbit is produced. It is an ellipse with the Sun at one of the foci, though on such a scale it appears nearly circular.

The Barycenter

The elliptical orbit of Earth is not traced out by the center of Earth, as one might expect. Rather, the center of mass (*barycenter*) of the Earth-Moon, which may be considered a double planet, scribes out the path.

The barycenter might be thought of as the center of balance of the Earth-Moon system. Because Earth is much more massive than the Moon, the barycenter is much closer to Earth; in fact, it is within Earth. As the barycenter traces out the orbit, the Moon swings wide around it, and Earth swings in a smaller circle.

Sometimes Earth is ahead of its center, sometimes behind; therefore when the Sun is observed, it will sometimes be ahead or behind its predicted position. These variations in the Sun's position work out to a location of the barycenter 4645 kilometers from the center of Earth. Its location, as well as the looping orbit of the Earth-Moon system, is shown in Figure 12–12.

Precession of the Equinox

The vernal equinox is a point on the celestial sphere (the apparent surface of the sky that surrounds us) where the celestial equator—an extension of Earth's equator to the sphere—and the *ecliptic*—the Sun's path across the sky—intersect. At the beginning of spring, the Sun is at the vernal equinox. However, the point is not fixed; it continually drifts westward among the stars. Apparently this movement was discovered by Hipparchus. At that time, second century B.C., the vernal equinox was in the constellation Aries. Hipparchus determined that in 3000 B.C. the vernal equinox was in the constellation Taurus. It is presently in Pisces. The equinox will continue its westward drift, reaching Taurus again in the year 23 000.

An explanation for this drift of the equinox was offered by Copernicus. He said it was a motion similar to that of a wobbling top. The motion is called precession, and it comes from precede, "to go before." It is wobbling, as Copernicus suggested, produced by the spinning of Earth, and the attraction of the Sun and Moon, mainly the Moon, on the equatorial bulge of Earth.

Earth is not spherical; the polar diameter is 12 578 kilometers, while the equatorial diameter (the bulge) is 12 683 kilometers. The

FIGURE 12–12   *The barycenter of the Earth-Moon system is located 2903 miles (about 4670 km) from Earth's center. As Earth and the Moon move around the barycenter, the barycenter traces the orbit of the system around the Sun.*

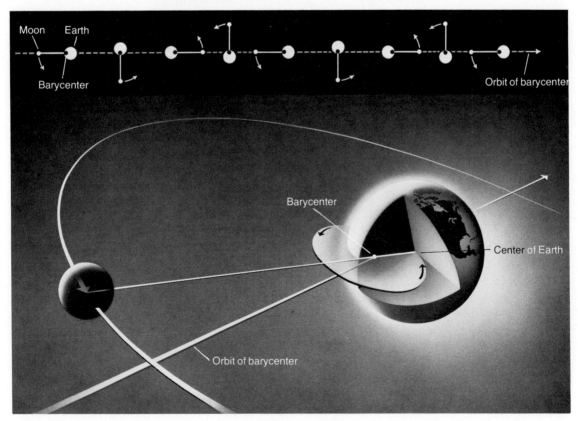

equator and the bulge are tilted 23½° from the plane of the ecliptic, while the Moon is on the ecliptic, or nearly so. Because of its nearness, the Moon exerts a strong attraction on Earth, stronger on the side toward the Moon than on the side away. This tends to put the axis of Earth's rotation at right angles to the ecliptic. Because Earth is spinning, it acts as a gyroscope and resists this tendency; it moves at right angles to the force. This results in Earth's axis scribing out a vast cone, the apex angle of which is 47°—twice 23½°—the angle of Earth's axis from the vertical. The motion of the axis is retrograde, opposite to the direction of rotation. Because the mass of Earth is so great, it precesses very slowly, taking about 26 000 years to complete one precession cycle. (Figure 12–13)

The inclination of Earth's axis remains 23½°; however, as time goes by, the axis points toward different regions of the sky. For example, our present pole star (the star nearest the celestial pole) is Polaris; but it was not always so. In the days of the Egyptians, the

FIGURE 12–13   *The gravitational attraction of the Moon upon Earth's bulge is the main cause of precession of the equinoxes, a 26 000-year cycle. Presently the vernal equinox is in Pisces; in about 500 years precession will cause it to be in Aquarius.*

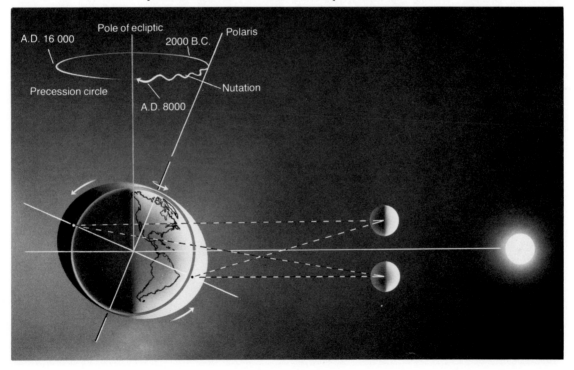

pole star was Thuban, which is in the constellation Draco. Some 6000 years from now our axis will point toward Alpha Cephei, and in A.D. 14 000 Vega will be the brightest star near the pole. Twenty-six thousand years from now, Polaris will once again have the distinction now given it.

Because of precession our sky-view changes. In 4000 B.C., the Southern Cross could be seen from Central Europe. Now one must be inside the Torrid Zone before it can be detected. Also, precession will eventually cause a shift of seasons. Summer will occur when Earth is in the part of its orbit opposite to where it occurs presently. Orion is now a winter constellation. But in the year A.D. 15 000 it will be seen on a summer's night—an alarming sight for today's observers, but no doubt perfectly normal for sky watchers of that far-off millennium.

The equinox drifts westward, moving about 50″ of arc in one year. However, the annual amount varies slightly because the forces that cause precession are not applied constantly. When the Sun and Moon are in line with the bulge, there is no force; this happens twice a year for the Sun and twice a month for the Moon. Therefore, there are slight fluctuations in precession called *nutations*, meaning "to nod."

The poles of Earth wander in other ways. Very careful calculations indicate Earth shifts under the poles in a circular orbit with a diameter of half a second of arc. This amounts to 18 or 20 meters. The irregularity may be caused by shifts of great masses of air, the flow of water masses of different densities, or perhaps the movements of plates—the great masses of mantle rock upon which continents float like islands.

Earth moves in a multitude of ways: it spins about 1600 kilometers an hour at the equator, revolves around the Sun some 105 000 kilometers an hour, drifts (with the Sun) toward Hercules 70 000 kilometers an hour, and goes around the center of the Galaxy a quarter of a million kilometers per hour. And that's not all; the Galaxy is moving with respect to other galaxies—how fast is hard to say.

Shape of Earth—Mass and Gravitation

At the time of the Renaissance, scholars did not question the roundness of Earth, but the majority of people believed in the adage "seeing is believing," and their senses told them Earth was flat. Millenniums before that period, however, there were isolated individuals who were convinced, on the basis of logic if nothing else, that Earth was spherical. In the sixth century B.C. Pythagoras taught that Earth was round, and two centuries later Aristotle supported the belief with observations. The shadow of Earth cast upon the Moon during a lunar eclipse was always the arc of a circle, he noted. Only a spherical body can cast a shadow that is the arc of a circle at all times. Also, the elevation of stars changed as one moved from one location to another. If Earth were flat, Polaris, for example, would always remain the same angular distance above the horizon. It did not, however; for each unit of distance traveled, the altitude of Polaris changed uniformly, and therefore the surface of the Earth had to be curved and the curve had to be uniform. Earth was spherical.

The roundness of Earth and its size were known to scholars of old. In the third century B.C., Eratosthenes, a Greek who was librarian at Alexandria, Egypt, during part of the long occupation by the Greeks, computed the size of Earth. At Syene, now Aswan, in southern Egypt and nearly on the Tropic of Cancer, Eratosthenes dug a well some 8 meters deep with spiral steps leading down to the water level. On the first day of summer, sunlight was reflected from the water—the Sun was directly overhead. On the same day at Alexandria, far to the north, the Sun was found to be 7°15′ from the vertical. Assuming that sunlight traveled along parallel lines, Eratosthenes reasoned that the surface of Earth between Syene and Alexandria must be curved in an arc of 7°15′.

A unit of distance used for measuring in those days was the *stadium*. This was probably 160 meters, the length of an actual stadium in Greece. The distance between Syene and Alexandria was believed to be 5000 stadia. The angle 7°15′ is $\frac{1}{50}$ of a complete circle; therefore, the distance around Earth should be some 250 000 stadia. Using 160 kilometers as the unit, we find that the procedure followed by Eratosthenes gives a circumference of 41 500 kilometers, remarkably close to the actual size of Earth (Figure 12–14).

The technique of Eratosthenes has been modified for modern use. Now careful measurements of the distance between degrees of latitude are made, then multiplied by 360. Repeatedly it is found that a degree of latitude is longer at the poles than it is at the equator, implying that the Earth-sphere is two different sizes. Obviously, the Earth cannot be two different spheres at the same time. It cannot be a sphere at all. Such measurements mean Earth is a flattened sphere, an *oblate spheroid*, or, more properly, an *ellipsoid*—the solid produced when an ellipse is rotated around its short axis.

By general agreement, the dimensions of Earth most widely used are those computed by J. F. Hayford of the U.S. Coast & Geodetic Survey in 1909—equatorial radius 6 378 880 meters and polar radius 6 356 912 meters, which gives a flattening of $\frac{1}{297}$. Flattening, or oblateness, is obtained by dividing the difference between equatorial and polar radius by the equatorial radius.

Precise measurements of satellite orbits indicate that the flattening is probably closer to $\frac{1}{298.2}$, representing a difference of about 75 meters.

In addition to being an ellipsoid, Earth appears to be unsymmetrical; that is, it would not appear the same upside down as rightside up. Measurements of satellite orbits indicate that the North Pole radius is about 15 meters greater than the South Pole radius.

Laser beams reflected from satellites and from reflectors placed upon the Moon are being used to refine measurements of the size of Earth as well as locations and drifts of continents. Answers within fractions of a centimeter accuracy are now possible. Even greater accuracy is possible using the light from quasars. The light is bright, and because of the great distance the positions of quasars in relation to the earth are fixed. Thus quasars are precise references against which the most exacting measurements can be made.

Gravitation. Precise measurements of gravitation at various locations on Earth's surface also indicate the ellipsoidal shape of the planet. Measurements vary with distance from Earth's center and allow the construction of a model.

FIGURE 12–14
Careful measurements of the angles of shadows at different locations on Earth enabled Eratosthenes to compute the size of Earth.

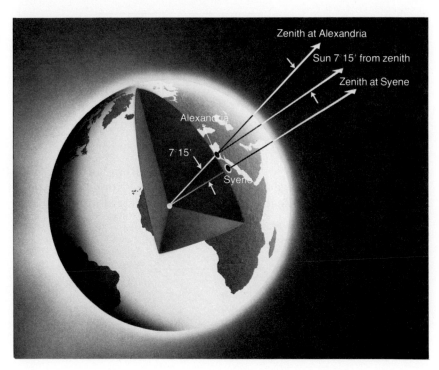

How fast an object would increase its rate of fall if it were falling in a vacuum provides a measure of gravity. At the surface of Earth, acceleration of gravity is approximately 980 centimeters per second, during each second of fall (980 cm/sec²). A mass would fall 490 centimeters the first second (it starts from 0 velocity), at the end of second two it would have fallen 1960 centimeters, and so on. During 5 seconds, the history of a fall would be as indicated below:

Time (seconds)	Velocity (in cm at end of each sec)	Acceleration (cm/sec²)	Distance (meters)	Total distance (meters)
1	980	980	4.9	4.9
2	1960	980	13.7	19.6
3	2940	980	24.5	44.1
4	3920	980	34.3	78.4
5	4900	980	44.1	122.5

Because centrifugal effect is diminished, and because the distance to Earth's center is less, gravitation increases with nearness to the poles, as shown in the tables.

Latitude	G (cm/sec²)
0°	978.049
30°	979.388
60°	981.924
90°	983.221

As one moves away from Earth's surface, the force of gravity is reduced because distance from the center of mass is increasing. Readings with altitude change are shown below:

Elevation (meters)	G (cm/sec²)
0	987.049
15 000	977.579
30 000	977.108
45 000	976.639
60 000	976.169

At the Moon's distance (384 000 kilometers), Earth's gravitational attraction is only $\frac{1}{3600}$ of what it is at the surface, because 384 000 kilometers is 60 times 6400 kilometers (Earth's radius). Gravitation is inversely proportional to the square of the distance: $\frac{1}{60}^2$, or $\frac{1}{3600}$. The acceleration of the Moon toward Earth is thus 0.272 centimeters a second ($\frac{1}{3600} \times 980$) each second. In one day, the Moon must fall 11 000 kilometers toward Earth.

Certainly the Moon is not falling directly toward us at any such rate. By "falling," we mean the Moon is deflected from a straight-line direction by Earth's gravitation. This is what keeps the Moon in orbit. Similarly, Earth's gravitation affects the orbits of artificial satellites.

Mass of Earth. In 1687 Sir Isaac Newton—the man who achieved greatness (as he said) by standing on the shoulders of giants—discovered the laws of gravitation. More than a hundred years were to pass before there was experimental proof of his statements. It came in the works of Henry Cavendish (1731–1810)—the last investigation he was to make in a lifetime of discoveries. In the experiment, small metal balls were connected to each end of a rod, and the rod was suspended at the center on a fine wire. Two massive lead spheres were brought close to the small ones, causing the wire to twist slightly. A force was applied to the small spheres to bring them back to position, and the force was measured. Repeated measurements gave a figure of 666×10^{-10} dynes—the constant of gravitation (G)—the force by which two 1-gram masses attract each other when placed 1 centimeter apart. By definition, a dyne is the force required to accelerate a 1-gram mass 1 centimeter per second per second.

Results of the Cavendish experiment can be applied to finding the mass of Earth. *Mass* is often confused with *weight*. Weight is really a measure of *force*—the force which a given mass exerts on another mass. A person's weight on Earth, for example, is the force which the mass of Earth exerts upon him. On Earth we might express weight in this manner:

W (weight) $= M$ (mass) $\times G$ (acceleration of gravity)

For example, the weight of a 1-gram mass could be expressed in force units as follows:

$W = 1 \times 980$
$W = 980$ dynes

The task of finding Earth's mass seems far removed from the above reasoning, but it follows logically. Suppose we put a 1-gram mass on a spring balance; the reading is 1 gram since Earth attracts the mass with a force of 1 gram, or 980 dynes. From Newton's law of universal gravitation, we have $F = (G \times Mm)/d^2$ where

$F = 980$ dynes
$G = 666 \times 10^{-10}$
$M = $ Mass of Earth
$m = $ Mass of 1 gram
$d = $ distance to center of Earth (640 000 000 cm)

Substituting these values in the equation:

$$980 = \frac{666 \times 10^{-10} \times M \times 1}{(640\ 000\ 000)^2}$$

When we solve for M, we find

$$M = 6 \times 10^{27} \text{ gm}$$

Though mass and weight are often confused, they can be differentiated readily. If a person stood on a spring scale on the Moon, he would weigh one-sixth of his Earth weight. But suppose he "weighed" himself with an arm balance. The mass of material needed to strike a balance would be exactly the same on the Moon as on Earth—or at any other location. Mass does not change with location.

Terrestrial Magnetism

Earth acts as a magnet. This fact was stated conclusively in 1600 when William Gilbert, an English physician, published the conclusion after experimenting with lodestones and making observations of the behavior of compasses and dip needles at various locations. Gilbert found no way to explain why Earth should be magnetic. Later on, it was believed that there was a metal bar extending through Earth and reaching the surface at the polar regions.

The idea had some merit. For one thing, it explained the existence of the magnetic poles. Actually, however, these poles do not coincide with the geographic poles—the magnetic axis does not pass through Earth's center. The magnetic north pole is at about 70° north latitude and 100° west longitude, while the magnetic south pole is at about 68° south latitude and 143° west longitude.

The north pole (actually the north-seeking pole) of a magnet points toward the magnetic north pole. The difference between that pole and geographic north is magnetic *declination*, a factor that changes continually. For example, in the vicinity of New York City, declination increases by about 1° in 60 years. Changes over a long time may be extreme. In 1576 in London, declination was 12° east, while by 1823 it had shifted to almost 25° west. Presently it is diminishing and will eventually shift to east of north (Figure 12–15).

From analysis of the arrangement of magnetized particles in rocks, geologists can infer shifts in Earth's magnetism that have occurred through millions of years. Apparently the magnetic north pole has shifted. Studies also indicate that reversals of polarity have happened many times during Earth's history.

Such observations indicate that Earth's magnetism must be produced internally, very likely by electric currents that result from movements of great masses of material—probably a dynamo effect involving the core of Earth and the mantle rock.

The central solid core of Earth is surrounded by a liquid layer. Being liquid, this layer can move with respect to the mantle rock which is solid. The core is probably metallic, a nickel-iron alloy, and so is able to conduct electricity; therefore a magnetic field is estab-

FIGURE 12–15
Lines of equal
magnetic declination.

lished. The system becomes self-supporting; that is, electric currents produce a magnetic field, and the moving magnetic field generates electricity. The structure of Earth and its internal movements create a dynamo with all the characteristics of the one first proposed by Michael Faraday in the early part of the nineteenth century.

The material in the core of Earth probably moves about a few hundredths of a centimeter a second. The source of energy for the motions is not clear. Convection currents due to temperature difference may explain the energy, or perhaps slight differences in density may set up motions. Whatever the cause, eddies of limited extent are no doubt established in the process. One cannot say where they will appear, but the general magnetic field seems to be made up of them.

Magnetism and Radiation Belts

Magnetic lines of force may extend hundreds of thousands of kilometers into space, becoming weaker as distance increases and finally fading away altogether. The volume surrounding Earth and containing the active part of the magnetic field is the magnetosphere. It appears to extend some 80 000 kilometers beyond Earth—compressed on the Sun side and extending in a long sweeping "tail" on the side away from the Sun (see page 280).

The magnetosphere is three-dimensional, having a diameter of tens of thousands of kilometers at right angles to a line joining Earth and the Sun. Were it to be seen from outer space, the magnetosphere would be shaped very much like a comet. The compression on the Sun side is caused by the solar wind—particles ejected from the Sun and entering the region with considerable force. Low-energy protons and electrons are captured throughout the magnetosphere. There are also many local regions that trap high-energy particles.

Some 3000 kilometers above the magnetic equator, there is a belt of high-energy protons. Twelve thousand kilometers beyond, high-energy electrons girdle the equator. The particles, trapped by the magnetic field, spiral around the lines of force, bouncing from one magnetic pole to the other in only a few seconds. As they bounce back and forth, the particles drift around Earth—electrons to the east, protons to the west. There is little, if any, trapping of particles beyond 75° latitude, the regions where the auroras occur. Space above the magnetic poles is quite devoid of particles. In the early days of manned space explorations, these regions were suggested as corridors of safety, free of radiation particles that might prove disastrous should men be exposed to them. Actually, passage through the particles was so rapid, no detrimental effects were noted.

Particles that comprise the inner parts of the radiation belts (often called the Van Allen belts after James Van Allen, the American physicist who discovered them) may originate from cosmic radiation, while particles in the outer regions may be more closely related to the Sun. Changes in the energy levels of the belts and composition of the belts are of constant interest to investigators. They are concerned with how these particles affect conditions here on Earth, and how we can adjust to them should the effects be detrimental.

REVIEW

1 The origin, age (about 4.6 billion years), and evolution of Earth began to be revealed in the eighteenth century.

2 The air ocean, which is layered, probably evolved from the interior of Earth during early stages of formation.

3 The amount of water in Earth remains essentially unchanged, though the shape and extent of the ocean, and ocean bottoms, change constantly.

4 The crust of Earth is in great plates which separate and move together, producing earthquakes, volcanic activity, ridges, chasms, and mountain chains.

5 Proofs and reasons for the many motions of Earth were produced by Kepler, Newton, Foucault, and Bradley, among others.

6 Knowledge of gravitation enables the determination of Earth's mass—6×10^{27} g.

7 Terrestrial magnetism probably results from Earth's rotation, and the presence of a semimolten layer.

8 Radiation particles that form belts around Earth are trapped and held by Earth's magnetic field.

QUESTIONS

1 How did Bradley's investigations prove that Earth revolves?

2 What is the barycenter of the Earth-Moon system? How is its location determined?

3 What evidence is there for believing that the age of Earth is 4.6 billion years?

4 What is the observational evidence to support belief that planets are being created presently?

5 What are distinctive characteristics of the various layers of the atmosphere?

6 What are the temperature variations with altitude? How can the changes be explained?

7 What evidence is there to support the plate theory of Earth's crust?

8 What evidence is there to support present conceptions of the internal structure of Earth?

9 What was the historic explanation for the rising and setting of the Sun?

10 Describe Jean Foucault's experiment, its results and implications.

11 What are some effects of Earth's rotation?

12 How can it be established that Earth's orbit is elliptical?

13 The vernal equinox, also called the first point of Aries, is located in the constellation Pisces, and not Aries. Why is this so?

14 How did Eratosthenes determine the size of Earth?

15 What is the acceleration of gravity? How is it determined? How does it vary from place to place?

16 How can the mass of Earth be determined?

17 What is the probable explanation for Earth's magnetic field?

READINGS

Calder, Nigel. *The Restless Earth*. New York: The Viking Press, 1972.

Gamow, George. *A Planet Called Earth*. New York: The Viking Press, Inc., 1948.

Marsden, B. S., and A. G. Cameron, eds. *Earth-Moon System*. New York: Plenum Publishing Corporation, 1971.

Matthews, Samuel W. "The Changing Earth." *National Geographic*, January 1973.

Scientific American. *The Planet Earth*. New York: Simon & Schuster, Inc., 1957.

THE MOON

CHAPTER 13

The Moon is our nearest neighbor in space, less than half a million kilometers from us—a distance covered in a few hours of coasting in an Apollo vehicle. It was called the goddess Selene by the Greeks, Luna and Diana by the Romans, revered by men of prehistoric times and always as a beneficent goddess that destroyed the blackness of night and so protected men from unseen dangers. Its soft light inspired poets and lovers. Its proximity prompted careful study of its features and the drawing of maps. When the telescope became available, the Moon became an even more intriguing object, for the telescope revealed just enough to pique man's curiosity, and to initiate debate and discussion about the tantalizing glimpses made possible.

During the Apollo 17 mission, astronaut Harrison H. Schmidt is shown next to a huge lunar boulder split at the left. (NASA)

In the mid-twentieth century, the Moon continued to be an object of fascination when landing a man on the Moon by 1970 and returning him safely to Earth became a national objective. On July 20, 1969, the goal was achieved when Neil A. Armstrong set foot on the Sea of Tranquility—the first man to walk upon the Moon.

THE MOON HOAX

Firsthand investigations revealed that the Moon is a barren world having no atmosphere, no water—a cratered, channeled, and mountainous surface that has never supported life of any kind. The lack of any signs of any kind of living things was disappointing to many, those who believed life to be a universal condition. The desire to find creatures elsewhere in the universe did not begin in recent times. Indeed, it has persisted through centuries. In 1835 there was such strong belief in the existence of life on the Moon that people avidly read and believed reports of the discovery of Moon-people supposedly written by Sir John Herschel. According to articles in the *New York Sun*, Herschel had recently set up a telescope with a 200-foot lens—so powerful he could see flowers on the Moon, Moon creatures eating melons, and mountains embedded with sapphires, rubies, and amethysts. It was true that John Herschel was surveying southern skies with a large telescope. But the stories were a hoax engineered and written by Richard Locke, editor of the *Sun*, who was in danger of losing his job if circulation did not improve. (Circulation picked up. Locke left the *Sun*, founded his own newspaper, and became a wealthy man.)

DISTANCE TO THE MOON

The Moon is our only natural satellite. Its average distance from Earth—center to center—is 384 393 kilometers. Precise determinations of distance (within a few centimeters) are now possible through the use of lasers. Beams sent from Earth are bounced back by reflectors placed upon the Moon during Apollo missions. Each month the Moon passes through perigee, some 355 000 kilometers from Earth, and apogee, some 400 000 kilometers. When the Moon is near us it appears much larger than when it is farther away, as shown in Figure 13–1.

The Earth and Moon are more nearly the same size than any other satellite-planet combinations in the solar system. The Moon's

FIGURE 13–1
When the Moon is
close to Earth
(perigee), it appears
considerably larger
than when it is farther
away (apogee).
(Willard MacCalla, in
Sky and Telescope)

diameter is 3456 kilometers, one-fourth that of Earth; the diameter of Ganymede, the largest satellite in the solar system, is 4990 kilometers; however, Ganymede is only one-twenty-eighth the diameter of its planet, Jupiter.

THE MOON ILLUSION

The Moon appears to change its size as it moves from one part of the sky to another. At moonrise, for example, the Moon seems much larger than it does when it is riding higher in the sky. This is the so-called "Moon illusion." And that's what it is, for if one were to photograph the Moon, taking exposures every few minutes with a stationary camera, one would see that the sizes of the images are uniform (Figure 13–2).

The Moon illusion may result because when the Moon is rising we see it in relation to nearby objects. Intellectually we know that the nearer something is to us, the larger it appears to be. Perhaps this factor is operating.

Or, as some suggest, the illusion may result from the fact that we think of the sphere of stars and sky as being round while the

actual appearance is more like a bowl—the rim being at an infinite distance.

The Moon surely does not change size as it changes position, so we know we are observing an illusion, even though there may be lack of agreement concerning the explanation. However, there can be no disagreement about motions of the Moon, some of which affect Earth considerably.

MOTIONS

Every object in the universe is affected by every other object. When bodies are relatively close together—as are the Earth, Moon, and Sun—the effects are significant. Earth pulls upon the Moon first one way and then another; so does the Sun. Because all three bodies are in motion, their movements become so complicated that it is questionable whether a given arrangement of the three bodies is ever reproduced precisely. It has been computed that there are probably 800 different motions of the Moon, too many for us to consider here. We will discuss only the major motions, those readily observable. One of these is revolution.

Revolution

The Moon revolves around Earth; actually, the two bodies revolve around each other, around the barycenter of the Earth-Moon system. The barycenter is the center of mass of the two objects. If we were to line up the Moon with a distant star, the time required for the Moon to return to the same position would be 27⅓ days (27^d 7^h 43^m 11.27^s). This is called the *sidereal* month—derived from the Latin *sidus*, meaning constellation or star group.

The time that elapses between repetitions of a given phase of

FIGURE 13–3   *New Moon occurs when the Earth, Moon, and Sun are in line—in syzygy. One sidereal period (27⅓ days) elapses between successive alignments of Earth, the Moon, and a distant star. A synodic period (29½ days) is the time between successive alignments of the Earth, Moon, and Sun—new Moon to new Moon.*

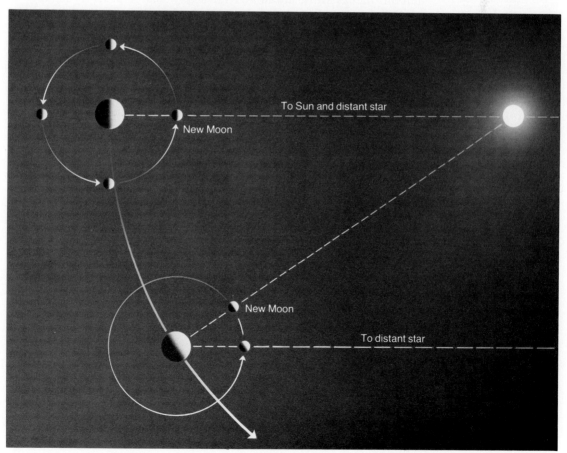

the Moon is 29½ days. This is slightly longer than the sidereal period because Moon phases depend upon the positioning of the Sun, Earth, and Moon. For example, the "new Moon" occurs when the three bodies are in line, when they are in *syzygy*, as it is called. While the Moon goes around Earth, Earth goes part way around the Sun. The Moon takes about two days to catch up, to return to the new-Moon position. This is called a *synodic month*, after *synod*, which means "meeting" or "assembly." The average length of a synodic month is 29½ days ($29^d\ 12^h\ 44^m\ 2.78^s$). (Figure 13–3.)

In our sky we see the revolution of the Moon around Earth as a motion from west to east. Since the journey is completed in 27⅓ days (the Moon moves through 360 degrees in 27⅓ days), it must move through about 13 degrees in 24 hours, or about its own diameter ever hour. At a given time if we mark the position of the Moon relative to nearby stars, we will be able to observe the amount of eastward motion. After a day has elapsed the change in position is considerable. Measurement of 13 degrees can be checked by keeping in mind that the distance between the pointer stars of the Big Dipper (the stars at the end of the bowl) is 5 degrees.

As the Moon revolves around Earth, it falls toward Earth 1.3 millimeters each second. Also, it moves 1000 meters along its orbit. The Moon stays in orbit because of Earth's gravitation. If no force were exerted upon the Moon, or if whatever forces exerted upon it were equal, the Moon would move in a straight line. (One of Newton's laws of motion states that a body in motion continues to move, and to travel in a straight line, unless a force is exerted upon it.) Earth's gravitation pulls upon the Moon—the force is not balanced by any other force, so the Moon must accelerate (change direction). The Moon's acceleration of 1.3 millimeters each second, together with its forward motion of 1000 meters a second, results in its following a curved path around Earth.

Moon Phases

The appearance of the Moon changes as it moves eastward, revolving around Earth. We cannot see the new Moon, for the lighted half of the Moon is turned away from us. About one day later a small sliver of the lighted half can be seen, however. This is the new crescent, and it is seen low in the western sky at sunset. As days go

FIGURE 13–4 (opposite page)   A cycle of lunar phases as seen from Earth.

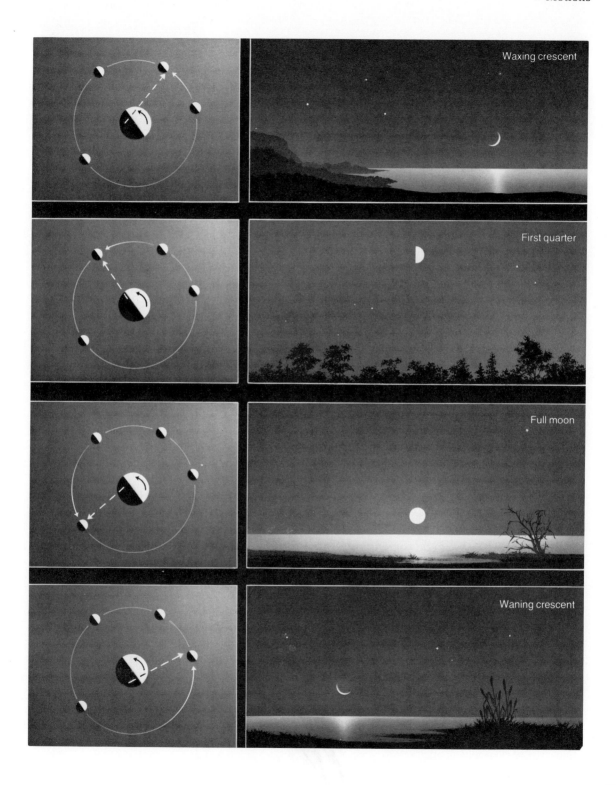

Waxing crescent

First quarter

Full moon

Waning crescent

by, the Moon appears farther eastward, and the crescent grows, or waxes. A week into the cycle, first quarter occurs. The Moon continues to "grow." It becomes a *gibbous* (from an Italian word meaning "humpbacked") Moon—larger than quarter but less than full. In two weeks, the Moon rises in the east as the Sun sets in the west. All of the lighted half of the Moon can be seen from Earth: the Moon is full.

Now the Moon appears to decrease in size; it is a waning moon. Each night it rises later and later. It precedes the Sun and can be seen in the daylight skies of morning. After 29½ days, the Moon is new once more; a lunar cycle is completed.

The changing appearances of the Moon are the phases. They appear to us only because of our location—the changing configuration of the Earth-Moon-Sun pattern. An observer in space looking down from high above would not see any phases. Nor would he see the Moon moving around Earth. Rather, he would see Earth and Moon swinging from side to side of Earth's orbit—the line scribed out in space by the barycenter of the Earth-Moon system—somewhat as shown in Figure 12–12 on page 321.

Rotation

Another basic motion of the Moon is rotation. At the equator, Earth rotates about 1600 kilometers an hour. The Moon moves only about 16 kilometers an hour at its equator, taking exactly one sidereal month to complete a single rotation. Because the periods of rotation and revolution are identical, the same half of the Moon is always turned toward Earth. Until 1959, when Russia's Lunik III photographed it, we had never seen the far side of the Moon. Actually, about 59 percent of the total lunar surface had been observed because of *librations* (from *libra,* "to balance") of the Moon that enabled us to see around it slightly. There are three different librations to be considered. One is libration in latitude.

Latitude is distance north or south of the equator of Earth, the Moon, or any other body. Just as Earth's axis is tilted to the plane of its path through space, so is the axis of the Moon. (On Earth the tilt is 23½ degrees, on the Moon it is 6½ degrees. As a result, we can see slightly (6½ degrees) beyond the north pole of the Moon when the axis is tilted toward us, and slightly beyond the south pole when it is tilted away from us. Over a period of time the Moon seems to rock from pole to pole (Figure 13–5).)

Libration in Longitude

Also, we are able to see slightly around to the east and west (the longitude) of the Moon. Longitude on Earth is angular distance

measured east and west of the Prime Meridian, an imaginary line
which passes from pole to pole and through Greenwich, England, a
short boat ride on the Thames from metropolitan London. On the
Moon, longitude has a similar meaning. It is an angle east and west
of a prime meridian which, by agreement, is the line going from

FIGURE 13–5   Libration in latitude.

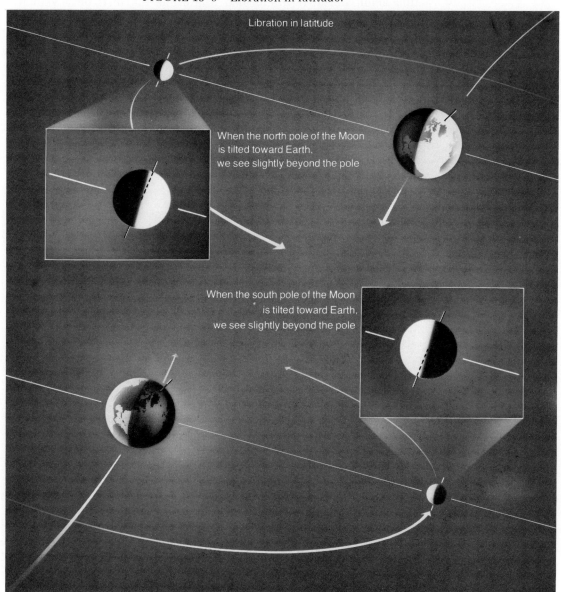

Libration in latitude

When the north pole of the Moon
is tilted toward Earth,
we see slightly beyond the pole

When the south pole of the Moon
is tilted toward Earth,
we see slightly beyond the pole

pole to pole and passing through the eastern edge of Mare Imbrium (the Sea of Showers) and Mare Nubium (the Sea of Clouds).

The Moon's orbit around Earth is an ellipse. When the Moon is close to Earth (at perigee) as it is once each month, it revolves faster than when it is at apogee. However, the rotational speed of the Moon does not change; therefore, revolution and rotation do not keep pace. When the Moon moves through one-fourth of its revolution about Earth, it would also rotate one-fourth if both velocities were constant. Sometimes the Moon rotates a bit more than one-fourth during a quarter revolution, so we can see a bit farther around the Moon, about 6 degrees east or west. The Moon seems to rock from east to west (Figure 13–6).

FIGURE 13–6   *Libration in longitude. (1) Arrow erected on Earth-Moon line appears at the center of the lunar disk. (2) Rotation has changed direction of the arrow by 90°. But the Moon has revolved more than 90°, so we can see a little of the western side of the Moon. (3) The Moon has rotated, and revolved, 180°, so the arrow appears once again at the center of the lunar disk. (4) The Moon has rotated 270°; however, it has revolved only 264°. We see more of the eastern part.*

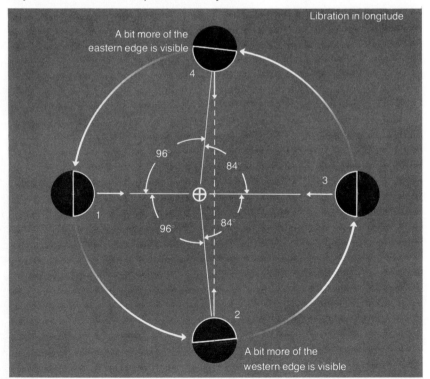

Daily Libration

Another factor, daily libration, is caused by the rotation of Earth. When the Moon is rising, we can see partly beyond the top of it (the western edge) because we are slightly above a line connecting the center of Earth and the center of the Moon. At moonset, we can see partly over the eastern edge of the Moon (Figure 13–7).

Because of these libration effects, 41 percent of the lunar surface has never been seen from Earth, another 41 percent is always seen during a full lunar cycle, and 18 percent is alternately visible and invisible.

Retardation of Moonrise

The apparent motion of the Moon is from east to west; it rises above the eastern horizon, moves westward, and disappears below the western horizon. However, the actual motion around Earth, its direction of revolution, is just the opposite, from west to east. As men-

FIGURE 13–7 *At moonrise and moonset we can see partially over the "top" of the Moon.*

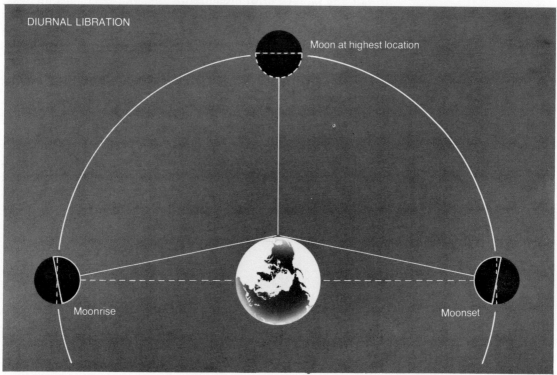

DIURNAL LIBRATION

Moon at highest location

Moonrise

Moonset

tioned earlier, during 24 hours, the Moon moves about 13 degrees eastward. Therefore, Earth must catch up with the Moon; it must move 13 degrees more than 360 degrees before the Moon rises the following night. The average time required for this is 50 minutes—the retardation of moonrise.

The actual delay in the time of moonrise varies a great deal throughout the year. It can occur as little as 22 minutes later, and as much as 80 minutes. We will explain why.

The path the Moon travels across the sky is essentially the same as that followed by the Sun—the ecliptic—the line along which eclipses occur. When the Moon is new, the Sun and Moon are on the same side of Earth, and they are at essentially the same location on the ecliptic. If the location were identical and the Moon were close enough to us, there would be a solar eclipse.

When the Moon is full, the Sun is on one side of Earth (and on the ecliptic) while the Moon is on the opposite side. We see the full Moon in the part of the sky that is opposite the Sun. When the Sun is low, as during the northern winter, the full Moon will be high; and the full Moon will be to the south and low in the summer when the Sun is high and to the north.

In fall, the Moon's orbit (essentially the ecliptic) makes the least angle with the horizon. At this time, the motion of the Moon has minimum effect on delaying the time of moonrise. Because of this the full Moons following the autumnal equinox are called the Harvest Moon and Hunters Moon. In spring, the opposite is true. The angle of the ecliptic to the horizon is much greater, and the delaying effect is increased (Figure 13–8).

In a sidereal month the Moon travels some 2.23 million kilometers; at perigee it speeds up, at apogee it slows down, but its average velocity is 3659 kilometers an hour. The speed of the Moon in its orbit varies because of Kepler's second law of motion, which, when applied to our problem, says that a line joining Earth and Moon will sweep through equal areas in equal periods of time, as shown in Figure 15–3 on page 399.

ECLIPSES

When the Moon and Sun are at the same location on the ecliptic (in conjunction), there is an eclipse of the Sun, as discussed in Chapter 12. When they are both on the ecliptic, or nearly so, but at opposite locations (in opposition), there is a lunar eclipse; Earth's shadow falls upon the Moon (Figure 13–9).

FIGURE 13–8
In fall the Moon rises at position 3. Twenty four hours later it is at A below the horizon. It takes about 22 minutes for the horizon to move down to that position. The next night the sequence is repeated —the Moon is at position B and it rises at 1. In spring the Moon moves from position 3 to C, and it takes about 50 minutes for it to appear at 4. The next night the sequence is repeated.

FIGURE 13–9 (below) Both solar and lunar eclipses occur only when the Sun, Moon, and Earth are on the line of nodes, or nearly so.

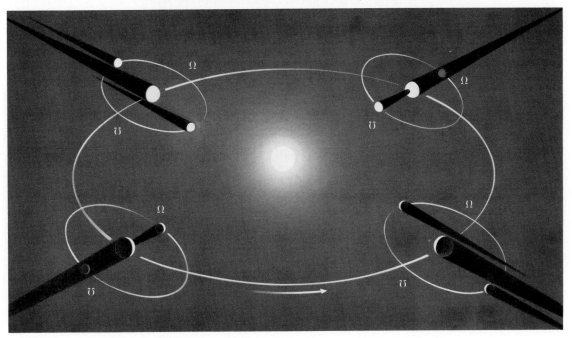

During the next few years, lunar eclipses will occur on the following dates:

1976	May 13	Partial
1977	April 4	Partial
1978	March 24	Total
	September 16	Total
1979	March 13	Partial
	September 6	Total
1981	July 17	Partial
1982	January 9	Total
	July 6	Total
	December 30	Total

If interplanetary space contained sufficient dust to reflect sunlight, an observer out in space would be able to see Earth's dark shadow against a brighter background. Although it is black against black and so is invisible, Earth casts a shadow of some 1.37 million kilometers. At the Moon's average distance of 381 000 kilometers, the shadow is some 9120 kilometers in diameter—more than enough to cover the entire Moon (Figure 13–10).

Because Earth is lighted by a broad light source, its shadow is made of two parts, called the *umbra* and *penumbra*. The passage of the Moon into the penumbra of Earth's shadow cannot be noted

FIGURE 13–10   *At the distance of the Moon, Earth's shadow is more than twice the Moon's diameter.*

readily, for the density of the shadow is not great enough to dim moonlight very much. The entrance of the Moon into the umbra is quite apparent, however. A darker region with a curved edge notches the eastern side of the Moon; the Moon moves into the shadow from the west. The notch grows larger, and after about an hour the Moon is in total eclipse. It is not totally obscured, however.

During totality, the Moon takes on a coppery-red color. This is because the longer wavelengths of sunlight still reach the Moon. They are bent by Earth's atmosphere around Earth and mix with the shadow.

As time passes, the Moon slowly moves out of Earth's shadow and the events mentioned above occur in reverse order.

Lunar eclipses are of limited value to modern astronomy. However, the shape of Earth's shadow on the Moon was an observation made by the early Greeks, and one that enabled certain of them to reach the conclusion that Earth is round; a sphere is the only object that always produces a shadow that is an arc of a circle. Also, since the dates when there were lunar eclipses can be determined, eclipses are valuable in fixing the dates when events of historical importance occurred. For example, it is known there was a partial eclipse of the Moon on the eve of King Herod's death. Knowing this, historians could fix the date of his death, and so the time for the beginning of the Christian Era.

TIDES

According to some people, the Moon affects the growth of crops. Above ground crops will not flourish if they are sown in the dark of the Moon, they say, this being the time when potatoes and other below-ground crops should be planted. Also, there are those who agree that the Moon has considerable influence upon the weather.

Such lunar effects may be debated. However, there are others that are not debatable; for example, the tides are very closely related to the Moon—of that we can be sure. People of ancient days were aware that the Moon affected tides, for they observed the relationship between them and lunar positions. No explanation could be given, however, until Newton's discovery of gravitation. The primary cause of tides is differences in the gravitational force that the Moon exerts on different parts of Earth.

The Moon is so close to Earth that the Moon side of Earth is much more strongly affected by lunar gravitation than the center of Earth—6400 kilometers farther away. The tidal force affects

Earth's atmosphere and land surface as well as the seas. Although continental tides may reach 15 to 22 centimeters, we are not nearly as aware of them as we are of water tides, where a change of a meter is common, and the variation from high to low reaches 15 meters in specific areas.

Gravitationally, the Moon acts as though all its mass were at its center. The gravitation forces exerted on Earth at the sublunar point (where the Moon is at the zenith) are greatest; at the center of Earth they are less, and at the far side of Earth (Moon at nadir) they are least (Figure 13–11).

Locations at right angles to the line connecting zenith and nadir positions have a force the same as at Earth's center, but in different directions.

If from each vector we subtract the force exerted at Earth's center, the differential tidal forces remain. A stretching occurs along the zenith-nadir line, and a squeezing at those locations where the Moon would appear on the horizon. At intermediate locations, a combination of the two effects occurs. The resultant of the forces causes the water to flow as indicated. Earth is deformed as shown, and, were Earth liquid, high and low tides would occur because of the flow of water toward the "zenith" and "nadir" locations.

As Earth turns, the high-tide location moves into low tide, and some 12 hours, 25 minutes later it experiences another high tide. During 24 hours, 50 minutes, the location would experience two high tides and two low tides.

If Earth were completely water, the tidal change would be 60 to 90 centimeters. In many places that is what it amounts to, but in others the land configurations produce wide departures. In a wide-mouth bay, for example, there may be a funnel effect, a piling up of the water to 15 meters above low water, as in the Bay of Fundy.

FIGURE 13–11   *Differential effects on Earth of lunar gravitation produce the tide-raising forces.*

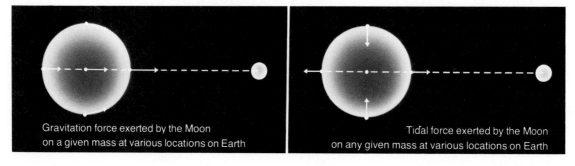

Gravitation force exerted by the Moon
on a given mass at various locations on Earth

Tidal force exerted by the Moon
on any given mass at various locations on Earth

Also, the tidal change will most often lag behind the Moon. The water will take considerable time to flow in through a narrow opening, and once Earth has turned and the effect of the Moon's gravitation has modified, water held in the bay will be retained longer because of the narrow outlet. Nevertheless, such regions will experience periodic changes from high to low water which are related to the positioning of the Moon (Figure 13–12).

FIGURE 13–12
At certain locations there are tidal changes that average 50 feet (about 15 meters). These photos show the variation at Cutler, Maine. (Maine Department of Sea and Shore Fisheries)

Although the Sun is much more massive than the Moon, its effect on Earth tides is secondary to that of the Moon. The Moon is 400 times closer. The tidal force varies as the inverse cube of distance. The tidal force of the Sun is only $\frac{5}{11}$ that of the Moon.

When the Sun and Moon are in line with Earth (when they are in syzygy), tidal effects are added together. At such times there are the greatest extremes in tides—the highest highs, the lowest lows. These are the spring tides that occur twice monthly, during new and full Moon. Here the word *spring* refers to a leap or jump, and not to the season. When the Sun-Earth line is at right angles to the Moon-Earth line, there prevails the least variation between tidal extremes. These are the neap tides (from an Anglo-Saxon word meaning "scanty"), and they occur during the quarter phases of the Moon; the Moon is at quadrature. Especially high tides are experienced when the new Moon is at perigee, only some 355 000 kilometers away.

MASS AND VOLUME

The mass of the Moon tells us how much matter it contains, while its density tells us how tightly the material is packed together. Density values are related to water. A liter of water weighs 1000 grams and the same volume of iron weighs 7800 grams—7.8 times as much; the density, more often called specific gravity, of the iron would be 7.8.

We know the mass of Earth is 6×10^{27} grams. To find its density, we have to work with the volume—1.083×10^{27} cubic centimeters. When mass is divided by volume, the answer is 5.5—the average density of our planet. The weight of Earth (if you could weigh it) is 5.5 times what it would weigh if the planet were entirely water.

We now have all the information needed to set up a ratio of masses and distances in the Earth-Moon system. From observations of solar motions, we obtain the barycenter—the center of balance, if you will, of the Earth-Moon system, as discussed in Chapter 12.

(The barycenter is 4645 kilometers from the center of Earth. This distance is $\frac{1}{81}$ of the distance from the barycenter to the center of the Moon. Therefore, the mass of the Moon must be $\frac{1}{81}$ that of Earth.)

The volume of the Moon is easily computed $(V = \frac{4}{3}\pi r^3)$ since the diameter is known to be 3456 kilometers. It turns out to be about ⅟₅₀ of Earth's volume—and its average density is 0.6 that of Earth, or 3.3, which, incidentally, is about equal to the density of rock layers about 40 kilometers below Earth's surface. This fact has led many investigators to believe the entire Moon may be made of rocks very similar to the basaltic underlayers in Earth's mantle.

SURFACE AND NOMENCLATURE

Meticulous analysis of Moon rocks and soil by researchers around the world indicates that basalt is common on the Moon, and some have inferred that this substance together with olivine comprises the bulk of the lunar structure.

Millions of years ago volcanic activity prevailed over the entire lunar surface. And it was not unusual to have active volcanoes inside craters of older ones. Many of the lunar minerals exhibit a structure that would result from extreme shock, the kind that would be produced by severe meteoritic bombardment.

(The types of minerals found upon the Moon are essentially the same as those which occur on Earth. However, they are extremely clear, indicating the absence of water in their structure. Also, percentages of materials in certain crystals are different on the Moon from those found on Earth. For example, the percentage of ilmenite, an ore of titanium, is much higher in Moon rocks (about 20 percent) than in Earth rocks where 5 percent would be normal. Ilmenite occurs as dark, shining crystals in otherwise dull basaltic rocks) This ore may be responsible for the generally dark appearance of lunar maria, the flat plains.

The entire lunar surface is shown in the maps on pages 354–357. The surface of the near side of the Moon is much more varied than that of the far side, craters are fewer and many are larger, and the maria are much more extensive. However, just as many features of the near side defy adequate explanation, so too do many features of the far side. Of special interest is Mare Orientale (the Eastern Sea) at 95°W and 17°S. It appears to be a huge crater (some 1100 kilometers across) having smaller craters inside it that are concentric with the outer rim (Figure 13–13b).

Nomenclature of the Moon began in the early part of the seventeenth century when Galileo called the flat plains *maria*, or seas, probably believing that they contained water. In 1650, Riccioli, an Italian astronomer, compiled the first map of the Moon. He needed some method of identifying features and so named moun-

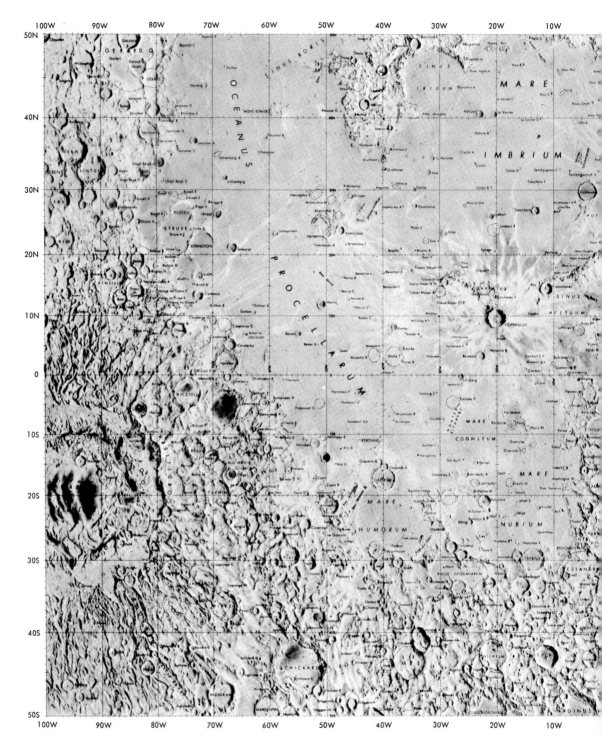

FIGURE 13–13a *U.S. Air Force map of near side of the Moon.*

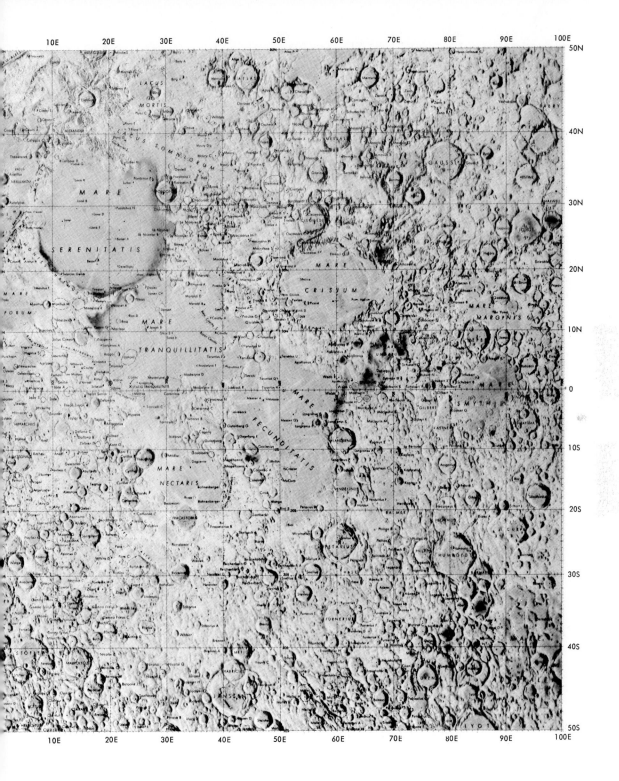

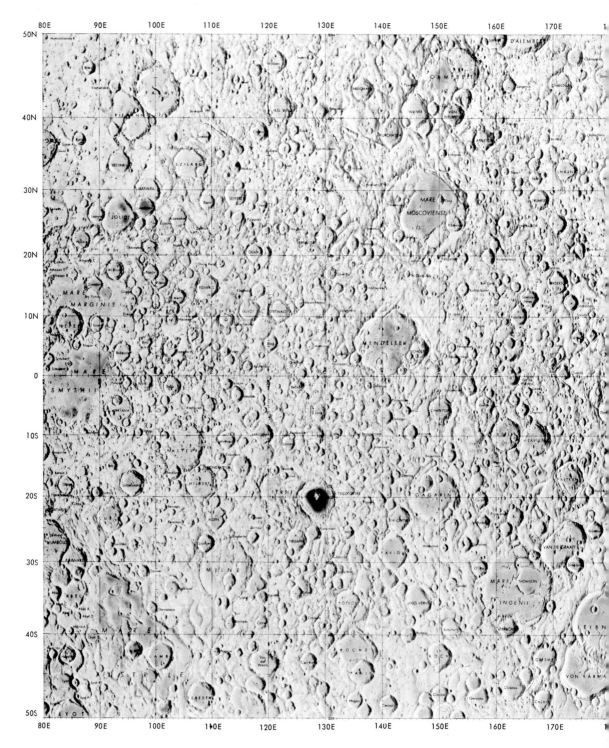

FIGURE 13–13b   U.S. Air Force map of far side of the Moon.

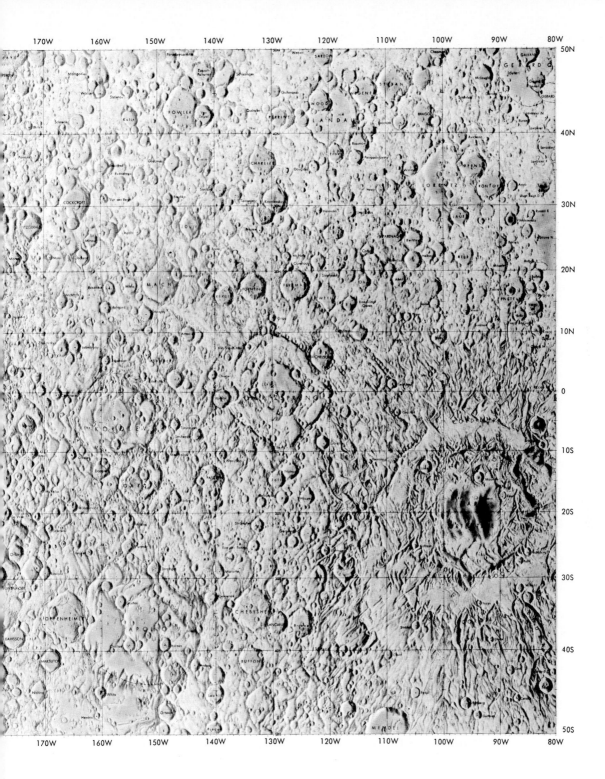

tains of the Moon after those of Earth—Alps, Appenines, Caucasus. However, a few are named after astronomers—Doerfel and Leibnitz. Craters are named after scientists and explorers—Archimedes, Copernicus, Tycho, Kepler, Tsiolkovski, Einstein.

By mutual agreement, modern naming of lunar features, and features of planets also, is done by committees appointed by the International Astronomical Union. All interested parties propose names they feel should be included; these are usually names of scientists, historians, or people who have been instrumental in lunar

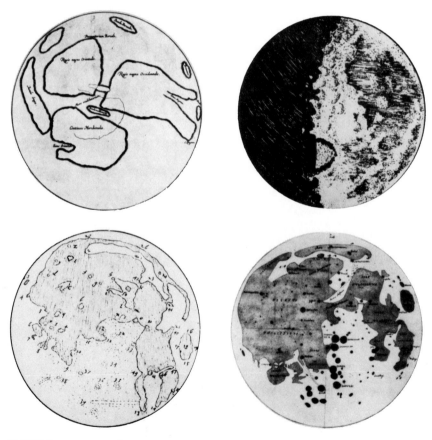

FIGURE 13–14 *Early maps of the Moon, as seen by the unaided eye of William Gilbert, sometime before 1603 (top left); from Galileo's* Sidereus Nuncius, *1610 (top right); first telescopic map of the full Moon by Thomas Harriott, sometime after 1610 (bottom left); and map containing proper names for features by Langren, circa 1628 (bottom right).*

exploration. Selections are made by the committee; Gagarin, Fermi, Hertzsprung, Michelson, Apollo, and Wan-Hoo are the names of some of the features of the far side.)

While there are thousands of formations on the Moon that are really craters, many features that go by that name are more like walled plains. Some are 80 to 90 kilometers across, and Clavius is 250 kilometers in diameter.

The inner walls of many craters are very high, often reaching 1500 meters and even reaching 8850 meters, as in the case of Newton, the deepest crater. The outer walls are usually not as high, indicating interior depression and probably pile-up of material around the exterior of the formation. This supports the meteoritic theory of crater origin, since volcanic craters have inner craters higher than surrounding regions.

Careful observation of the Moon with tripod-supported binoculars or a low-power telescope is most interesting. The best time to observe is when the Moon is around quarter phase. Sunlight is then coming at an angle, and deep shadows cast by mountains and crater walls put the features in bold relief.

During the period when the Moon is waxing, the "pleasant sounding" features become apparent. Many of the lunar formations, especially the maria, are named after the kinds of events people related to Moon phases. During waxing of the Moon, it was believed there would be good weather; it was a time when crops flourished and all was well with the world. So we have the Sea of Tranquility, the Sea of Nectar, the Lake of Dreams. When the Moon was on the wane, weather would be severe; it was a cold and barren time, one to be feared because soon there would be no light in the sky to ward off the dangers of the night. So we have the Sea of Rains, the Ocean of Storms, the Sea of Moisture.

Each area of the Moon has many interesting features. We might consider just one of them—Mare Imbrium, the Sea of Rains. We can see it with binoculars. It is a plain some 1200 kilometers across. Close inspection during the Apollo mission showed it to be not nearly as smooth as had been believed, but like most of the maria, the floor is dotted with craterlets and with the remains of craters that have been probably worn down by meteorite bombardment or have been filled in with accumulations of meteoritic dust. The plain is surrounded by mountains—the Carpathians in the southeast, the Appenines toward the southwest, the Caucasians, the Alps, and the Jura Mountains to the north. The Jura Mountains wall in Sinus Iridum, the Bay of Rainbows. Its walls are little more than remnants of what they were long ago.

In the northwest is Plato, a smooth-floored symmetrical crater surrounded for the most part by the Alps. In the southwest is a cluster of craters, the largest of which is Archimedes.

Toward the east Mare Imbrium blends into Oceanus Procellarium, the Ocean of Storms, the most extensive flat area. It contains scores of craters, the most dominant being Copernicus. It is readily discernible when lighted from the side. Even when the Moon is full it is intriguing because of the light-colored rays that radiate from it (Figures 13–15, 13–16, and 13–17).

AGE

The Moon's age appears to be about the same as that of Earth. Many of the rocks and soil samples indicate ages of more than 3 billion years, and specimens up to 4.6 billion years old have been found. The Moon was probably formed at approximately the same time that Earth was formed, and from essentially the same basic substances.

The Moon is probably solid throughout. If there were a semi-

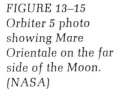

FIGURE 13–15 Orbiter 5 photo showing Mare Orientale on the far side of the Moon. (NASA)

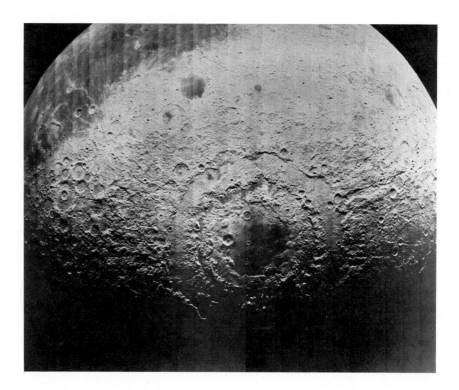

molten region, the Moon would have a magnetic field, much as Earth has. Also, *mascons*, mass concentrations, have been detected under some of the maria. They would have to be supported by solid material. Various theories attempt to explain lunar structure and its origin. One of them grew out of the study of tides in general that was followed by George Darwin (1845–1912), second son of the more famous Charles Darwin. He proposed that at one time during its early history, Earth rotated in only about four hours. Rhythmic tides of enormous size on both land and sea set up stresses and strains so great that Earth was torn apart. One great mass—the material originally in the Pacific Ocean basin—ultimately became the Moon. The theory is little more than an interesting idea, for it breaks down when investigated carefully. So does the so-called "capture theory," which supposes that the Moon was an asteroid–like mass. At some time, so the theory teaches, the mass passed close enough to Earth to be captured and held by Earth's gravitation.

The capture theory provides a way of explaining differences in mineral structure and composition between the Moon and Earth, but practically it appears to be more an explanation of convenience than one based upon cold scientific information. However, one cannot

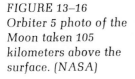

FIGURE 13–16
Orbiter 5 photo of the Moon taken 105 kilometers above the surface. (NASA)

FIGURE 13–17
Orbiter 5 photo of the Moon showing the crater Kepler in Oceanus Procellarum, at an altitude of 730 kilometers. (NASA)

completely dismiss the possibility that the Moon was at one time a deep space wanderer.

It appears more likely that Earth and the Moon, and the Sun and all the other planets, had a common origin from a spherical mass of gases that had a diameter of some 15 billion kilometers. Many of the lighter materials escaped from the system, and the remainder contracted to about one-half the original size. As the mass contracted, it spun faster, collapsed, flattened, and ultimately became disk-shaped. The diameter diminished to some 10 billion kilometers, and its thickness to about 2 billion. The large central mass became the Sun, and smaller concentrations evolved into planets and satellites.

Another fascinating theory of lunar formation supposes that at one time Earth was surrounded by a ring (much like the rings of

Saturn) and that the Moon evolved out of a consolidation of the particles comprising that ring. According to the theory, the ring may have been 200 000 kilometers in diameter and 40 000 kilometers wide. Particles making the ring may have been the remains of a primitive, dense atmosphere that enveloped Earth during late stages of its formation. There may have been sudden flare-ups in solar activity that pushed away less dense materials. Those of greater density became the Moon.

The theory may be another of convenience, for it explains some of the observations made when lunar rocks were investigated. For example, the Moon's density is much less than the density of Earth. A primitive Earth atmosphere would be made of less dense materials; those of greater density would be left behind. Also, the Moon appears to be essentially the same at various depths as would be the case with a formation growing out of ring masses. Metals are relatively rare on the Moon; there appears to be no metal core. We would expect a primitive atmosphere to contain no metals, or at least they would be extremely rare.

The manner in which the Moon came into being is still not well understood. Further study of lunar rocks and soil may provide additional clues.

FATE OF THE MOON

In his study of tides, George Darwin calculated the past and the future of the Earth-Moon system. At one time Earth and the Moon were only 16 000 kilometers apart, Darwin stated. At that time Earth turned on its axis once in 5 hours, and the Moon went around our planet in just over 5 hours. Because the two bodies were so close together, they must have raised tremendous tides on each other.

Whenever a tide is produced, there is friction; on Earth this is the drag of the water as it moves over the surface. When the Moon was close to Earth, the friction would have been tremendous, causing a slow-down in rotation of Earth. Gradually the Moon moved farther from Earth and the month became longer.

Darwin believed the process had been going on for billions of years. Indeed, it continues today. Earth is turning more slowly; the day is getting longer, about $\frac{1}{1000}$ of a second every hundred years.

The process will go on, Earth will continue to slow down, the Moon will move farther away, and the months will become longer. After billions of years Earth will take 47 days to make a single rotation, and the Moon will require 47 days to go around Earth. Friction of the water against land, caused by lunar tides, will become zero.

Then solar tides will become effective in producing changes in the system. Tides produced by the Sun's attraction will slow down Earth's rotation. The day will become longer than the month. Once more lunar tides will become effective, but now the result will be to speed up Earth. The Moon will lose energy to Earth, and so the Moon will move toward Earth. It will get closer and closer.

(There probably will never be an Earth-Moon collision. As the Moon nears Earth, gigantic tides of the lunar surface, produced by Earth's gravitational attraction, will erupt. These tides would probably pull the Moon apart, shattering it into small pieces that would form into a ring, much like the rings of Saturn.) This would be in accordance with the theory propounded by the French astronomer Edouard Roche in 1850. He found that if a satellite having the same density as its planet moved within *Roche's limit*, a distance 2.44 times the planet's radius, the gravitational force would pull the satellite apart. The theory appears to have validity, for the ring system of Saturn is inside Roche's limit.

Should this happen eventually to our satellite, we have nothing to fear. The process would take billions of years, perhaps longer than the Sun will shine.

REVIEW

1 The Moon, an arid world lacking an atmosphere and life of any kind, is our natural satellite.

2 Apparent motion of the Moon is east to west; real motion is west to east around Earth, covering about 0.5 degrees in one hour.

3 Lunar eclipses, which can be seen at any location where the Moon is visible, occur when the Moon enters Earth's shadow.

4 Differentials in the gravitational force of the Moon on Earth produce land and sea tides in Earth—spring tides when the Moon is new and full, and neap tides when the Moon is in a quarter phase (at quadrature).

5 The Moon is made of essentially the same materials that comprise Earth, except that percentages of certain elements may vary, as well as the type of crystalline structure.

6 Both Earth and the Moon appear to have been formed at approximately the same time—some 4.6 billion years ago.

QUESTIONS

1   What are the various theories that have been proposed to explain the origin of the Moon? Which seems most reasonable to you and why?

2   Who were some of the early mappers of the Moon? What similarities can you see in their creations?

3   Explain why Earth-bound observers can see more than 50 percent of the lunar surface.

4   During what parts of the lunar month can the Moon be seen in the daytime?

5   In which part of the 24-hour day can the various phases be observed?

6   What is the explanation of the Harvest Moon?

7   What are some similarities and differences among lunar rocks and terrestrial rocks?

8   How might lunar features have been formed? What are reasons for your conclusions?

9   What might be the ultimate fate of the Moon?

10   What is the mass of the Moon? How is it determined?

11   What is the principal cause of tides? Why do they vary at different locations?

12   If there were no Moon, what would be some of the effects on Earth?

13   What have been the effects of Project Apollo upon man's outlook and perspectives?

14   Discuss the suggestions often made that someday there will be astronomical observatories on the Moon, and also industrial plants to achieve certain processes best done under vacuum.

READINGS

Anderson, Don L. "The Interior of the Moon." *Physics Today*, March 1974.

Cherrington, Ernest H. *Exploring the Moon through Binoculars*. New York: McGraw-Hill Book Company, 1969.

Cooper, Henry S. *Moon Rocks*. New York: The Dial Press, 1970.

Mason, Brian, and William G. Melson. *The Lunar Rocks*. New York: Wiley—Interscience Publishers, 1970.

Scott, David R. "What Is It Like To Walk on Another World?" *National Geographic*, September 1973.

Weaver, Kenneth F. "Have We Solved the Mysteries of the Moon?" *National Geographic*, September 1973.

Whipple, Fred. L. *Earth, Moon and Planets*. 3d Ed. Cambridge, Mass.: Harvard University Press, 1972.

THE SKY
AS SEEN FROM
EARTH

CHAPTER 14

Except for an occasional man-made satellite that looks like a fast-moving planet, the sky appears to us today just as it has to men down through the centuries. Our interpretations of observations are quite different, however, from those of our ancestors.

An early scientific explanation of sky observations, especially the motions of the Sun, Moon, and the five visible planets, was made by Eudoxus of Cnidus (408–355 B.C.). He proposed that each of the planets is fastened to a transparent sphere which, while it rotates, is inside another sphere, and so on. To account for the observed motions, each planet required four spheres, each rotating in a different direction and each performing a different function.

For locating objects, the sky is divided into segments with lines running along the "surface" from "pole to pole," and also around Earth parallel to the equator. Were one able to "see" these lines from a location on the "surface," the vista would appear as shown.

The model that Eudoxus originated required 26 spheres to account for motions of the components of the solar system. Later on, as additional observations were made, more spheres were needed.

The theory was elaborated by many individuals, including Aristotle. Apollonius of Perga (fl. 247–205 B.C.), the great geometer who was educated at Alexandria under Euclid's successors, sought a simpler explanation. He eliminated the spheres of Eudoxus, replacing them with epicycles and deferents; the planets moved in small circles (epicycles), the centers of which were on great circles (deferents) which encircled Earth, or the eccentric, a point removed from the Earth. Later on, the theory was described and written down by Ptolemy (second century A.D.) in the *Almagest*. The theory served quite well to explain not only direct motions of the planets, but also the retrograde motions. It held sway until Copernicus modified it drastically, but it was not eliminated entirely until Kepler discovered the laws of motion that bear his name.

Eudoxus enclosed his transparent spheres inside one that encompassed all—the star sphere. In our enlightened age we still find it convenient to use the "star-sphere" idea. Celestial objects are located by supposing they are equidistant from Earth and are somewhere on that star sphere, the celestial sphere.

Certain locations on this sphere belong to the observer alone. For example, the point in the sky directly over a person's head is his *zenith*. Obviously, the zenith changes with the person's location. (His feet point toward the *nadir*—the location directly opposite his zenith.) The line passing through the zenith and from the North Pole to the South Pole is his *meridian*. This also changes with his location on Earth. It is of considerable use as a reference; for example, solar noon occurs when the Sun is on a person's meridian, midnight when it is 180 degrees removed. When a star is on the meridian, we say it *culminates*. Objects can be located in a general way by giving positions so many degrees east or west of one's meridian or how long ago it crossed the meridian. It is the "M" in A.M. and P.M.

MAPPING THE CELESTIAL SPHERE

Altitude—Azimuth

At first glance, the night sky appears to be uniform throughout, not providing us with any means of identifying positions or locations; however, the experienced observer sees guideposts that enable him to find his way among the stars and to locate precisely a given star, planet, or other celestial object. One way we can find our way is to start from the north position on our horizon. North can be determined by using the Big Dipper as a guide. The two stars at the end of the bowl—Merak and Dubhe—are the pointers. In our imaginations we join Merak and Dubhe, and extend the line five times the distance between the stars. This will take us almost directly to Polaris, our present north star. Then we drop a line vertically from Polaris to the horizon to obtain geographic north. We measure in degrees eastward along the horizon—east being 90°, south 180°, west 270°, and north 0° or 360°. East is the fundamental direction; it is where sky objects rise. When we find position, we refer to east by saying we "orient" ourselves (Figure 14–1).

The angular measure along the horizon gives us the *azimuth* of the stars (azimuth is from Arabic roots meaning "the way"). In order to locate the star we need to know how high the star is above the horizon, the *altitude* of the star. Altitude is measured in degrees

FIGURE 14–1
A line dropped from Polaris to the horizon indicates the location of terrestrial north. The azimuth-altitude system for indicating locations of celestial objects is good for only a certain place at a certain time.

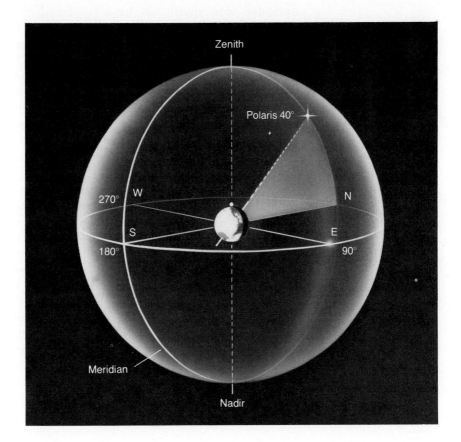

from the horizon to our zenith—from 0° at the horizon to 90° at the zenith.

Earth rotates from west to east, so the entire celestial sphere appears to move the other way—east to west. Since Earth rotates, one's relationship to the sphere changes constantly; therefore, the azimuth-altitude reading a person determines for a given star is valid only for him, and only for a specified time. The system has only limited value in astronomy.

Right Ascension—Declination

A better coordinate system uses measurements from a given location on the celestial sphere itself—something that moves with the sphere—rather than from a moving Earth. The equatorial coordinate system is the system that has been developed. It is similar to the latitude-longitude system for location on Earth.

Measurement in an east-west direction around the sky (*right ascension*) is made from the vernal equinox. If Earth's equator is extended into space, we have produced the *celestial equator*. It divides the sky into two hemispheres just as Earth's equator divides the planet. Another great circle is the Sun's path through the sky at an angle of 23½° to the celestial equator. This is the *ecliptic*. The ecliptic and celestial equator intersect at the *vernal equinox* and, 180° from that, at the *autumnal equinox*. The vernal equinox is the ascending node, the point of intersection where the Sun "moves" north of the celestial equator. The autumnal equinox is the descending node (Figure 14–2).

The poles of Earth are "extended" into space. Where they meet the celestial sphere we have the *north* and *south celestial poles*. Lines, *hour circles*, run from pole to pole. The hour circle that passes through the vernal equinox, by agreement, is the zero hour circle.

A star's east-west position is expressed as its angular distance eastward from the vernal equinox—so many hours, minutes, and seconds; each hour equals 15 degrees. This parameter is called *right ascension*. "Right" may come about because the measurement is eastward, and eastward is the real, or right, direction of motion of the Sun, Moon, and planets. (The apparent westward motion of these objects, rising and setting, results from Earth's west to east rotation.) Some maintain that the term "right ascension" arises because at the equator the celestial equator, along which right ascension is measured, is at right angles to the horizon, or, perhaps because if one is facing south, objects move to the right.

Right ascension tells us where the star is "around" the sky, just as longitude tells us where a location is "around" Earth. But in

FIGURE 14–2
The celestial sphere.

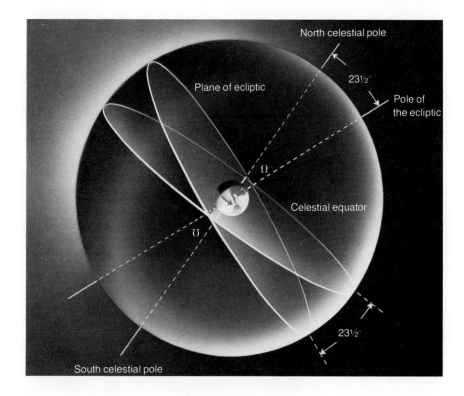

order to pinpoint the star, another coordinate is needed. We must know where the star is north and south, just as we must when finding location on Earth. To specify this, the position is given in degrees north or south of the celestial equator. This coordinate, called *declination*, is positive when the star is north of the equator, and negative when it is south of the equator. And so we say that a star is located in R.A. $2^h30^m10^s$ and Declination $+60°2'$ (Figure 14–3).

The position of a star on the celestial sphere changes very little from year to year. Over an extended period, however, its position will change because of precession of the equinox. The vernal equinox moves westward about 50″ a year; it is now in Pisces and moving toward Aquarius. Therefore, a star's position is given for a particular epoch—1900, 1950, 1975, or whatever—and adjustments are made accordingly. In early days of observing, the vernal equinox was located in the constellation Aries and was called the "first point of Aries." The term is still used, even though the intersection of the celestial equator and ecliptic (the vernal equinox) occurs in a different constellation.

Astronomers use this coordinate system extensively; however, even the casual viewer will find his involvement in sky watching more significant when he is able to imagine the locations of the celestial equator and the ecliptic. The celestial equator passes through the east and west points on the observer's horizon. The elevation of the celestial equator above the southern horizon in degrees is equal to the difference between his latitude and 90°. For example, if his latitude is 41°, the maximum altitude of the celestial equator above the southern horizon will be 49°.

The ecliptic or apparent path that the Sun travels in the course of a year is at an angle of 23½° to the celestial equator. In spring and fall the Sun will be on the celestial equator; it will be at the vernal or autumnal equinox. In summer and winter, the Sun will be 23½° north or south of the celestial equator at the time of the summer or winter solstice respectively.

It follows that the maximum elevation the Sun reaches at any given location is equal to 90° minus the latitude plus 23½°, and to obtain the minimum elevation, 23½° is subtracted from 90° minus

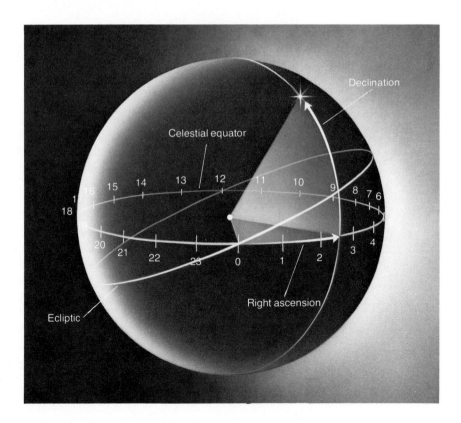

FIGURE 14–3
Locations of celestial objects by right ascension and declination.

the latitude. In order for the Sun to appear directly overhead, at the zenith, at some time during the year, one must be somewhere in the Torrid Zone, within the Tropics of Cancer and Capricorn.

RETROGRADE MOTION

The solar system is essentially a flat plane; the planets, except for Pluto, appear on or very near the Sun's path—the ecliptic. To see the planets, therefore, one looks along the ecliptic. If the planet is not directly on the ecliptic, it will be within the *zodiacal belt*. This is the band around the sky extending 8° either side of the ecliptic. The band passes through the 12 constellations of the zodiac (the word comes from the Greek root *zoön* that also gives us zoology)—all of them representing animals except Libra, the scales. The constellations of the zodiac eastward around the sky are Aries, Taurus, Gemini, Cancer, Leo, Virgo, Libra, Scorpius, Sagittarius, Capricornus, Aquarius, and Pisces (Figure 14–4).

FIGURE 14–4
The constellations of the zodiac eastward from the vernal equinox. Presently the "First Point of Aries" is in the constellation Pisces.

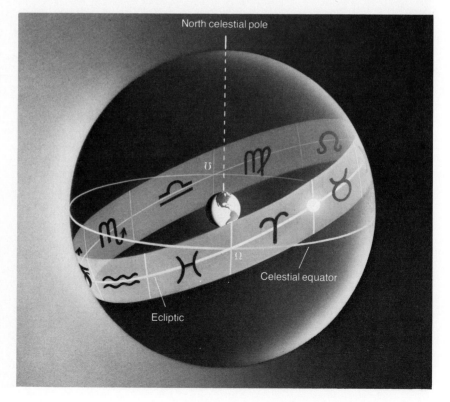

During a year, the Sun moves from one constellation to another, and the planets do also. Since the stars do not "wander" in the fashion of the planets, the planets were considered to possess mystical qualities. Ancient men associated them with the gods and believed that their positions affected significantly the course of men's affairs. Early astronomers were responsible for advising rulers of the portents of the "stars" to enable them to make decisions. Belief in astrology, or that the positions of the planets at the time of conception and birth affect our entire lives, persists to the present day, as witness the popularity of magazines devoted to the subject.

While the planets do move within the zodiacal belt, unlike the Sun they do not move always in the same direction. At times they

FIGURE 14–5   *The cause of retrograde motion of a planet.*

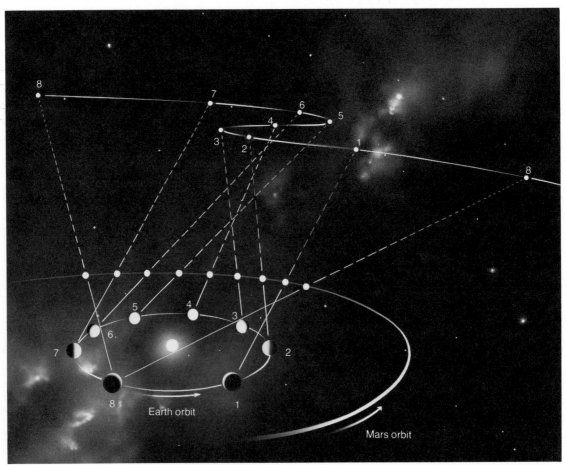

appear to stop their eastward motion, make a loop westward, and then resume motion toward the east. This retrograde motion confounded ancient observers, and it was the single observation that made it necessary for Eudoxus to propound his spheres, and for Hipparchus and Ptolemy after him to devise the complex systems of epicycles, deferents, and eccentrics. One can understand the consternation that "backward" motion would have caused.

We know today that planets cannot in any way reverse themselves. The observation is an illusion, an effect that results from the fact that each planet moves around the Sun at a different velocity, some moving much faster than others (Figure 14–5).

In the illustration we show a retrograde loop of Mars and explain why it is produced. The inner circle represents the orbit of Earth and its positions at seven monthly intervals. The outer circle represents the orbit of Mars and its positions during the same intervals. When an Earth-located observer looks at Mars from positions 1, 2, and 3, Mars appears to move from west to east against the background stars. When Earth overtakes Mars (its velocity is 29.6 kilometers per second while that of Mars is 24 kilometers per second), the observer sees Mars from positions 4 and 5. Since Earth is passing Mars, the red planet appears to be moving backward—retrograding, from east to west. When Mars is seen from positions 6 and 7, the planet appears to have resumed its west to east motion.

THE MILKY WAY

The Milky Way is another phenomenon readily observed from Earth. It can be seen throughout the year, providing the sky is dark and visibility conditions are good. However, because in summer the night view is toward the center of the Galaxy, the Milky Way is most apparent during that season. It is the cloudlike band that arches across the sky from horizon to horizon and through Scorpius, Sagittarius, Aquila, Cygnus, and Cepheus. See page 204 in Chapter 9.

Until Galileo observed the Milky Way through his 32-power telescope, the hazy belt was not understood at all. It was sometimes thought to be a band of weather clouds, but its persistence caused consternation and doubt. The Greeks called it *galaxias kyklos*, or "milky circle," so obviously they had little understanding of its true composition. The Romans called it *via lactea*, which translates literally to "milky way."

In the early seventeenth century Galileo discovered its true nature. After viewing it, he wrote: "The Galaxy is nothing else but

FIGURE 14–6
The plane of the
Galaxy is tilted 62°
to the celestial
equator.

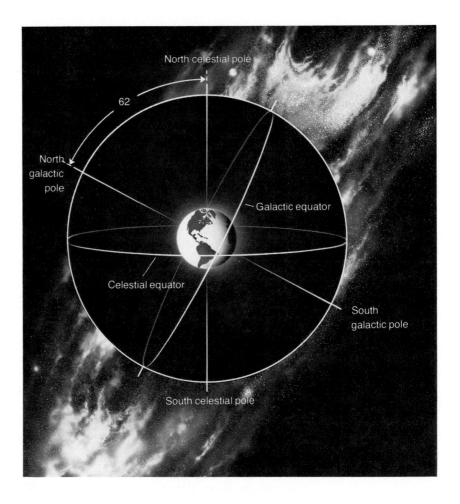

a vast conjury of stars planted together in clusters . . . the number is quite beyond determination."

When you observe the Milky Way, keep in mind the location of the celestial equator. The center line of the belt makes an angle of 62° with the celestial equator. The relationship of the two is shown in Figure 14–6.

THE MOON

As discussed in Chapter 13, we can observe the phase changes of the Moon, keeping in mind the relative positions of the Sun, Moon, and Earth to better understand how each of the phases is produced.

The points of the crescent Moon are called the *cusps*. To determine the location of the Sun when the Moon is crescent-shaped, in our imaginations we pretend that the crescent is an arrowhead with the cusps trailing. The arrow points toward the Sun. It will be immediately apparent that the *waxing crescent* Moon appears in the western sky and becomes most visible after the Sun has set and the sky has darkened. It follows the Sun. The *waning crescent* Moon will be seen in the early hours before sunrise. The line we constructed will point to the Sun and the position on the horizon where the Sun will soon appear. The Moon is chased by the Sun.

Now that we have some orientation to the sky and a feeling for the location of the celestial equator and ecliptic, we can relate the Moon's position to these lines. The Moon's orbit is tilted about 5° to the plane of the ecliptic, so it appears somewhere along the ecliptic, although it may be as much as 5° above or below. When the Moon is new, it is close to the Sun and near, or on, the ecliptic. Two weeks later, the Moon is opposite the Sun. During summer, therefore, when the Sun is high, the full Moon will appear low in the night sky (Figure 14–7). During winter, when the Sun is low in the sky, the full Moon will appear at its highest.

The most casual sky observer has seen that the Moon seems to change shape. It also changes in appearance: sometimes it is brilliant white, more often yellowish, but on occasion reddish or orange-red —infrequently a blue Moon is reported. Gemini and Apollo missions to the Moon reveal that the Moon lacks any color intrinsically; it is a dull, drab grey-brown, rather nondescript. The distinct colors that we see must therefore be phenomena that arise here on Earth. Moon colors are most apparent at moonrise and moonset, at those times when the Moon is near the horizon. The moonlight must shine through dust particles and water droplets suspended in the atmosphere. The particles scatter the light, allowing those of longer wavelength (the reds) to come through, and dissipating those of shorter wavelength (the blues). The particles are more dense in the lower atmosphere, and so the color changes are most extreme when the moonlight is coming through the greatest amount of atmosphere, moonrise and moonset. A blue Moon must be produced in some similar fashion. It has been suggested that the phenomenon, which is rare indeed, may result when there is a very thin layer of ice crystals at a high altitude which scatters red light, allowing only the blue to pass. Or the color may be produced by an overabundance of some particular substance, for example, when there is a large amount of sulfur in the atmosphere which might result from extensive forest fires or volcanic eruptions.

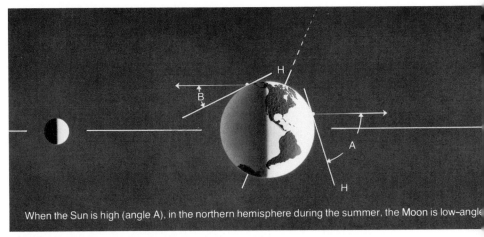

When the Sun is high (angle A), in the northern hemisphere during the summer, the Moon is low–angle

FIGURE 14–7   When the Sun is high in the sky (angle A at the left), the full Moon appears low. When the Sun is low in the sky (angle A at the right), the full Moon appears high.

STAR FINDING

The celestial sphere serves astronomers because it is convenient to locate upon it stars, constellations, planets, nebulae, and galaxies. As late as the seventeenth century it was believed that the sphere was very real and that it was 65 million miles away. The sphere does not exist actually, so obviously no objects are really on it. When we say, for example, that there is a bright nebula in Orion, or a great galaxy in Andromeda, we mean that the formation appears in that particular region of the celestial sphere. The nebula may be much closer to us or much farther away than the stars in that region. In the case of galaxies, they are most certainly farther away. The *constellations* themselves are phenomena that really do not exist. The familiar configurations of stars within constellations, those that compose Orion let's say, arise only because of our particular location in the Galaxy and because of our direction of viewing. Suppose one could be removed to some distant location in the Galaxy. The stars of Orion would form an entirely different configuration (Figure 14–8).

Historically, people were very aware of the sky. All objects were considered to be stars: planets were wandering stars; comets were long-haired stars; meteors were shooting stars. Because ancient people failed to comprehend their observations, it is understandable that they attached mystical significance to the stars. They felt it

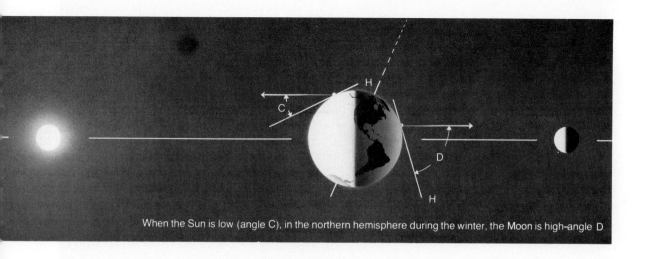

When the Sun is low (angle C), in the northern hemisphere during the winter, the Moon is high–angle D

FIGURE 14–8   On the left of the cube is Orion as it appears to us.
Were we able to view the stars from the right face, we would see an
entirely different configuration.

Distance in light-years

(1) Betelgeuse 520

(2) Bellatrix 470

(3) Saiph 2000

(4) Rigel 900

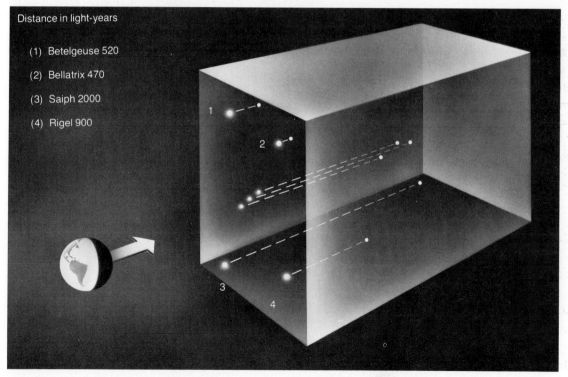

GEMINI

Pleiades

TAURUS

Aldebaran

Betelguese

Bellatrix

ORION

Great nebula

Rigel

Saiph

LEPUS

ERIDANUS

Sirius

CANIS MAJOR

only right that their gods and goddesses should have their own regions. So the sky was divided into sections, each relegated to a particular deity. The procedure was followed by the Greeks, Romans, Babylonians, Arabs, and Persians. The mythology associated with the gods has come down to us today, as have the names of the characters.

Various astronomers through the ages have attempted to represent the gods and other creatures, relating the configurations to the stars themselves. While they have produced impressive works of art, the procedure has not made it any easier for an observer to "see" Andromeda, let us say, or Perseus or Pegasus. Also, astronomers often introduced new constellations, or removed those they felt were rather unimportant. Ptolemy's *Almagest* contained 48 constellations; Hevelius added Canes Venatici, Leo Minor, and half a dozen others; Johann Bayer in 1603 added 13 new ones (it was he who invented the practice of giving bright stars names of Greek letters plus the constellation name); other contributors were Bertschius and Lacaille. Depending upon his stature, a contributor's ideas were either accepted or rejected. No matter how they were received, the changes invariably caused confusion rather than simplifying identification. Some constellations overlapped, and some areas of the sky did not "belong" to any constellation. In 1925 a commission appointed by the International Astronomical Union studied the problem and settled upon 88 constellations covering the entire sky. These are recognized today and form the basis for divisions of the celestial sphere. Among these 88 constellations there are 14 men and women, 1 head of hair, 9 birds, 2 insects, 19 land animals, 10 water creatures, 1 serpent and 1 dragon, 2 centaurs, 1 flying horse and 1 unicorn, 1 river, and 29 other inanimate objects. (This doesn't add to 88 because some of the constellations have more than one creature.) Don't try to find them all, for many of them are obscure. However, certain stars and constellations dominate the skies of certain seasons, and the casual observer finds familiarity with them quite satisfying. The charts we have drawn show the sky looking south about 9:00 P.M. in midseason. For example, the Orion group would be seen on or near your meridian about 9:00 P.M. in mid-February. Accompanying each of the four maps for the seasons, and the map of circumpolar stars, we've listed information about some of the stars and about objects in the vicinity which you may find interesting. Try finding them for yourself.

FIGURE 14–9 (opposite page) The winter sky showing Orion, Taurus, the Pleiades—and the Great nebula (M42).

Winter Skies—The Orion Region

Winter offers a spectacular sky, one that is dominated by Orion (see Figure 14–9), all the main stars of which are considered first or second magnitude. Ancient cultures picture Orion as a giant, a king, a great hunter. The Egyptians thought of him as Osiris; the Chinese call this constellation the Three Kings because of the three stars that line up to make the belt of Orion. The image most often given is that held by the Greeks and recorded by Johann Bayer in his *Uranometria* in the early thirteenth century—a hunter holding a shield before him on his left arm and flourishing a club high over his head with his right arm.

The brightest star in Orion is Alpha (α) Orionis (Betelgeuse). Stars are often referred to after the system devised by Bayer. He assigned small letters of the Greek alphabet to the brighter stars of the constellation, alpha being the brightest, and followed this with the possessive of the Latin name for the constellation. When there are many stars of almost the same brightness, the stars will be named in order of position in the figure, beginning at the head. Betelgeuse means "the armpit of the white-belted sheep" and probably was devised by Ptolemy to indicate the position of the star in the conventional figure used in those days.

Betelgeuse is a red supergiant star (a term given to those stars having diameters greater than 100 times the Sun's diameter). The diameter of Betelgeuse changes, however, at times being 530 times that of the Sun, and on other occasions shrinking so that its diameter is only 360 times greater. The color indicates a relatively cool temperature, around 3000°C. The color also indicates that the density of the star is very low. Even though it is a gigantic star, many times larger than the Sun, a cubic meter of it would weigh only 1 gram. (A cubic meter of air weighs more than 900 grams.) Betelgeuse is nearly a red-hot vacuum.

The belt stars, all of essentially the same brightness and in a straight line, make a distinctive feature of the constellation. In order from east to west, they are Alnitak, Alnilam, and Mintaka, Arabic words that refer to belts or strings. Mintaka is nearly on the celestial equator.

Below the belt is a configuration of stars referred to as Orion's sword. If we look carefully we may see a slight fuzziness toward the center of the sword. This is the Great nebula of Orion, a cloud of gases some 16 light-years in diameter and at a distance of 1500 light-years.

If you cannot detect the nebula with the unaided eye, mount a pair of 7×50 binoculars on a tripod. On a clear dark night you'll be able to see the nebulosity. (When you use binoculars for astronomy, you should try to mount them on a steady stand. Slight body motions distort the image too greatly.)

Another notable object in Orion is the "Horsehead" nebula. Its location is just a bit below Alnitak, the easternmost star in the belt. It is a tremendous cloud of cool gases that obscures the light from stars that lie beyond it. It was named by the Irish astronomer William Parsons, the third Earl of Rosse, who was struck by its resemblance to a horse's head.

The Orion region is rich with prominent stars and constellations. On the map we indicate those which stand out prominently. Aldebaran and the Taurus configuration can be located by using the belt stars as pointers. If we follow the line westward we should find Aldebaran in the V-shaped formation.

If our line of sight is directed eastward along and beyond the belt stars, we will come to Sirius. This is the brightest star in the nighttime sky, having a magnitude of −1.42. It is Alpha Canis Majoris, the brightest star in Canis Major, the hunting dog of Orion. The word Sirius probably comes from Siris, the Egyptian name for the Nile River. The all-important annual flooding of the river began when that star appeared in the morning twilight just before sunrise. The Egyptians probably associated the star with the flooding, and so with life itself.

Above Sirius is Procyon, the brightest star in Canis Minor. And proceeding farther we see Castor and Pollux, the brightest stars in Gemini, Alpha and Beta (β) Geminorum. Summer begins when the Sun is in Gemini.

Under good seeing conditions we should have little trouble spotting all the stars mentioned above.

Spring Skies—The Leo Region

About the middle of April, Leo, the lion, is on our meridian at 9:00 P.M. (see Figure 14–10). It is among the oldest of all the constellations, being found in all astronomical records and with essentially the same boundaries it has today. When men first related the stars and Sun, the Sun reached its highest elevation (summer began) when it was in Leo. In those days the Sun was worshiped, and Leo was given a regal position. The brightest star in the constellation is Regulus, meaning "the prince." It marks the base of a sickle-like formation often referred to as the head of the lion.

Regulus is actually a double star, one that is 84 light-years away and 100 times brighter than the Sun. It is located almost directly on the ecliptic.

Beta Leonis (Denebola) lies east of Regulus. It is the easternmost of the three stars that form a right triangle, the hypotenuse of which is uppermost. Denebola is derived from Arabic words meaning

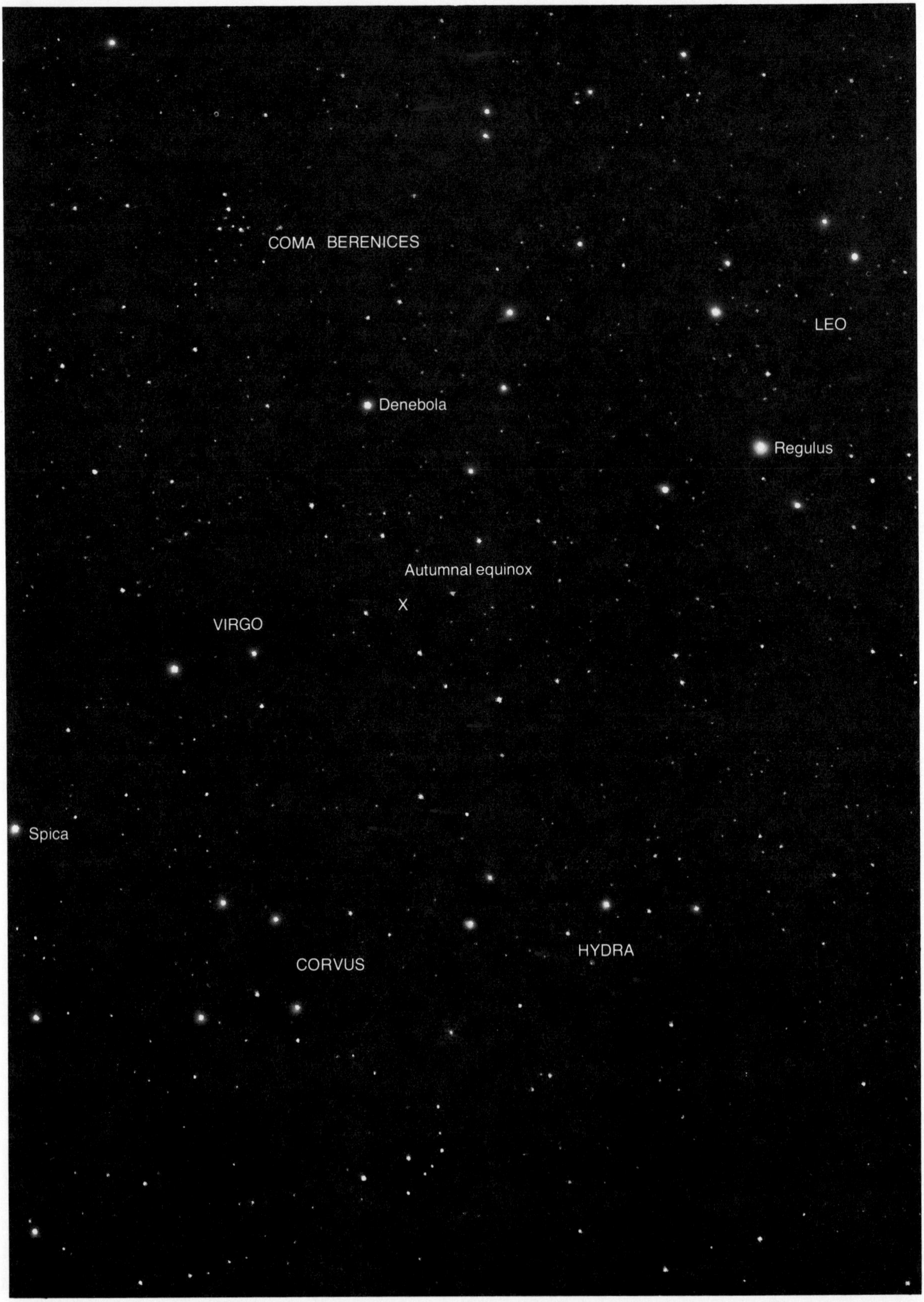

COMA BERENICES

LEO

Denebola

Regulus

Autumnal equinox

X

VIRGO

Spica

CORVUS

HYDRA

"the lion's tail."

The autumnal equinox, the intersection of the ecliptic and the celestial equator, lies just east of Denebola and 15° below it.

The north pole of the galaxy (R.A., 13°50′, Decl. 28°) is east of Denebola about 15° and north about 12°.

The region east of Leo contains vast numbers of galaxies. It has been suggested that our galaxy is one of a cluster of galaxies that comprise a metagalaxy—a family of galaxies. Perhaps this galaxy of galaxies is shaped somewhat like the Milky Way Galaxy. We may see in this region the central region of the metagalaxy.

Directly above Leo, high in the sky, we see the Big Dipper. The three stars in the handle make a curve. If we follow this curve, we come to Arcturus—off the page to the left in our illustration—the brightest star in Bootes. Continuing along the curve, we reach Spica, the brightest star in Virgo, and then the curve takes us to Corvus where four fairly bright stars stand out, the principal ones of the constellation.

Arcturus is 36 light-years away. In the 1930s it was believed to be 40 light-years away. Light from the star was used to activate a switch that turned on the lights at the World's Fair in Chicago in 1933. The light had left the star 40 years before, according to knowledge at the time, which would have been the time of the previous Chicago World's Fair held in 1893.

Arcturus (Alpha Bootes) is extremely bright, having a magnitude −0.06; it is one of our close neighbors.

Spica (Alpha Virginis) is an eclipsing binary star system. An eclipse occurs every four days, but the magnitude changes are slight, only about one-tenth of a magnitude. Spica is very hot and bright, some 220 light-years away.

Summer Skies: Sagittarius

Sagittarius, the archer, is on our meridian at 9:00 P.M. in about the middle of August (see Figure 14–11). In some versions, the western stars represent a bow and the eastern stars a hand gripping an arrow. Supposedly the arrow, marked by a star just west of the bow, is being shot at Taurus, which is far to the west when Sagittarius is rising in the east. The constellation is often called "the Teapot" because the configuration reminds one of a pot complete with handle and pouring spout.

The Milky Way sweeps across the skies of summer. The Sagittarius region is especially brilliant because the center of the Galaxy is in that direction; we are looking toward the greatest concentration of stars. Northward, the galactic equator stretches through Aquila,

FIGURE 14–10
(opposite page)
The spring sky showing Leo the Lion, Coma Berenices, and some galaxies that appear in that region.

Deneb
Sadr
CYGNUS
Vega
Gienah
LYRA
Albireo
Altair
AQUILA
SCUTUM
CAPRICORNUS
Algedi
Kous Borealis
Nunki
SAGITTARIUS

the eagle, and Cygnus, the swan. These two constellations, together with Lyra, comprise the summer triangle—a most beautiful region of the skies in that season. Altair is the brightest star in Aquila (Alpha Aquilae), and it is a nearby object, being only 16.5 light-years away. Just above Altair is Tarazed, and just below it to the east is Alshain. Both of these names come from the name the Persians gave the constellation—*Shahini Tarazad,* "the soaring falcon."

Deneb (Alpha Cygni) marks the tail of the swan. Often the constellation is considered a cross; then Deneb becomes the uppermost star in the upright of the northern cross. It is some 1600 light-years away and is probably 30 000 times brighter than the Sun, placing it among the brightest of all stars—a supergiant star giving off stupendous amounts of energy.

West and south of Deneb is Vega, the third star in the triangle. It is Alpha Lyrae, the third brightest star in the Northern Hemisphere—magnitude 0.05. It is a striking, blue-white star that dominates summer skies.

The Sun and the entire solar system with it are moving toward the region of Vega, R.A. 18°4′, Dec. +30°, at a speed of some 20 kilometers per second. This is our motion with respect to the nearby stars. At the same time, Vega is moving toward us at some 14 kilometers per second. Vega, the Sun, and the nearby stars are all moving in the direction toward Cygnus as a result of the rotation of the Galaxy.

Just below Vega are four stars that make a parallelogram. The lower one to the east, Shelyak (Beta Lyrae), is actually two stars going around one another. The brightness of the system drops a full magnitude every 13 days. Apparently the two stars are separated by only about 5 million kilometers, yet both are much larger than the Sun. The two stars must be considerably misshapen by the strong gravitational forces exerted upon them. The gases composing them must be strung out, perhaps even exchanging from one star to the other (see page 132).

Autumn Skies—The Pegasus Region

About the middle of October, Pegasus will be on our meridian at 9:00 P.M. (Figure 14–12). The constellation is easy to pick out because three of its stars combine with Alpha Andromedae (Alpheratz) to produce a well-defined square, the square of Pegasus. The square is often thought of as the forequarters of an inverted flying horse.

The three corner stars of the square, clockwise from Alpheratz, are Scheat (Beta Pegasi), Markab (Alpha Pegasi), and Algenib [Gamma (γ) Pegasi]. Scheat is a red giant, its diameter some 100

FIGURE 14–11 (opposite page) The summer sky showing Sagittarius, Aquila, Cygnus, Lyra, and Capricornus.

387

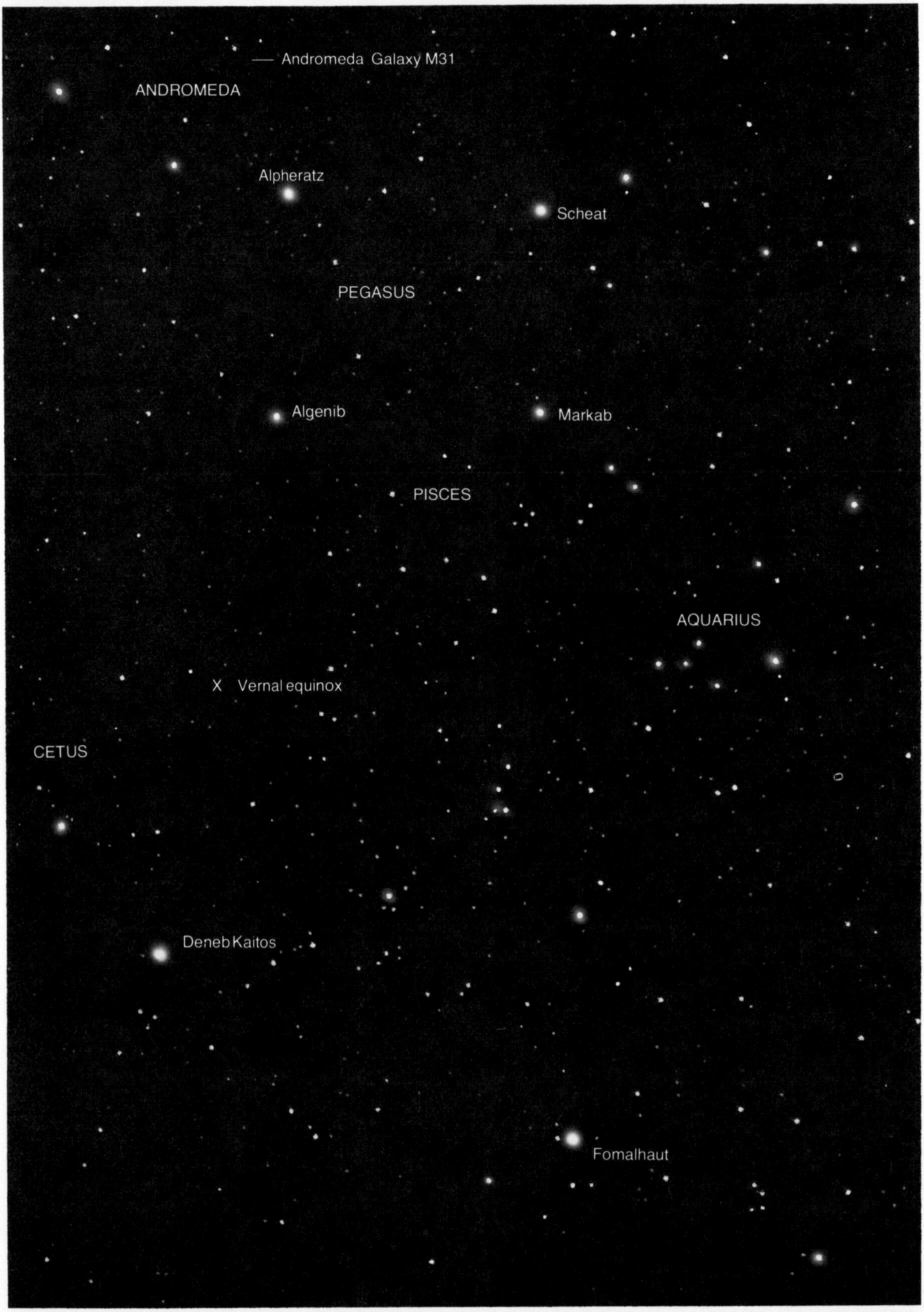

—— Andromeda Galaxy M31

ANDROMEDA

Alpheratz

Scheat

PEGASUS

Algenib   Markab

PISCES

AQUARIUS

X   Vernal equinox

CETUS

Deneb Kaitos

Fomalhaut

times that of the Sun; Markab is a hot white star; and Algenib is a short-period variable, going through a slight change—a fraction of a magnitude in a bit over three hours.

The fourth corner star is Alpheratz (Alpha Andromedae), the westernmost of three bright stars in a slightly curved line. The next star east of Alpheratz is Mirach, and then Almach (Beta and Gamma Andromedae).

Just above Mirach about 5 degrees is the great galaxy in Andromeda. Through good binoculars mounted on a tripod it can be seen as a hazy patch of light. Looking at it is always a thrill, realizing as one does that the light he is seeing left that galaxy some 2 million years ago.

If we look below Algenib just about as far as the distance between Alpheratz and Algenib, and slightly westward, we will see the location of the vernal equinox—the intersection of the celestial equator and the ecliptic. Spring begins when the Sun is at this location. The vernal equinox is now in the constellation Pisces, the fishes (one of the fish below Pegasus, the other vertical and to the east). Because of precession, the vernal equinox drifts westward, and in about 600 years it will be in Aquarius.

Stars of the North

Perhaps the most well known of all the constellations is Ursa Major, the Great Bear (see Figure 14–13). Certainly people around the world have recognized the region for generations. Most people regard the Big Dipper as being the constellation. Actually, the Bear covers a much broader region, and the Dipper outlines only the hindquarters of the bear and its tail. The early Greeks knew about the Great Bear. They called it Arktos Megale, the "greater bear." (Incidentally, it is from the Greek *Arktos* that we get our word arctic—the northern part of the world—the country of the Great Bear.)

There are seven stars in the Dipper; the one having the lowest declination (50°) is Alkaid. This means that for regions at and above 40° north latitude, the entire Dipper is above the horizon year-around. During winter evenings it is below Polaris for the most part, and difficult to see because it is near the horizon. In summer it is high in the sky; the water spills out, as they say. This is a case where the stars are identified not in order of brightness, but in order of position, beginning at the "head" of the figure. Alpha Ursae Majoris (Dubhe) is one of the "pointer" stars at the end of the bowl. The other pointer is Beta Ursae Majoris (Merak). They provide a line of direction for locating Polaris. The other stars, Gamma (γ), Delta (δ), Epsilon (ε), Zeta (ζ), Eta (η), are Phecda, Megrez, Alioth, Mizar, and

*FIGURE 14–12
(opposite page)
The autumn sky
showing Pegasus, the
Andromeda galaxy
(M31), and Pisces.*

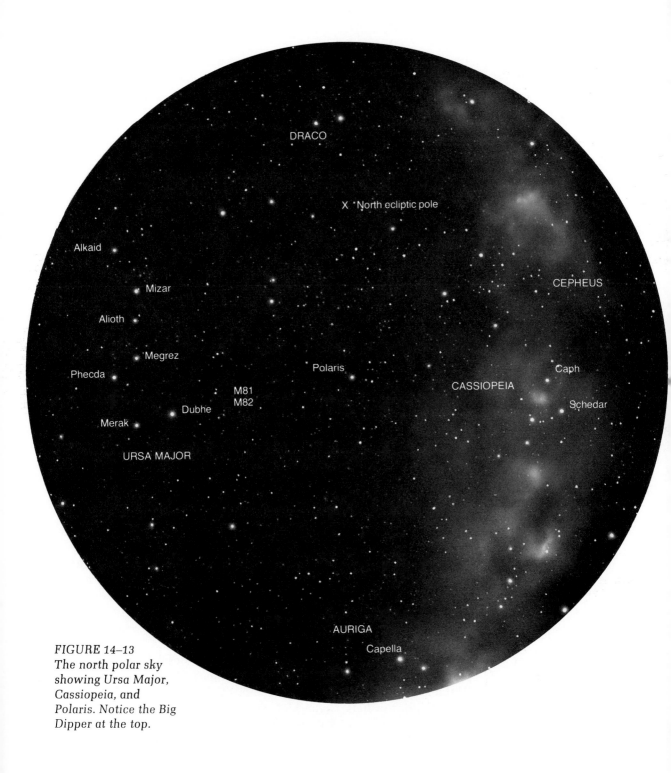

DRACO

X •North ecliptic pole

Alkaid

Mizar

Alioth

Megrez

Phecda

M81
M82

Dubhe

Merak

URSA MAJOR

Polaris

CEPHEUS

CASSIOPEIA

Caph

Schedar

AURIGA

Capella

FIGURE 14–13
The north polar sky
showing Ursa Major,
Cassiopeia, and
Polaris. Notice the Big
Dipper at the top.

Alkaid—Arabic words that refer to the back, the loins, and the thigh, or various parts of the bear. Mizar, the star next to the end of the handle, has a companion, Alcor, which can be seen with the unaided eye.

Six Messier objects are located in Ursa Major. The most impressive are M81 and M82 which are close together and just beyond the pointer stars. With binoculars one can see them in the same field. One galaxy is edge on, the other tilted at an angle of about 30°. With larger telescopes, hundreds of extragalactic objects can be discerned in the Big Dipper region.

Across the sky from the Big Dipper, beyond Polaris, one sees the W- or M-shaped formation of stars that composes part of Cassiopeia. In winter it is high in the sky. A line from Alpha Cassiopeiae (Schedar), going between a line joining Beta and Gamma Cassiopeiae and continuing, goes rather close to Polaris.

Close by Caph, which is Beta Cassiopeiae, is the location where Tycho Brahe saw his star, the supernova which appeared in 1572 and which he studied until its disappearance some 18 months later. A paper on the subject brought him considerable fame as an astronomer.

The north pole of the ecliptic lies at right angles to a line joining Beta Cassiopeiae and Gamma Ursae Majoris, 23.5° from Polaris and in the constellation of Draco. This location 90° from the ecliptic provides us with another help toward sky orientation.

REVIEW

1 The celestial sphere, an imaginary sphere upon which stars are placed, is helpful in determining stellar positions by reference to various parts and locations on the sphere.

2 Stars are usually located by their right ascension and declination, measured from the vernal equinox and the celestial equator.

3 The Sun and planets move through the zodiacal region in a generally west to east direction; however, the planets occasionally appear to stop and reverse their motion, a phenomenon known as retrograde motion.

4 There are presently 88 constellations recognized by the International Astronomical Union.

5 Star study is simplified if one constellation is keyed to each season, and departures to other regions of the sky are made from that base.

6 Stars are usually given Greek letters followed by the Latin possessive of the constellation's name, in order of brightness: for example, Alpha Orionis. Sometimes, however, the stars in a constellation may be named in order of position.

QUESTIONS

1 What is the horizon system of locating celestial objects? What are its advantages and disadvantages?

2 What is the equatorial system of position finding? What are its advantages and disadvantages?

3 What is the relationship between the celestial pole and the ecliptic pole? What are the coordinates of each?

4 When there are good seeing conditions and the season is right, trace out the celestial equator in the sky; the ecliptic; the vernal equinox; the autumnal equinox.

5 What is the system for identifying individual stars in a given constellation?

6 What are the altitude and azimuth of a star at each of these locations: east and halfway to the zenith, northwest and a third of the way to the zenith?

7 What are the right ascension and declination of the Sun on the first day of spring, the beginning of summer, and the beginning of winter?

8 Why are the coordinates of a star given for a certain epoch?

9 The Sun moves eastward along the ecliptic. How can this be reconciled with the westward movement of the Sun? What are the explanations for each motion?

10 What is the maximum elevation of the Sun at your location?

11 Next time the Moon is full, notice its location. Where is it with reference to the Sun? to the ecliptic?

12 What is the zodiac? Where is it located? Of what value is it?

13 In summer the Milky Way is more apparent than it is in winter. Why is this so? What is its relation to the ecliptic? The celestial equator?

14 What is the actual color of the Moon? How can its changing apparent colors be explained?

15 Throughout the year, be aware of the stars. Use the guides in this chapter to identify a few stars and constellations each season.

READINGS

Bowditch, Nathanial. *American Practical Navigator*. Washington, D.C.: U.S. Government Printing Office, 1962.

North, J. D. "The Astrolobe." *Scientific American*, January 1974.

Muirden, James. *Astronomy with Binoculars*. Princeton, N.J.: Van Nostrand Reinhold Company, 1963.

Norton, Arthur P., and J. Gall Inglis. *Norton's Star Atlas*. Cambridge, Mass.: Sky Publishing Corp.

Nourse, Alan E. *The Backyard Astronomer*. New York: Franklin Watts, Inc., 1973.

EARTH
AS A CLOCK

CHAPTER 15

Time is elusive. We are aware of present, past, and future. We also realize that the present is instantaneous, always changing; that the past continually grows larger; that the future grows shorter. A dictionary defines time as "indefinite continuous duration regarded as that in which events succeed one another" —obscure, to say the least.

Astronomy offers a much more concrete definition. Time is the hour angle of an object: the Sun, the Moon, a star, a point in the sky.

When an object such as a star is on our meridian, the day begins: the hour angle is zero. As Earth rotates eastward, the object appears to move westward. Its hour angle increases 15° in

To Earth-bound observers all points in the sky appear equi-distant—Earth seems to be at the center of a celestial sphere as shown in this view from outer space. The gnomon of a sundial approximates the axis of Earth.

one hour (an angle of 15° equals 1 hour of time, since 24 hours are required for Earth to turn through 360°). If a star were 30° west of our meridian, the time would be $2^h0^m0^s$ (Figure 15–1).

The star, of course, does not move around Earth. Rather, Earth rotates and so the star appears to move westward. The rotating Earth is the basis for time reckoning. It is good for this purpose because the rotation is fairly uniform over both long and short periods of time, and it can be observed readily.

We are not aware of Earth's rotation directly, but we can observe it by relating the motion to some object outside Earth. For example, we might use a star. However, stars move—in some cases, slightly east or west. A more firmly fixed sky location is the vernal

FIGURE 15–1 *Time is the hour angle of an object, in this case the Sun, past the local meridian (the line from the celestial pole through the zenith of the observer to the celestial equator). The angle is measured westward along the celestial equator.*

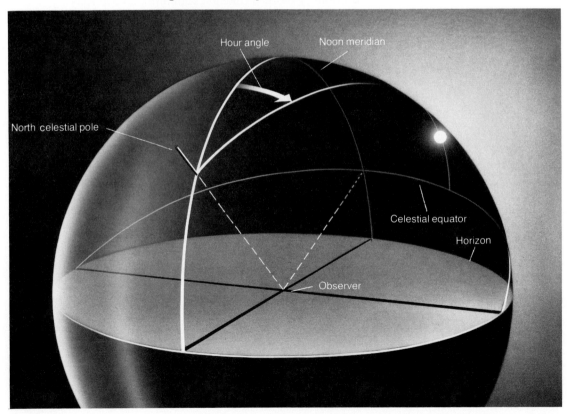

equinox. This, too, drifts westward, but its motion is precisely predictable, and it is very slight, so it is the basis for determining sidereal time—time by the stars.

SIDEREAL TIME

The vernal equinox, we recall, is the point where the celestial equator and the ecliptic intersect. When it is on the local meridian, the time is $0^h0^m0^s$—the hour angle is zero degrees. When the hour angle of the vernal equinox is 6^h, it will be setting; when it is 18^h, it will be rising. The 360° are divided into 24 hours (15° per hour), and the 24-hour clock system is used.

Right ascension and declination are measured from the celestial equator and the vernal equinox. Star positions are therefore related to sidereal time. This makes sidereal time a great convenience to the astronomer. Laymen, however, have little use for sidereal time. A sidereal day can begin at any time of day or night. During a year, it would begin at solar noon on the first day of spring, at 6:00 in the evening on the first day of summer, at midnight on the first day of fall, and at 6:00 in the morning on the first day of winter.

Our lives are regulated by the Sun, not by the stars, and so it is more logical that our time keeping should be related to the Sun. Just as sidereal time is the hour angle of a star (or the vernal equinox), solar time is related to the hour angle of the Sun.

SOLAR TIME

A *solar day* begins when the Sun is on our meridian—when the hour angle of the Sun is $0^h0^m0^s$, or when it is solar noon. It would be awkward, however, to run our civil affairs on the basis of such a solar-day system. Our mornings would be one day, our afternoons another. Since our affairs are for the most part pursued during daylight hours, it was decided to begin a day at midnight. Therefore, solar time becomes the hour angle of the Sun plus 12 hours. On the 24-hour system, then, 3:00 P.M. becomes 15^h—the actual angle of the sun (3^h) plus 12^h. Also, 15 hours is 3:00 P.M., the P.M. coming from *post meridiem*, meaning "after midday," or "after noon." In astronomy, the term would be more appropriate were it *post meridian*—"after the meridian." However, the connotations are the same. In the same system, 3:00 A.M. (*ante meridiem*), which is an hour angle of $15^h0^m0^s$, is derived by adding 12 hours to give 27^h. Since this is more than one day, 24^h, we subtract 24^h and get 3:00 A.M.

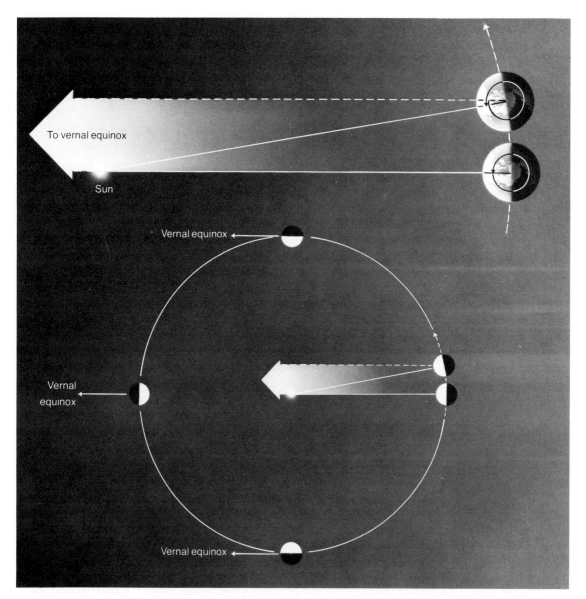

FIGURE 15–2   A location on Earth requires 23^h 56^m to realign with the vernal equinox (sidereal day), but 24 hours to realign with the Sun (apparent solar day). Note: Because the Sun and stars are so distant, light is assumed to come from them in parallel lines. During a year the vernal equinox may be on the meridian (the start of the sidereal day) any time during daylight or darkness.

The solar time we have been discussing is referred to as *apparent solar time*—apparent because it depends on where the Sun appears in the sky.

We recall that sidereal time is defined as the hour angle of the vernal equinox, whereas apparent solar time is the hour angle of the apparent Sun (plus 12 hours). Because Earth revolves around the Sun, and the Sun is relatively close by, the solar day is longer than a sidereal day (Figure 15–2).

Earth takes one year (365 days) to go around the Sun, or to travel through 360°. Therefore, in one day Earth moves through about one degree in its orbit. If the meridian of a location on Earth lines up with the Sun and with a distant star, that observer's meridian will line up again with that distant star after 23^h56^m ($23^h56^m4^s.09$). However, an additional 3^m56^s are needed before that meridian realigns with the Sun. A clock day is 24 hours long; a sidereal day is

FIGURE 15–3
When Earth is closest to the Sun (during winter in the Northern Hemisphere), it moves most rapidly in its orbit. When it is farthest from the Sun (during northern summers), Earth moves most slowly.

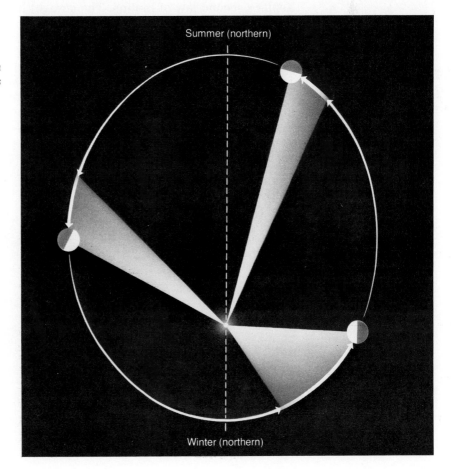

23^h56^m of solar time. A sidereal clock dial is marked with 24 sidereal hours; however, the gearing is designed so that a complete turn of the hands is made in 23^h56^m of solar time.

Mean Solar Time

Sidereal time keeping has both advantages and disadvantages. So also has apparent solar time. For example, the time that the Sun crosses the meridian may vary by 15 or 16 minutes from the time that a uniformly moving Sun would. This is the result of two factors, the first of which is the annual variation in the motion of the Sun. As Earth goes around the Sun, the Sun seems to move west to east around the sky. If the orbit of Earth were circular, the motion of the Sun would be the same at all times. In a given period of time, the Sun would always cover the same distance. But the orbit of Earth is elliptical, and so the speed of the planet is variable, as expressed in Kepler's second law of motion. The straight line joining a planet and the Sun sweeps out equal areas in equal periods of time. In January,

FIGURE 15–4 *Because the Sun moves along the ecliptic, it usually moves north or south of east. True west to east motion occurs only at time of the solstices. Therefore, distance along the ecliptic covered by the real Sun in a given time period varies from distance along the celestial equator covered by the fictitious, or mean, sun (see inserts).*

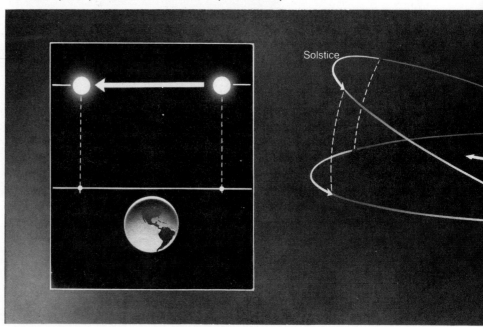

we are closest to the Sun and so Earth moves most rapidly (and from swift-moving Earth, the Sun seems to move most rapidly). In July, when Earth is at the greatest distance from the Sun, Earth's revolution is slower and so is the Sun's motion (Figure 15–3).

Another condition that affects the length of an apparent solar day is the fact that the Sun moves along the ecliptic and not along the celestial equator. Therefore, the Sun usually does not move directly eastward in its annual motion. Rather, it moves either north of east or south of east, depending upon the season. The only time the Sun moves directly eastward is at the beginning of summer and the beginning of winter (Figure 15–4).

A sundial indicates apparent solar time. The shadow cast by the gnomon is always shortest when the Sun is due south—directly on the meridian. Until the clock—a device that measures time and against which Earth's rotation can be checked—was invented, sundial time keeping was quite satisfactory. However, it became necessary to devise a method of time keeping that would smooth out discrepancies that appeared as the days were measured. This was done by inventing a fictitious Sun, a "mean" Sun that moved uniformly along the celestial equator. It served to average out apparent solar days and is known as *mean solar time*, time that is kept by a clock.

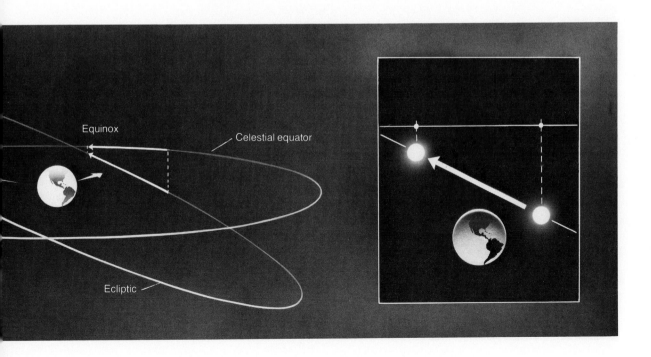

Equinox

Celestial equator

Ecliptic

The difference between apparent solar time (sundial time) and mean solar time (clock time) is called the *equation of time*. In our illustration, the dotted line shows effects caused by the fact that the ecliptic and the celestial equator do not coincide. The dashed line shows variations that result from the fact that the Sun's motion around the celestial sphere is not uniform. When the two factors are added together, the total effect on sundial time is indicated by the solid line. To adjust sundial time to clock time, minutes are added to the sundial reading when the equation of time indicates plus, and subtracted when the indication is minus. For example, about November 4th, 17 minutes must be subtracted from sundial time to obtain clock time (Figures 15–5 and 15–6).

But that does not give us the proper correction either. It would, if mean solar time were reckoned from the local meridian, as it used to be. When people did little traveling, it made small difference whether a place had one time or another. The situation became awkward as soon as there was rapid movement from one longitude to another. A place east of a given location had a later time, and whenever a person moved only a few kilometers east or west he had to set his watch a few minutes ahead or behind.

FIGURE 15–5   *The equation of time shown in the solid line results from two causes: (1) changes in velocity of the Sun in its orbit (dashed line), and motion north and south of the celestial equator (dotted line).*

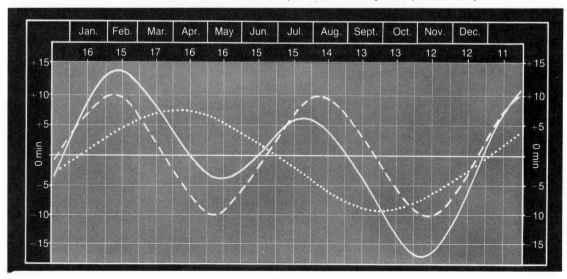

FIGURE 15–6
The analemma is another representation of the equation of time. It shows the locations of the Sun at civil noon during each day of the year. The analemma is shown at the bottom of the illustration.

STANDARD TIME

In 1883, it was decided that instead of each hamlet's using its own meridian as the "starting place for the day," noon in the United States would be reckoned from only four meridians—75°, 90°, 105°, and 120°. These would be the central meridians of four time zones, 15° wide. Time throughout the zone would be the same regardless of the position of the Sun. A year later, at an international conference held in Washington, D.C., and attended by 26 nations, it was decided to divide the world into 24 standard time zones, numbered

FIGURE 15–7   *The standard time zones of the world.*

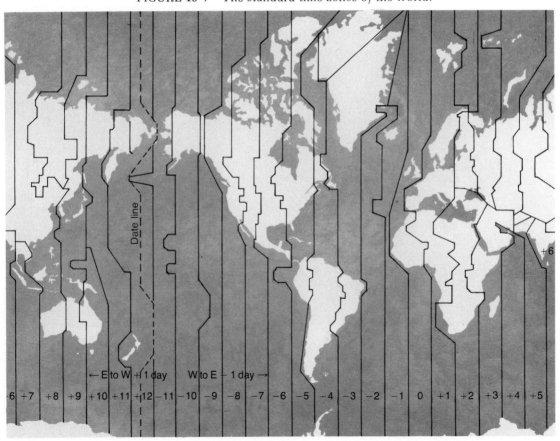

in order from the Greenwich meridian. The zones east of Greenwich were given times later than Greenwich, while those west of Greenwich were given earlier times. The Central Standard time zone, for example, is 6 hours earlier than Greenwich time (Figure 15-7).

Under such an arrangement, travelers' clocks need be changed only in units of whole hours. Sundial time must be corrected to account for the position of the sundial relative to the location of the meridian for that particular time zone. If the sundial is east of the meridian, it will be ahead of clock time—the amount being 4 minutes for each degree of longitude; west of the meridian, the sundial will be behind clock time. To convert apparent solar time (sundial time) to mean solar time we must apply the equation of time, and to convert this to standard time we must make the adjustment for longitude.

Daylight saving time is a convenience now adopted generally. On the last Sunday in April, clocks are pushed ahead one hour (spring-ahead) and remain so until the last Sunday in October, when they are set back an hour (fall-behind). The plan provides an additional hour of daylight in the evening during the summer months, when the weather is pleasant. There is one more hour of darkness in the morning, but the adverse effect of this hour is not nearly as great as the benefits of the additional hour of daylight in the evening.

INTERNATIONAL DATE LINE

Because Earth rotates west to east, locations east of us see the Sun before we do; their time is ahead of ours. When it is 12:00 noon on Monday in California, the time in New York is 3:00 P.M. In London, the time is 8:00 P.M. Around the world opposite to California, the time is 12 hours earlier; it is 12:00 midnight on Monday. If one were to continue beyond that location, the time in the next zone would be 1:00 A.M. on Tuesday. Continuing through 12 zones to California, and figuring an hour for each zone, the time at California would turn out to be 12:00 noon on Tuesday. It is patently impossible to have two different days at the same time at one location. The problem is avoided with the International Date Line, a line running north and south through the Pacific Ocean and following roughly the 180th meridian. By agreement, travelers moving eastward across the International Date Line drop back one day; they repeat the day. If it is 3:00 A.M. Tuesday when they cross the line, the time becomes 3:00 A.M. Monday. Westward travelers skip an entire day when they cross the International Date Line. If it is 1:00 A.M. Sunday the time immediately becomes 1:00 A.M. Monday; a day drops out.

TIME MEASUREMENT

We determine time by rotation of Earth. Time is measured and variations in the rotation of Earth are detected by using various devices such as the pendulum. In 1581, Galileo discovered that the period (the swing time) of a pendulum depends only upon the length of the pendulum. After Christiaan Huyghens (1629–1695) invented the escapement, the pendulum was used to regulate clocks. For centuries it was the basis for accurate time keeping. Improvements of many kinds were made; for example, temperature was kept constant to prevent expansion and contraction, and the apparatus was operated in a vacuum chamber. Nevertheless, the very best pendulums had errors amounting to one part in ten million, the equivalent of a second or two each year.

Quartz Crystal

A great improvement in time keeping resulted from the use of vibrations of quartz crystals to regulate electric clocks. An electric clock operates on a 60-cycle frequency that is maintained at the generating plant. But the frequency wanders, and so does the time kept by the clock. In a quartz-electric clock, a quartz crystal (which vibrates at a steady frequency of many millions of Hertz) controls the clock mechanism. Such a clock is accurate to one part in a hundred million, or, as is sometimes stated, it loses only one second in many years. With the quartz clock, it became possible to detect slight variations in Earth's rotation.

Ammonia Clock

Even greater precision can be obtained when the vibrations of individual molecules, or atoms, are used rather than those of crystals. In the ammonia clock, radiation is tuned to a certain fixed frequency by a quartz crystal. The radiation is then introduced to a tube of ammonia gas that absorbs that frequency. Should the frequency change, ever so slightly, it is not absorbed but passes through the ammonia. The radiation retunes the quartz crystal, correcting the drift and maintaining its time-keeping accuracy. In 50 years such a clock will vary by only one second.

Cesium Clocks

Cesium clocks employ vibrations of single atoms rather than molecules and provide an accuracy so high that only a second is lost in a thousand years. When first developed, these superclocks were large, cumbersome, and permanent installations. Technological improvement reduced size to the point where they could be moved about and carried in aircraft. This has made possible interesting experiments in which clocks have been carried both east and west around Earth, enabling corroboration of certain of Einstein's conclusions concerning time and velocity.

Presently it is possible to detect slight variations in the X-ray energy (one part in a trillion) emitted during the radioactive decay of cobalt. Should efforts to apply the idea to time keeping be successful, it will permit an accuracy to one second in 30 000 years.

Better time-measuring devices enable us to check Earth's rotation, our basic standard for determining time. There are variations in rotation; for one thing, the period of rotation is getting longer by about $\frac{1}{1000}$ of a second in a century. This is the result of tidal drag— —the interaction of the water and land masses. This drag is cumulative and, after millenniums of time, will increase Earth's period considerably. In addition, there are seasonal variations that amount to two- or three-hundredths of a second. The changes are probably produced by shifts in huge air masses, by the melting and freezing of the polar caps, and by shifts in crustal plates. Relative to a uniform time, Earth seems to be slow in the spring and fast in the autumn.

THE CALENDAR

Rotation of Earth is the standard for determining the day—a short period of time. The periods of revolution of Earth about the Sun, and of the Moon about Earth, are used for determining larger units—the year and the month. When we use more than one of these units, there is confusion: the number of days in a year is not a whole number; the number of days in a month is $27\frac{1}{3}$; the number of lunations in a year is not an even number. Through the centuries, various calendars have been used in an attempt to reconcile these three units, but none of them has been completely successful.

The Lunar Calendar

The lunar calendar was the first to be employed. Its division into weeks was not based upon astronomy, except that a week is the approximate time required for the Moon to pass from one phase to the next. The week is an arbitrary time division that probably originated out of astrology, which assigned certain hours of the day to the seven celestial objects that moved "around Earth"—the five planets known to the ancients, the Sun, and the Moon. It was assumed that the fastest moving objects were closest to Earth. So, according to their speeds, the objects were placed in the following order: Saturn (the slowest moving) was the most distant, then Jupiter, Mars, Sun, Venus, Mercury, and the Moon. The days were named for the planet that ruled the first hour of the day after sunrise. The second hour was ruled by the next celestial body in the order above, and so on throughout the day.

For example, if Saturn ruled the first hour after sunrise, the day would be Saturday. The second hour would be dominated by Jupiter, the third hour by Mars. If one continues in this manner through the 24 hours of the day, he finds that the 25th hour is ruled by the Sun: Sunday follows Saturday. The same procedure will reveal that the 25th hour after Sunday will be ruled by the Moon.

Repeating the process, we come to the days for Mars, Mercury, and so on. The names of the days now in use are Saxon modifications of the Latin versions; they are based upon the names of the Saxon gods, not on the Latin words for the planets as was done originally.

The English versions together with the Saxon and Latin are shown below.

English	Saxon	Roman
Sunday	Sun's day	Dies Solis
Monday	Moon's day	Dies Lunae
Tuesday	Tiw's day	Dies Martis
Wednesday	Woden's day	Dies Mercuri
Thursday	Thor's day	Dies Jovis
Friday	Frigg's day	Dies Veneris
Saturday	Saterne's day	Dies Saturni

The month was not arbitrary. It was the length of a synodic period, beginning with the appearance of the crescent Moon after sunset and about one day after new Moon. A lunation was a regularly occurring event, and the time that elapsed between successive appearances of the new crescent was a convenient longer interval. Eventually, the lunar calendar was regulated. Rather than relying on priests to announce the months as they sighted the crescent Moon, the organizers of the calendar made the 12 months of the lunar year alternately 30 and 29 days long, bringing the total number of days in the year to 354. Obviously, in this calendar there is no attempt to relate to the solar year discussed below. Mohammedans, for religious purposes at least, still use this ancient lunar calendar.

The Lunisolar Calendar

The lunisolar calendar represents an effort to reconcile the cycle of the Moon with the annual appearance of the Sun at a particular location. To keep the two in step, a month is added every third year. Essentially this is the procedure still employed in the Jewish religious calendar.

The Solar Calendar

The solar calendar disregards the phases of the Moon, placing emphasis on annual motion of the Sun and its position in the sky. There are 12 months in the calendar, generally longer than the lunar month. Before the solar calendar was adopted in Rome, the Egyptians had used it for centuries, though their calendar had only 360 days. The additional five were days of celebration.

The Gregorian calendar, used today throughout the world (but not by Mohammedans, the Jewish religionists, and the Russian Orthodox church), is men's attempt to devise a logical reckoning device. While it has desirable features, many aspects of it are illogical, as becomes apparent when one reviews the history of its development.

The Ancient Roman Calendar

The calendar of the early Romans began in the year 753 A.U.C. (anno urbis conditae—"from the founding of the city"). The first of the 10 months of the calendar was March, a reasonable time for a year to begin since it marked the end of winter and the beginning of a revival of living things. March (Martius) was named for Mars, which at that time was in charge of growth of crops and harvest. April (Aprilus) was named after asperire—"to open."

Time was not counted between December and March; the year was discontinuous. Very likely this calendar came out of the cold barbaric region of central Europe where this period was an inactive time of the year: plants were dormant; it was the dead season. Eventually, it became necessary to reckon that interval, and so January (after Janus, the two-faced god of entrance) was added, and February (after the god Februus, associated with atonement and purification). In 153 B.C., January 1 replaced March 1 as the first day of the year. It seems an especially inappropriate season to celebrate a reawakening, but this is only one of the incongruities one finds in the calendar. Incidentally, the word "calendar" comes from the Latin kalendae, or "callings." Officials charged with the calendar watched for the appearance of the crescent Moon. When they saw it, they called out the beginning of the month: the first day was known as Kalendae, or Kalends.

The 10 months alternated in length between 29 and 30 days, to bring the average length to 29½ days, the synodic period of the Moon. But the year ended with only 354 days. Every three years or so, at the discretion of the priests and officials, an extra month (an intercalary month) was put into the calendar. If the politicians were powerful, they would convince the priests to lengthen the year; those less powerful had to be content with less.

The Julian Calendar

In the first century B.C. Julius Caesar became ruler of Rome. Discontent with the Roman calendar had increased. Caesar knew that the Egyptian calendar avoided many of the pitfalls that had plagued the Romans. At that time, Sosigenes, a Greek astronomer who probably came out of Alexandria, Egypt, was employed by Julius Caesar to reform the Roman calendar.

Months were eliminated as reckoners of the year. The Egyptians knew there were 365¼ days in the year. The Julian calendar became 365 days long, with 366 days every fourth year to take care of that bothersome quarter day.

In 46 B.C. when Sosigenes reformed the calendar, the seasons were so woefully out of step with the date that two months were added. The year 708 A.U.C. became the last "year of confusion." The new Julian calendar, as it was called, became effective January 1, 45 B.C.

The first month of the year was given 31 days, and so were all odd-numbered months. The even-numbered months were allotted 30 days each, except February, which had 29.

To do honor to Julius Caesar as well as to Augustus Caesar, and probably also to fill less honorable purposes, the Roman senate later changed the name Quintilis (the name for month number five had been retained even after it had become number seven) to July, and Sextilus became August. In addition, August was given 31 days, the extra day being taken from February. To avoid having "long" months together, September and November were made into 30-day months, and October and December were each given 31 days.

The 366th day in the Julian calendar (the intercalary day) was called *bissextus*. It followed February 24, which was the sixth day before the calends (the first day) of March. It counted, therefore, as a second sixth day—thus the name.

Today the extra day is February 29 (leap day). The name comes about because after that date the calendar "leaps over" a day. For example, in successive regular years a given date falls upon the next day of the week: in 1970, April 1 was on a Wednesday, in 1971 on a

Thursday; in 1972, it was not on Friday, but on Saturday—we leaped over one day.

The Julian calendar was a tremendous improvement over the system previously used, but the Moon was still a cause of confusion. Many of the dates of religious festivals were determined by the Moon. Passover was the day of the first full Moon after the vernal equinox. Easter occurred about the same time. However, most Christians desired to have Easter on a Sunday, and so they decreed that it should fall upon the first Sunday after the fourteenth day of the Moon. Those who persisted in placing Easter on the fourteenth day were heretics, and were called *quartodecimans*. The problem was not settled until A.D. 325 when a council of church officials called by the emperor Constantine met at Nicaea in Bithynia, close by the emperor's summer palace. Among other decisions, they decided that Easter should be the first Sunday after the fourteenth day of the Moon (reckoned from new Moon) that occurs on or after March 21. At that time, the vernal equinox occurred on March 21. Today, the same rule applies, except the counting must be from the time of the vernal equinox, which may be March 20, 21, or 22.

This Easter problem indicates the kind of complications that were not reconciled by the Julian calendar. There were others. The extra day helped to keep the dates in step with the seasons—but not completely, for the Julian year was 11^m14^s too long. It is a small annual interval, but in 128 years the intervals add up to an entire day. Over the centuries, therefore, dates and seasons got out of step once more. In 1200, Roger Bacon figured there was an error of eight days. In 1474, Pope Sixtus IV realized the calendar needed fixing, and he called in Regiomontanus (the German astronomer Johann Müller became known as Regiomontanus, which was a modification of the name of the town where he was born) to study the problem. Nothing happened, however, because Pope Sixtus died.

The Gregorian Calendar

In the sixteenth century—1582 to be exact—Pope Gregory XIII was aware that those 11^m14^s had added up to ten days. The first day of spring fell on March 11, and, if the trend continued, spring festivals would be occurring in winter.

The first thing to do was to drop ten days out of the calendar. This would put the vernal equinox on March 21—the date it occupied in A.D. 325 when the Council of Nicaea was assembled. Pope Gregory simply issued a proclamation saying that the day following October 4, 1582, would be October 15, 1582. There was great confusion; what happened to all events—social, business, ecclesiastical, political—that were scheduled for October 5 through 14? And certainly no adult relished the idea of becoming ten days older overnight. Chaos or not, the decree held.

The second thing to do was to arrange the calendar in such a way that the problem would not arise again. The real error in the Julian calendar was that the year was 11^m14^s too long, which amounted to a full day in 128 years, three days in 400 years. Pope Gregory decreed that century years should be leap years only when divisible by 400.

At the present time the intercalation procedure is as follows: all years divisible by 4 are leap years, except century years which are leap years only when divisible by 400. Exceptions are the years 4000, 8000, 12 000, and so on, which are not leap years. When this procedure is followed, the location of Earth in its orbit at the beginning of the year would vary no more than a day from its present place in some 200 centuries. No system of intercalation will ever work completely; only approximations can be made because there is no simple ratio between Earth's periods of rotation and revolution.

Countries in which the dominant religion was Catholic adopted the Gregorian calendar at once. Other countries were slow to act. Denmark and Protestant states of Germany did not use the Gregorian calendar until 1700, over a hundred years later. England (and her American colonies) adopted it in 1752, Japan in 1873, China in 1912, Russia in 1917 (Russia dropped it almost immediately until 1940), Turkey in 1927.

When England and the American colonies adopted the Gregorian calendar, the year began for the first time on January 1 instead of March 25. The discrepancy between the Julian and Gregorian calendars now amounted to 11 days, so Parliament decreed that the day after September 2, 1752, should be September 14, 1752. Unscrupulous landlords attempted to collect a full month's rent for September, and the populace in general did not agree at all with the decision. They felt they had been cheated out of the days, and government had no right to make such a decision.

Up to that time, the beginning of the year in England (and her colonies) was on March 25—March was the first month of the year. It became necessary to change this, to begin the year in January. To this end, January and February 1751 were dropped, as also were 24 days of March 1752. The Gregorian calendar was adopted January 1, 1752.

Eleven days were dropped out of the calendar: no history was made between midnight September 2, 1752, and midnight September 13. Also dates of celebration had to be adjusted. For example, our history books tell us that George Washington was born February 22, 1732. That's according to the Gregorian calendar, and with the corrections that were made when England adopted it. The calendar in use at the time of Washington's birth would have read February 11, 1731.

Astronomers' Calendar

To avoid the pitfalls produced by calendars—by the attempt to find a ratio between days and years—astronomers use no conventional calendar. They count only the days.

In 1582, Joseph Justus Scaliger (1540–1609) proposed the system that is now employed. Scaliger was an eminent scholar interested in, among other pursuits, tracing back the early history of Greeks, Romans, and Babylonians. To keep track of chronological sequences, he proposed that days alone should be counted. He arbitrarily proposed that they should be numbered from 12:00 noon on January 1, 4713 B.C. The date probably was suggested by his research. Incidentally, it is considerably earlier than 3928 B.C., the date Earth began, according to a statement made in 1642 by John Lightfoot, a scholar at Cambridge University.

To differentiate these days from all others, Scaliger called them "Julian days" in honor of his scholar-father Julius Caesar Scaliger. The name has no connection with the Julian calendar.

Julian day (J.D.) 2 440 576.5, corresponds to January 1, 1970. Greenwich mean midnight. Julian days are divided decimally, which makes computation simpler; also, since they begin at noon, a whole night's observing will occur on a given Julian date. Astronomers have circumvented the calendar problems that have plagued man since first he felt the need to keep his activities in step with motions of Earth.

New Calendar?

Historical records, contracts, and legal documents of all kinds in most countries are drawn according to the calendar which is in use today and which, it is presumed, will be in use for the next several centuries. Only a few nations do not use the Gregorian calendar. If the calendar were changed, confusion would result.

In spite of this, proponents of calendar reform emphasize weaknesses of the present calendar: it cannot be divided evenly into quarter periods; the length of the months ranges between 28 and 31 days; the months begin and end on all days of the week; weeks at the start and close of months are split between months.

To correct these flaws, various plans have been suggested, the most popular being the *world calendar*. This plan calls for 364 days divided into four equal quarters which would always remain the same, each beginning on Sunday and ending on Saturday. The first month in each quarter would contain 31 days, and the other two 30 days each. Each year one day would be added at the end of the last quarter—called Year-End-Day. It would be an extra Saturday. Every fourth year (with exceptions already practiced), a day would also be added at the end of the second quarter. This would be another extra Saturday, and might be called Leap-Year Day.

In 1956, the plan was proposed to the United Nations. It was vetoed. Established procedures are difficult to change. We expect it will be many, many decades, if not centuries, before the world calendar (or any other calendar) displaces the calendar we use presently—one that is basically the Egyptian calendar of some 3000 years ago, modified by Sosigenes for Julius Caesar, and then improved upon by Pope Gregory.

REVIEW

1   A day is the interval between successive passages of a star, the vernal equinox, or the Sun across one's meridian as determined by Earth's rotation.

2   To compute long periods the astronomer uses the Julian day, a system devised by Joseph Scaliger and reckoned from 4713 B.C.

3   Since time by the Sun (sundial time) requires adjustments, the world uses a fictitious Sun for reckoning time—one that always moves the same interval each day along the celestial equator.

4   Standard time zones divide the world into 15° zones in each of which the time is the same.

5   Precise clocks that hold to certain frequencies enable detection of slight variations in Earth's rotation and also enable split-second timing of various events, an essential requirement in many aspects of space exploration.

6   Calendars are attempts to reconcile days, weeks, months, and years, a situation that is awkward because there is not an even number of days in a year.

7   Most of the world now uses the Gregorian calendar, one that has evolved from its beginnings in ancient Egypt.

QUESTIONS

1   Define time, both in layman's terms and in astronomical terms.

2   Why is the vernal equinox used in reckoning sidereal time?

3   What is the value of sidereal time to the astronomer; and why is it of little value to the layman?

4   Construct a sundial, using plans in the references below or devising your own design. Calibrate it by marking shadow lines.

5   Note the changing lengths and directions of shadows during different hours of a day—from week to week and month to month.

6   How would the day change if the Moon were used as a time reckoner rather than a star?

7   How would a clock designed for keeping time on the Moon be different from a terrestrial clock?

8   On what date do sidereal and solar time agree? When is there a difference of 12 hours betwen them?

9   If a star rises at 9:00 P.M., what time will it rise two months later?

10   Suppose there were exactly 365^d2^h in a year. How would this change our method of keeping the date in step with the seasons?

11   Explain the factors that produce the equation of time.

12   If a place has a time 6 hours ahead of Universal time (Greenwich mean time—GMT), what is its longitude? $5\frac{1}{2}$ hours behind?

13   What are the features of the suggested world calendar?

14   How do astronomers circumvent the time-keeping problems of a calendar?

15   Devise clocks and calendars for Jupiter, Venus, or another of the planets.

READINGS

Cowan, Harrison J. *Time and Its Measurement.* New York: World Publishing Company, 1973.

Hood, Peter. *How Time Is Measured.* 2d Ed. New York: Oxford University Press, 1968.

Mayall, R. Newton, and Margaret Mayall. *Sundials: How To Know, Use and Make Them.* Cambridge, Mass.: Sky Publishing Corp., 1973.

Rohr, Rene R. *Sundials, History, Theory, and Practice.* Toronto: University of Toronto Press, 1970.

Waugh, Albert. *Sundials.* New York: Dover Publications, Inc., 1973.

THE PLANETS-
MOTIONS
AND MASSES

CHAPTER 16

Ancient observers of the sky saw the Sun of daytime, and the Moon and stars of night wheeling about Earth. Earth was indeed the center of the universe. Observations corroborated the belief. Milleniums of cogitation and conjecture were to pass before man was able to imagine himself outside Earth looking back— "seeing" it from an entirely different perspective. Copernicus was to provide the most extensive modern "outside" view not only of Earth but of the solar system—a view that very probably still lacks many details for it to be a true representation.

Although Copernicus conceived of the solar system as being Sun-centered, he believed orbits of the planets were circular. Kepler's ellipses explained variable motions, retrograde loops, and changes in brightness.

MOTIONS

To early man all celestial objects, except the Sun and Moon, were thought to be stars. And so there were long-haired stars (comets), shooting stars (meteors), and wandering stars, called planets (wanderers) by the early Greeks. Failing to understand the nature of these bodies, and their motions among the stars, the ancients would be expected to consider them as gods. Since all gods, regardless of their abode, regulated the lives of people, it was logical that earthly events should be related to the motions of the planets and the locations they occupied at the time of an event. Early astronomers were responsible for observing planet positions, for predicting where the planets would be at future dates—for keeping records of the "celestial gods" so that kings and princes, and common people, too, could know whether the time was right for war, business, love affairs, or political maneuvers.

Aristarchus

Among early men there were individuals not content with the traditional mythological explanations of the planets and their motions. In the third century B.C., Aristarchus of Samos believed that planets were Earthlike, and he was able to explain observed motions of the stars and planets by assuming the Sun stood still while Earth and the planets moved around it.

The brilliant insight of Aristarchus came before its time. People were not ready to dispense with the gods that populated the sky. Astrology was deeply entrenched, and to influential persons it was important because it affected their decision making; it was also considered a respectable science that merited much study. Since astrological discoveries were based upon a geocentric viewpoint, no one was inclined to accept any idea that required a different orientation. Eudoxus (Eudoxus of Cnides, 408–355 B.C.) and Callippus (Callippus of Cyzicus, c. 370–300 B.C.) taught that the planets were attached to spheres that moved within one another in a most complicated fashion.

Ptolemy

For 500 years the theory of spheres was accepted. In the second century A.D. Ptolemy, rejecting the idea of spheres, devised a system of circles to explain planetary motions. He wrote his ideas in the *Almagest*, the title being an Arabic translation made around A.D. 800. It was a monumental work of 13 sections including, among other information, a catalog of over 100 stars, discussion of the sizes

FIGURE 16–1
Ptolemy believed that the planets moved on deferents and epicycles centered upon the equant, a theoretical point on the oposite side of the eccentric from Earth. All objects had to move in perfect circles, it was believed; the circular epicycle explained retrograde motion, and the circular deferent (offcenter from Earth) explained variable velocity of the planets.

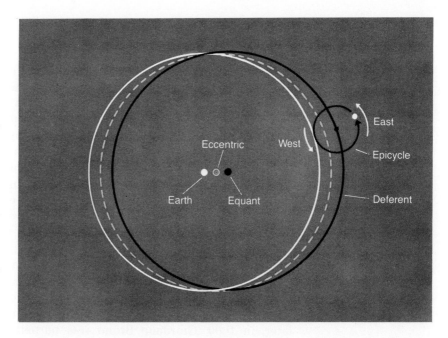

and distances of the Sun and Moon, eclipses, precession, and motions of the planets. Planets did not move on spheres, he said, but on circles. Each planet moved on an *epicycle*, a small circle the center of which revolved around Earth along the *deferent*. The center of the deferent was Earth. Some observations could not be explained by such a geometry, and so the deferent had to be eccentric—located a bit off center (Figure 16–1). Also in order to make observations agree with the geometry, Ptolemy had the center of the epicycle move around the *equant* the same angular distance in a given period of time. The equant was a point in space at a distance from the eccentric and on the side opposite Earth. The system of epicycles enabled planet predictions to be made with reasonable accuracy.

Copernicus

The teachings of Ptolemy contained in his *Almagest* remained man's guide to the stars and planets, the Moon, and the Sun for an incredible 1500 years. There were detractors here and there, but the *Almagest* held firm until the sixteenth century, when Nicholaus Copernicus (1473–1543) published *De revolutionibus orbium coelestium* (*Concerning the Revolutions of the Heavenly Spheres*), often shortened to *De revolutionibus*. It resulted from observations and interpretations he had made during his lifetime, and which he had hesitated to publish, fearing the ridicule of his peers and the wrath of the Church, which supported the ancient Ptolemaic viewpoint. It was finally published as Copernicus lay on his deathbed, making this an early example of "publish and perish."

Copernicus said Earth was not stationary. Sunrise and sunset resulted from Earth's spinning motion, and not from the movement of the Sun around Earth. Also, Earth was a planet not unlike the other planets—all moved around the stationary Sun.

There was little new in such a belief, for it had been proposed 2000 years earlier by Hipparchus. Copernicus knew this, but curiously did not acknowledge it in his publications (Figure 16–2).

Retrograde motions of the planets could now be explained. Those planets closest to the Sun move most rapidly, and so planets catch up and pass one another, producing the illusion of backward motion. However, Copernicus still believed the planets moved in circular orbits, and so most of the epicycles, deferents, and eccentrics of Ptolemy had to be retained to explain observed planetary motion. Copernicus set Earth in motion. His teachings were revolutionary, and spread rapidly among the intellectual world of the day. There were few immediate repercussions. But some 50 years later, in 1600, Giordano Bruno was burned at the stake because, among other things, he advocated the teaching of *De revolutionibus*. Galileo Galilei (1564–1642) was to be excommunicated for his support of an idea deemed heretical. Under pressure, Galileo recanted before the Inquisition his belief in the heliocentric universe and his teachings that Earth moved. Nevertheless, a famous statement of Galileos, one reported to have been made after his confession, is *"E pur si muove!"* (*"Nevertheless it moves!"*).

Galileo died at Arcetri, Italy, on January 8, early in the year 1642. Toward the end of that same year, on December 25, Isaac Newton, who was to provide proof positive of the Copernican theory, was born in Woolsthorpe, England.

Before considering the work of Newton, we should first review the contributions of Johannes Kepler and Tycho Brahe, which preceded those of Newton.

Tycho Brahe

Refinement of previous observations is an occupation of astronomers today, and the effort has persisted through the centuries. The most painstaking and precise observing which led to revisions of all catalogs that had gone before was done by the Danish astronomer Tycho Brahe (1546–1601), or simply Tycho. He assembled volumes of observations of planetary positions so accurate that they later allowed Kepler to discover his "laws" which were even later explained by Newton's law of gravitation as it applied to planetary motion. While

Tycho was a skilled observer, and an admirer of the work of Copernicus, he found it impossible to accept the heliocentric universe because he could detect no evidence of the Earth's motion. He compromised by combining the Ptolemaic and Copernican viewpoints. Earth was the center of the solar system, he said. The Sun went around the Earth and, at the same time, the other planets went around the Sun (Figure 16–3).

Johannes Kepler

Shortly before the death of Tycho, Johannes Kepler (1571–1630) became his assistant. Kepler continued to observe, but his main interest was the study of the tables compiled by Tycho, a project that was to occupy the next 25 years of his life.

For about ten years Kepler tried to fit the motions of Mars (the

FIGURE 16–2 (left)   *The Copernican system retained many of Ptolemy's epicycles and deferents; it had to because the belief persisted that planets moved in circular orbits and there was no other way to explain retrograde motion and variable velocity. However, Copernicus placed the Sun at the center of the solar system as Hipparchus had taught 2000 years earlier.*

FIGURE 16–3 (right)   *Tycho Brahe could not accept fully the Copernican model. He believed that the planets went around the Sun while the Sun went around Earth.*

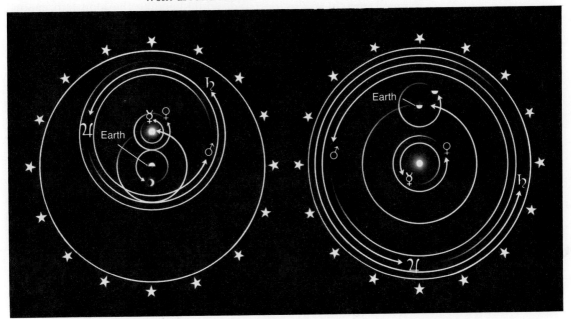

planet for which Tycho had prepared the most extensive data) to the deferents, epicycles, eccentrics, and equants which were believed essential. But no approach would explain the data completely. One of his hypotheses was accurate to within 8 minutes of arc (about a quarter of the diameter of the full Moon). Kepler could not accept that Tycho could have been in error by such an amount (which he wasn't). Kepler's thinking had been hampered by the circle. It was believed then, as it had been for thousands of years, that planets had to be fitted to this conception. Kepler finally broke away from this restriction. He experimented with an oval and ultimately came to the elliptical orbit, the true path of all the planets.

The ellipse. Circles and ellipses, as well as parabolas and hyperbolas, belong to the family of curves known as conic sections. If a cone made of clay, let us say, is cut along a line parallel to one side, a parabola is produced. If it is cut along a line intersecting both sides of the cone, an ellipse results: the greater the slant, the flatter the ellipse. A cut perpendicular to the axis of the cone produces a circle. A hyperbola is produced when the cut is made at a slant greater than that of the side of the cone (Figure 16–4).

An ellipse is not an oval, nor is it eggshaped, as one often sees in the literature. One way of producing an ellipse is to cut a cone, as explained above. Another way is to stick two pins vertically into a piece of cardboard. Drop a loop of string over the pins. Place a pencil upright in the loop and, while keeping the loop taut against both pins, scribe out the curve. It will be an ellipse. The two pins are the *foci* of the ellipse; the long dimension of the ellipse is the *major axis*; half of it the *semimajor axis.*

The shape of the ellipse depends upon the distance between the pins—if they are together, a circle is produced. The eccentricity of an ellipse is obtained by dividing the distance between the pins (between the foci) by the major axis. A circle is an ellipse with zero eccentricity. A parabola has an eccentricity of 1, the location of the second focus being at infinity, and the major axis being of infinite length (Figure 16–5).

Intense application of Tycho's Mars data to the ellipse hypothesis revealed that the orbit of the planet is elliptical, and that the center of the Sun is one of the foci of the ellipse. The other focus is empty. It is an extremely "fat" ellipse, so nearly a circle that on any reasonable scale the difference cannot be discerned. (The eccentricity of the orbit is 0.093.)

While working with the ellipse idea, Kepler determined that the distance between the Sun and Mars changed as the planet moved

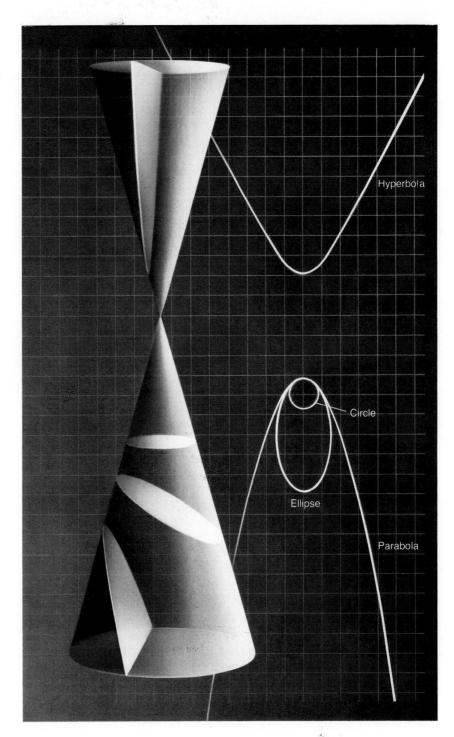

FIGURE 16–4
Circles, ellipses,
parabolas, hyper-
bolas, are all sections
of a cone.

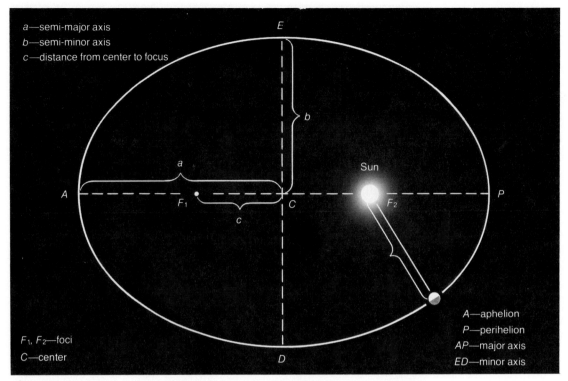

FIGURE 16–5   Parts of an ellipse.

in its orbit. He used Tycho's tables of positions to compute the changing distances to Mars in terms of the distance between the Sun and Earth.

It was known that Mars takes 687 days to go around the Sun, and Earth takes 365 days. Therefore, while Mars completes one revolution, Earth does also, plus $^{322}/_{365}$ of a second revolution, or $^{322}/_{365}$ of 360°.

Distance to Mars. By referring to Tycho's tables, Kepler knew that on a certain date Mars was separated from the Sun by an angle of 119°, and 687 days later the angle was 136°. A model can be drawn in which the position of Earth at the start of the period can be shown, as well as the position 687 days later. The two angles, determined by observation, can also be drawn. Just as the Sun's position is at the intersection of sides of the two angles, the position of Mars must be at the intersection of the other sides. The distance from the Sun to Mars can be measured on the scale model. On ours it turns

FIGURE 16–6
From Tycho's data of
the postions of Mars,
Kepler was able to
construct a model
showing relative
distances from the
Sun.

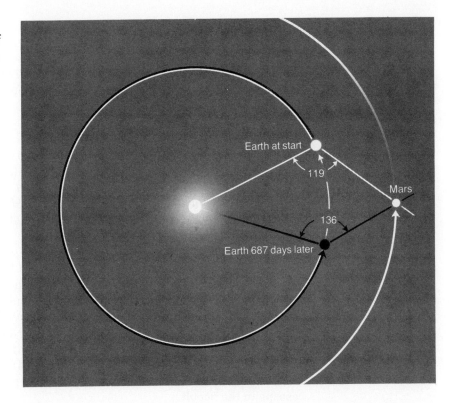

out to be 6 centimeters, or 1.5 times 4 centimeters, the length of the line connecting Sun and Earth. No matter what scale one were to use, the ratio of the distances would always be 1 to 1.5; the distance from the Sun to Mars is 1.5 times the distance from the Sun to Earth (Figure 16–6).

Equal areas. With care, Kepler made accurate measurements in this fashion for the distance to Mars throughout the many years for which Tycho had assembled data. Kepler not only discovered that the distance between the Sun and Mars changed, he also discovered that when Mars was nearest the Sun it covered more distance in its orbit during a given time interval than it covered when the planet was farther from the Sun. Scrutiny of the models Kepler constructed and the changes in planetary positions revealed a basic relationship between speed and distance. A line joining the Sun and Mars (or any planet) sweeps out equal areas in equal periods of time.

No longer was there any need to describe orbits as being circular, the problem that had led to the deferents and epicycles of Ptolemy and, later, of Copernicus. The planets moved in ellipses

(Kepler did not prove this for any planet other than Mars, but assumed it to be true for all) and therefore traveled at differing speeds. (When at aphelion, Mars moves 23.40 kilometers per second, and 21.76 kilometers per second at perihelion.) Therefore, Earth caught up and passed slower moving planets (the outer planets) and was passed by the faster moving inner planets.

The simplicity of Kepler's first two laws—elliptical movements of the planets and the equal-areas relationship—is deceptive. We quote them glibly today, but their discovery was achieved only through persevering dedication. Kepler's calculations were made without the use of computing devices, or even logarithms. They first appeared in Kepler's *Commentaries on the Motions of Mars*, published in 1609—the year that Galileo first used a telescope to make celestial observations.

Period-distance relationship. Kepler was beset by personal problems. At an early age, he was stricken with smallpox that left him with impaired eyesight and a generally weak condition. His father deserted the family. Later, Kepler lost two wives yet fathered 12 children of which only 2 survived. Twice he had to flee from religious persecution. In addition, he spent years to free his mother from condemnation for practicing witchcraft. Small wonder that Kepler became something of a mystic, finding solace and probably escape in the occult.

His facility with mathematics led him to search out mathematical relationships—harmonies and rhythms among celestial events. Nine years after his *Commentaries* appeared, Kepler published his *Harmony of the Worlds*. It was composed of largely far-fetched ideas that had no scientific value. However, the volume contained one nugget of great value: Kepler's third law of planetary motion, which is that the squares of the periods of the planets is in direct proportion to the cubes of their mean distance from the Sun:

$$P^2 = A^3$$

The law does not give one the actual distance to a planet. However, it does provide distance in some given unit, such as the astronomical unit—the mean distance between the Sun and Earth. For example, the period of Saturn is 29.46 years, as determined by observation; its square is close to 868. The cube root of that figure is very close to 9.5, the actual mean distance to Saturn in astronomical units.

Isaac Newton

Kepler died in 1630 at the age of 59. Twelve years later, Galileo died at the age of 78; and Isaac Newton was born. Kepler had shown relationships between the periods and distances of planets and had eliminated circular orbits. He told us *how* the planets moved, but was unable to explain *why* they moved as they did. It remained for Newton to carry on the work of Galileo with pendulums and falling objects to formulate the laws of motion and universal gravitation, which he did in his *Philosophiae Naturalis Principia Mathematica* (the *Principia*), published in 1687. Publication was achieved only after much acrimony caused by John Flamsteed, who was then Astronomer Royal and was loath to relinquish the planetary data that Newton required. However, encouraged by Edmund Halley, who was to succeed Flamsteed as Astronomer Royal, Newton persisted in his efforts. The relationship between bodies and their motions he expressed in three statements:

1   A body at rest remains at rest, and a body in motion continues to move and in a straight line unless some force is exerted upon it.

2   A force exerted upon a body causes it to accelerate (speed up, slow down, or change direction)—the amount and direction are proportional to the amount and direction of the force.

3   For every force exerted, there is an equal force exerted in the opposite direction.

According to the first statement, there had to be some force exerted upon the planets; otherwise they should be moving in straight lines, and so should the Moon. In his *Principia* Newton included a drawing of Earth with a cannon mounted upon it and the paths of projectiles shot from the cannon.

When a shot is fired, the cannonball moves forward and downward in a curved path—pulled by Earth's gravitation, just as Galileo had proved experimentally a half-century earlier. If the charge is increased, the ball moves farther before it reaches Earth, but still along a curved path. Theoretically, the curve of fall of the cannonball could be made to match the curve of Earth's surface. The cannonball would be in orbit around Earth (Figure 16–7).

Newton devised his theoretical experiment to explain the Moon's motion around Earth. However, it is also the explanation for the trajectories of artificial satellites, for the paths of planets around the Sun, or stars around stars.

If a satellite has insufficient speed, it will fall into Earth. If it has too much, it will escape from Earth and go into orbit around the Sun. Depending upon the thrust given a satellite, it will go into a

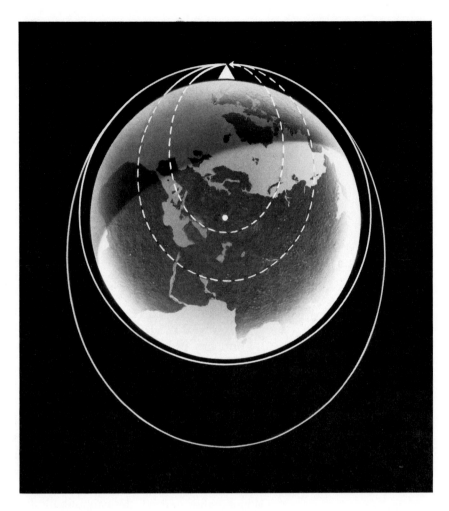

FIGURE 16–7
Newton explained the orbit of the Moon around Earth, and the planets around the Sun, by supposing a cannon located atop a high peak firing balls at sufficient velocity tangent to Earth's surface.

circular orbit or an elliptical orbit, in which case the distance to apogee is dependent upon the energy of the satellite. What is true here for Earth and a satellite also applies to the Sun and its planets.

Galileo had experimented with falling bodies, and so such things as rate of fall and the action of pendulums were known. Newton also was aware that a falling apple pulled upon Earth, just as Earth pulled upon the apple. The puzzlement in his mind concerned distance: could Earth's pull extend to the Moon? Could the pull of the Sun extend to the planets and thus answer the "why" of motion posed by Kepler's answer to how planets moved? Newton surmised an affirmative answer and was able to prove that the force F exerted by Earth upon the Moon (and vice versa) or the Sun upon the planets depended upon the distance, d, between the bodies and their masses (the amount of material they contained), M_1 and M_2.

$$F = G \times \frac{M_1 \times M_2}{d^2}$$

The G factor, the constant of gravitation, is obtained experimentally. When *dynes* for force are used (a dyne is the force needed to accelerate 1 gram mass 1 centimeter per second per second), grams for mass, and centimeters for distance, G has the value 6.67×10^{-8}.

The force that the Sun exerts upon Earth is equal to G times the mass of Earth multiplied by the mass of the Sun and divided by the square of the distance between them. As expressed in Newton's third statement, Earth exerts the same force upon the Sun. However, because the Sun is so massive, Earth's effect upon the Sun is barely noticeable.

Each of the planets exerts a force on each of the other planets, and upon satellites. Indeed, all objects have mass and therefore must exert a force on all other objects. However, because of the much greater mass of Earth we are not aware of the effects of a steel ball, let us say, on Earth. And the Sun is so massive (99.86 percent of the mass of the entire solar system is in the Sun) that the force of all the planets together exerted upon the Sun is inconsequential.

Newton's theoretical cannonball experiment and Galileo's experiments with falling bodies provided the clues to explain the orbit of the Moon and the planets.

A ball on Earth's surface, 6400 kilometers from the center, is accelerated by 9.80 meters a second each second, attracted by Earth's gravitational force. The Moon is about 384 000 kilometers away, 60 times farther. Therefore, the force that Earth exerts upon the Moon should be 60×60, or 3600, times less. Each second the Moon falls 0.0027 meters toward Earth while it moves ahead 1080 meters. In a distance of 1080 meters, Earth curves 0.0027 meter; therefore the Moon gets no closer to Earth (assuming a circular orbit). After $27\frac{1}{3}$ days, the Moon has fallen around Earth; an orbit has been completed. The same line of reasoning explains motions of artificial satellites around Earth, and planets around the Sun.

MASSES OF THE PLANETS

Mass and weight are often mistakenly interchanged. For example, we sometimes see the figure 5.98×10^{24} kilograms referred to as the weight of Earth. Actually, it is Earth's mass: the amount of material contained in the planet. Weight is a measure of force. When a person stands on a scale, the reading tells him the amount of force pulling him toward the center of Earth, and the force pushing upward on the platform—the force he feels on the soles of his feet. Were he to go farther from the center, his weight would decrease. But his mass would not change. Were he to go to the Moon, his weight would be a measure of the Moon's gravitational force exerted upon him. This would be one-sixth that of Earth, so his weight would be one-sixth that on Earth. His mass, however, would be unchanged; the total amount of matter would not be modified.

The mass of Earth can be determined experimentally. Once this figure is known, we can compute the mass of a planet relative to Earth's mass. One way to do this is by observing the effects of a planet's gravitation upon its satellite. We use the equation $P^2 = A^3$—the expression of Kepler's third law. The equation was modified by Newton to $P^2(M_1 + M_2) = A^3$ since the masses of the two bodies involved affect the results. Kepler was not aware that the planets affected the Sun, just as the Sun affected the planets. He could not be, for the effect of planets on the much more massive Sun is negligible. If the mass of the Sun is given a value of 1, the mass of Jupiter (the most massive of all the planets) is 0.001. So the sum of $M_1 + M_2$ became 1.001, only slightly greater. However, these deviations resulted in the slight variations between the matching of the squares of periods and cubes of distances for Jupiter and Saturn.

Newton's knowledge of the law of universal gravitation (the fact that the two bodies were revolving around each other) made it possible for him to introduce the mass factor.

Kepler's third law as modified by Newton became

$$P^2(M_1 + M_2) = A^3$$

When mass is the unknown factor, the equation becomes

$$(M_1 + M_2) = \frac{A^3}{P^2}$$

Suppose we wish to determine the mass of Jupiter (M_J). We select one of the planet's satellites (M_s) and consider the two mutually revolving. Practically, because the mass of the satellite is so small, it can be disregarded, and the equation then becomes

$$M_J = \frac{A^3}{P^2}$$

The distance between Jupiter and its satellite can be measured. It is expressed in terms of the distance of the Moon from Earth. For example, were we to use the satellite Ganymede, its distance would be 1 070 213 kilometers, the distance from Jupiter to Ganymede, divided by 384 472 kilometers, the distance between Earth and Moon, or 2.78.

The period of revolution of Ganymede around Jupiter would be expressed in sidereal months (1 sidereal month $= 27^{d}07^{h}43^{m}$). Ganymede revolves around Jupiter in 0.2614 sidereal months. Thus, in round numbers,

$$M_J = \frac{(2.78)^3}{(0.2614)^2} = \frac{21.4850}{0.0683} = 314$$

When more precise figures are used, the mass of Jupiter figures out to be 318 times greater than the mass of Earth. The masses of the planets are listed in the chart on pages 434–435.

This same line of reasoning—using Newton's modification of Kepler's third law—can be applied to finding the masses of those stars that are revolving around one another, and to estimate the mass of our galaxy as well as other galaxies.

OTHER PLANETS?

There are nine planets in the solar system, four of which are rather dense—Mercury, Venus, Earth, and Mars. These are called the *terrestrial planets*. The major planets—Jupiter, Saturn, Uranus, and Neptune—all have low densities. They are sometimes called "gas giants." Pluto is something of a maverick. It appears to have a density more nearly like that of the inner planets, yet it is the outermost planet of the solar system.

Traditionally, the solar system has been considered unique. However, modern astronomers consider that planetary systems may be common. Indeed, stars that exist as solitary objects without companions, or without bodies held in orbit by the stars' gravitation, may be the exception. There may be billions of other solar systems going around stars in our galaxy, or freely moving in space, not associated with a particular star. There also may be billions of galaxies beyond our own, so the possibilities for the existence of planets elsewhere in the universe are virtually numberless.

Vulcan

In the early part of the nineteenth century it was noted that Mercury was not moving as prediction said it should. Because of perturbations caused by other planets, the *line of apsides* of a planet's orbit (the line connecting the apses—aphelion and perihelion) rotates slowly. Therefore, the locations of aphelion and perihelion change. It was noted that the line of apsides was rotating slightly more rapidly than it should, an observation that no one could explain. Most of the motion is due to the effects of the other planets. However, the French astronomer Urbain Jean Joseph Leverrier (1811–1857), one of the co-discoverers of Neptune, suggested that the excess movement could be explained if there were a planet between Mercury and the Sun. If such a planet existed, it would be close to the Sun and very hot; therefore, it was called Vulcan, after the Roman god of fire.

Leverrier's statement was sufficient to start observers on a search for the elusive planet. In 1859, Dr. Lescarbault, a French physician and amateur astronomer living at Urgeres, reported that he saw a small, dark spot move across the face of the Sun. The supposition was that he had seen Vulcan, and so the search was given new impetus. Since the planet was supposedly close to the Sun, the only time it could be seen was during a total solar eclipse, when the region close to the Sun was dark. During scores of solar eclipses observers have scanned the space inside of Mercury, hoping for a glimpse of the elusive planet. It has not been found, and probably does not exist.

The presence of an intra-Mercurial planet remained the best explanation for the slight motion of the line of apsides until the beginning of the twentieth century. In his theory of relativity, Einstein said that one of the effects of the high velocity of Mercury (it moves over 170 000 kilometers an hour) would be for its line of apsides to rotate an extra 43 seconds of arc in a century. Observation shows it to be about that amount.

In spite of this entirely adequate, and no doubt correct, explanation, there are still those who persist in trying to prove that Leverrier was correct. Wherever total solar eclipses occur, staunch souls still set up their telescopes and cameras to scan the black sky near the solar corona, hoping to pick up the telltale light of Vulcan. When premature news stories reported that Mariner 10 had "seen" a satellite of Mercury, hope sprang anew in the hearts of Vulcan-searchers. (The "satellite" proved to be an image of a distant star.) During the 1970 solar eclipse, one team of observers reported an intra-Mercurial object they called "Zoe." It has not been confirmed since.

Planet Number Ten

Another planet was for a time thought to exist beyond the outer boundary of the solar system, rather than close to the Sun. In the early 1970s, an astronomer, Joseph L. Brady, checked records of 1700 years of observations of Halley's comet. He found that regularly the comet appeared four minutes earlier or four minutes later than predicted. Could this mean that there was an undiscovered planet out in that vicinity, one that was massive enough to cause the observed perturbations?

The possibility called to mind the early 1930s. Shortly after Pluto was discovered, observation of its motion revealed that the planet did not move as prediction said it should. Clyde Tombaugh, the discoverer of Pluto, spent several years searching for a trans-Plutonian planet. Brady's approach was different from Tombaugh's. Using computer technology, Brady built models of a ten-planet solar system—mathematical models of the masses of the planets, their locations, motions, and the perturbations caused by each of the planets. After several years of study, it became apparent that the observed motions of the planets (and Halley's comet) could be largely explained if there were a planet beyond Pluto. Planet Number Ten would have to be three times more massive than Saturn (about 300 times Earth's mass). The planet would have a period of 464 years, and its orbit would be tilted 60° to the plane of the solar system. It was also predicted that the planet would be 6 billion miles from the Sun, moving retrograde (opposite to the motion of the other planets), and in the Cassiopeia region. Such a planet would upset all our theories of the formation of the solar system.

After Brady's announcement, several astronomers in different parts of the world searched the sky region where the planet, if it exists, should be located. No sightings of it were made. Most astronomers now believe there is no Planet Number Ten. They believe there is at least one other possible explanation for the perturbations of Halley's comet. Frozen gases are the principal component of the nucleus of a comet. Occasionally, the solid materials change directly to gases which are spurted out of the core, causing the comet to roll and change course—the action being similar to that of altitude-control jets on a space vehicle. Those skeptical of the Planet Ten theory contend that a planet 300 times Earth's mass, and having a predicted magnitude of 14, would have been sighted sometime during the history of sky observations by telescope. They also point out that the gas-jet theory is quite adequate as an explanation for the unscheduled early and late arrivals of Halley's comet.

There may be planets beyond Pluto, but they would be so distant that the chances of discovery are slight. Pluto still is the outer boundary of our solar system.

Table 16-1   *Information about the planets*

	Mercury	Venus	Earth	Mars	Jupiter	Saturn	Uranus	Neptune	Pluto
Mean distance (million km)	57.8	108.2	149.6	227.9	778.3	1429.9	2869.6	4496.6	5911.7
Greatest distance from Sun (million km)	68.8	108.2	151.2	247	810.9	1500	2937	4507	7305
Least distance from Sun	46.4	106.7	146.4	207.2	735.7	1335.4	2718	4413	4426
Mean distance (a.u.)	0.387	0.723	1.000	1.524	5.203	9.539	19.18	30.06	39.44
Period of revolution	88.0d.	224.7	365.26	687.0	11.86y	29.46y	84.01y	164.8y	247.7y
Orbital velocity (km per sec.)	47.5	34.9	29.6	24	12.9	9.6	6.7	5.4	4.7
Inclination of axis	7°?	32°?	23°27′	23°59′	3°04′	26°44′	97°53′	28°48′	(?)
Inclination of orbit to Ecliptic	7.0°	3.4°	0.0	1.8°	1.3°	2.5°	0.8°	1.8°	17.2°
Eccentricity of orbit	0.206	0.007	0.017	0.093	0.048	0.056	0.047	0.009	0.249

	Mercury	Venus	Earth	Mars	Jupiter	Saturn	Uranus	Neptune	Pluto	Sun	Moon
Atmosphere (main elements)	Xe Ar	CO_2	N_2 O_2	CO_2	H_2 He	H_2 He	H_2 He	H_2 He	None detected	—	—
Temperature °C	349 day; −185 night	426	14	−68 to +28	2000	−156?	−180?	−198?	−226?	6000	−156 to +132
Volume (Earth = 1)	0.06	0.88	1.00	0.151	1318	736	50	43	0.2(?)	1 300 000	0.02
Albedo	0.056	0.76	0.36	0.16	0.73	0.76	0.93	0.62	0.14(?)	—	0.067
Mean apparent diameter of Sun (as seen from planet)	1°22′40″	44′15″	31′59″	21′00″	6′09″	3′22″	1′41″	1′04″	49″	—	31′59″
Magnitude (brightest)	−1.9	−4.4	—	−2.8	−2.5	−0.4	+5.7	+7.6	+12	−26.8	−12.6
Satellites	0	0	1	2	13	10	5	2	0	—	—
Synodic period (days)	116	584	—	780	399	378	370	367	367	—	27.3
Rotation period	58.65^d	244 (retrog.)	23^h 56^{m}04^s	24^h 37^{m}23^s	9^h 50^{m}30^s	10^h 14^m	10^h 49^m	16^h	6.387^d	25^d-35^d	27^d 07^{h}43^m
Equatorial diameter (kilometers)	4868	12 112	12 756	6787	142 830	119 330	47 200	50 000	5800	1 392 000	3476
Oblateness	0	0	1/298	1/192	1/16	1/10	1/16	1/50	(?)	0 (?)	0
Mass (Earth = 1)	0.055	0.815	1.000	0.107	318.0	95.2	14.6	17.3	0.06 (?)	332 958	0.0123
Masses contained in Sun	6 000 000	408 600	333 434	3 093 500	1047	3501	22 869	19 314	5 549 000 (?)	1	—
Density (water = 1)	5.50	5.23	5.52	3.93	1.33	0.69	1.56	1.54	4 (?)	1.41	3.36
Surface gravity (Earth = 1)	0.38	0.90	1.00	0.38	2.64	1.13	1.07	1.08	0.3 (?)	27.9	0.16
Escape velocity (km per sec.)	4.2	10.0	11.3	5.0	59.6	35.4	22.4	24.8	(?)	618.2	2.4
Symbols	☿	♀	⊕	♂	♃	♄	♅	♆	♇	☉	☾

REVIEW

1 Motion of the planets on concentric spheres, a belief held by ancient Greeks, was modified by Ptolemy to motion on circles, or epicycles, the centers of which were on deferents. The deferents were eccentric.

2 Ptolemy's *Almagest* contained information about the stars, Moon, Sun, and planets that was the basis of astronomical thought for some 1500 years.

3 In the fifteenth century, Copernicus argued in support of the Sun-centered universe.

4 Using data obtained by Tycho Brahe, Johann Kepler discovered three basic laws of planetary motion—the elliptical orbits, equal areas, and the distance-period relationship.

5 In the 1680s Sir Isaac Newton published his *Principia,* which, in addition to giving other information, described his three laws of motion.

6 Newton discovered that gravitation is universal and is dependent upon the masses involved and upon the distance between them.

7 There are nine planets in the solar system. Jupiter, Saturn, Uranus, and Neptune, all of low density, are usually classified as major planets, while Mercury, Venus, Earth, and Mars, all of higher density, are usually classified as terrestrial. Pluto does not quite fit into either category.

QUESTIONS

1 Assume a comet has a semimajor axis of 50 a.u. What would be its period?

2 Using the distances to the planets in the table on pages 434–435, calculate the sidereal periods of the planets. Check your answers against those in the table.

3 Why could Tycho Brahe not accept the Copernican "universe"?

4 How did Kepler explain the variable motions of the planets? How was he able to construct a scale model of the Sun-Earth-Mars geometry?

5 What are Newton's three laws of motion? Apply them to an explanation of the movement of an artificial satellite around Earth.

6 How do mass and weight change as one changes his location on Earth? in the solar system?

7 Describe how the mass of Earth can be determined.

8 Compute the mass of Saturn using the satellite Hyperion—mean distance 1 472 000 kilometers and sidereal period 21 days.

9 Why is Vulcan, supposedly a planet between the Sun and Mercury, no longer sought seriously?

10 The eccentricity of the orbit of Venus is 0.007. What does this mean?

READINGS

Adamczewski, Jan. *Nicholas Copernicus and His Epoch.* Philadelphia: Copernicus Society of America, 1973.

Asimov, Isaac. *The Universe.* New York: Walker & Company, 1966.

Dreyer, J. L. E. *A History of Astronomy from Thales to Kepler.* New York: Dover Publications, 1953.

Heuer, Kenneth. *City of the Star Gazers.* New York: Charles Scribner's Sons, 1972.

Pannekoek, A. *A History of Astronomy.* New York: Wiley-Interscience Publishers, 1961.

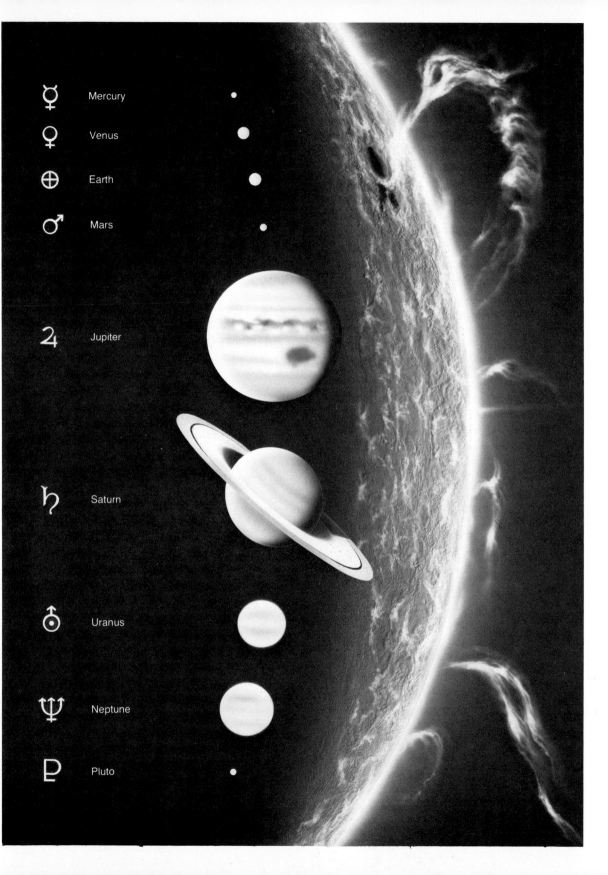

THE PLANETS-
MERCURY
TO PLUTO

CHAPTER 17

The history of astronomy is a history of man's evolving
perspective of himself from the narrow geocentric outlook
to the cosmic view.

To the Greeks Earth was the central planet around which the
universe turned. Herschel, Thomas Wright, and others of that
period believed our star was at the center of the Galaxy—
an idea that was changed only a few decades ago.

Modern man has his own delusions. It is hard for him to
comprehend that planets are insignificant among the stars and
that Earth is average among the planets. Our world is a mediocre
planet moving around an average star—one of billions
located in the outer portion of a wispy arm of an average
galaxy.

The planets vary
widely in size, mass,
and density. They are
impressive, yet all the
planets together
comprise less than
one percent of the
mass of the solar
system.

MERCURY

Mercury is the fastest of all the planets, the closest to the Sun, and the smallest. It is about the same size as Titan, the largest satellite of Saturn, and just a bit smaller than Ganymede, one of the four satellites of Jupiter that were discovered by Galileo.

The mean distance, semimajor axis, of Mercury from the Sun is 58 000 000 kilometers. However, its elliptical orbit is highly eccentric (the ellipse is flattened), so there is considerable difference between *aphelion* (69.5 million kilometers) and *perihelion* (45.7 million kilometers). In the case of Earth, the range is only about 4.8 million kilometers.

Mass

It is difficult to obtain the mass of Mercury. The planet has no satellite, so until 1974 when Mariner 10 made a close approach we had to rely upon observed effects of the planet's gravitation on objects such as comets and asteroids. Opportunities for such observations are rare; nevertheless, it was believed that the mass of Mercury was close to 0.055 that of Earth's mass. This belief was corroborated by Mariner 10. The density of the planet works out to be 5.46, very close to 5.52, the density of Earth. If the composition of the planet were the same as that of Earth, the density should be considerably lower, because the lower gravitation of Mercury would have packed the material less tightly together. Therefore the implication is that Mercury contains a larger proportion of heavy minerals than does our own planet.

Although Mercury is small and has an *albedo* of 0.05 (it reflects very little of the sunlight that falls upon it), the planet sometimes has a magnitude of −1.9, brighter than Sirius (−1.42). Because the planet is close to the Sun, it cannot appear more than about 28° from it. The sky does not become completely dark until the Sun is some 15° below the horizon; therefore, Mercury at its best position for observing against a dark sky is only 10° or 12° above the horizon. Atmospheric distortion is maximum at such a low elevation, so Mercury is an elusive object to observe. Most telescopic observing of Mercury is done in daytime when the planet is high in the sky. Obviously, there are drawbacks here, too. However, some observers report they have seen fixed dark markings on Mercury, and some 100 years ago both Schiaparelli and Antoniadi compiled maps of the surface. Those maps, and others of more recent origin, show only the haziest of markings. As revealed by Mariner, there are no markings as such. The observers must have been looking at light variations caused by crater walls and basins.

Anyone who has done any sky watching has probably seen Mercury, even though he may not have known so at the time. When it is near *eastern elongation*, the planet appears as a bright star just after sunset and before the sky is completely dark. It can be seen over a period of a few days. At *western elongation*, Mercury rises about an hour before sunrise and can be seen in morning twilight near the eastern horizon. The early Greeks did not realize the evening "star" and morning "star" were identical; they called the evening object Mercury, the morning object Apollo.

When observed telescopically, Mercury exhibits a complete cycle of phases during a synodic period of 116 days. When the planet is closest to Earth, we cannot see it (new phase or inferior conjunction), but it appears as a bright crescent on either side of inferior conjunction. When farthest from Earth (superior conjunction), Mercury is at full phase. Because of the greater distance, the planet appears to be much smaller at the full phase than at the crescent phase.

Rotation

When Mercury is at the most favorable position for viewing from Earth, the same face is always turned toward us. This fact led observers to the conclusion that the rotation and revolution were identical, 88 days. Also, the Sun-Mercury relationship was equated with the Earth-Moon situation; Earth's gravitation has slowed the Moon's rotation, so the same face is always toward Earth; and it was reasonable that the Sun's gravitation had affected Mercury in a similar fashion.

In 1965 astronomers at Arecibo, Puerto Rico, used the radio telescope to send signals to Mercury. The signals bounced off the planet's surface and returned to Earth. Signals from one edge of the planet showed a red shift (movement away) while those from the opposite edge showed a blue shift (movement toward). Analysis of these Doppler shifts revealed that Mercury was rotating in 59 days, not in 88 days as had been believed. In two years, Mercury makes three rotations, and thus the same face is again toward Earth when the planet is at the most favorable location for viewing (Figure 17–1).

If one were on Mercury, this combination of 88 days for a revolution and 59 days for a rotation would result in long periods of daylight. In one day, the drift of the Sun from the eastern horizon westward (which results from the rotation of Mercury) would be about 6 degrees. However, the eastward drift in one day caused by Mercury's revolution would be some 4 degrees. There would be about three months between sunrise and sunset. Of all the planets, Mercury moves fastest in its orbit around the Sun. Its mean speed

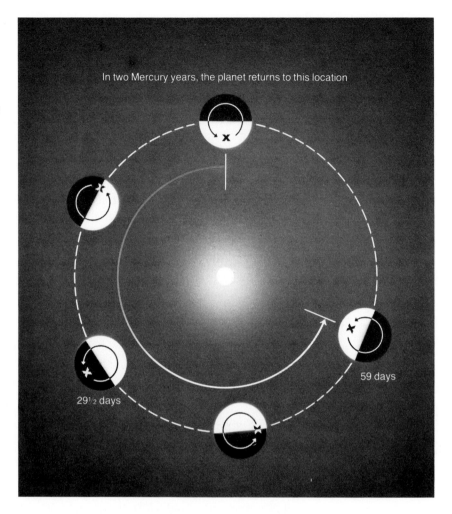

FIGURE 17–1
A year on Mercury (its revolution period) is 88 Earth-days long, and the planet rotates in 59 days. In two years, Mercury makes three rotations, thus again presenting the same face to Earth.

is 48 kilometers per second (170 000 kilometers per hour), while that of Pluto, the slowest planet, is less than 5 kilometers per second. According to Kepler's third law that relates the solar distances of the planets to their periods, one would expect this to be so: the nearer a planet is to the Sun, the shorter will be its period (the greater its velocity).

Temperature

Temperature was obtained by measuring infrared radiations from the planet. The bright side turns out to be about 330°C. More recent

FIGURE 17–2
Photomosaic of Mercury, constructed from Mariner 10 photos, from 700 km. The arc of mountains on the left forms the boundary of the largest basin in Mercury. It is presently called "Caloris" because of its location at the hot subsolar point when at perihelion. The mountains, up to 2 km high, form a ring that is 1300 km in diameter. The floor of the basin is a series of fractured plains. Similarity to the lunar surface is very close. Implications are that Mercury and the Moon are similar in chemical and mineral structure. (NASA)

investigations using radio astronomy and instruments aboard Mariner tend to corroborate this result. When the bright-side measurements were made, it was not possible to measure the dark-side temperature. However, it was believed that the planet rotated and revolved in the same period, and so the same face of Mercury was always turned away from the Sun. Therefore, the only heat on that surface would be that which traveled to that face by conduction from the hot side. Since this would be minimal, the temperature of the "dark side" of Mercury was thought to be at least −240°C.

When Mercury was scanned at different wavelengths by radio telescopes, the indicated temperature of the dark side was much warmer, around 2°C. Radio astronomers can determine surface temperatures because all heated objects radiate energy. They give it off in all wavelengths, although the greater part will be at one particular wavelength, as determined by temperature. A radio telescope "tunes

FIGURE 17–3   *Mercury's surface indicates an old planet; crater walls are smoothed down, and cratered plains partially filled with debris.* (NASA)

in" the radiation coming from a planet. When the signal is strongest, that wavelength reveals the temperature of the planet producing the signal. The initial explanation for the modified reading of Mercury was that apparently the planet had an atmosphere that served as an insulator. Over the centuries, heat had accumulated and was held under the blanket. Here was a contradiction, for the high temperature of the bright side, the low mass of the planet, and the low surface gravitation argued against an atmosphere. There wasn't enough gravitation to hold one, and, even if there had been, the high temperature would have accelerated gaseous particles to escape velocity.

In 1965, when it was discovered that the rotation and revolution were nonsynchronous, the apparently contradictory observations were reconciled. While the dark side radiated a great deal of heat into space, the rotation of the planet was sufficiently rapid to prevent the extreme loss previously surmised. Also, the planet appears to have a nebulous atmosphere of rare gases such as argon and xenon, probably solar wind particles captured by the planet's weak magnetic field.

FIGURE 17–4
A main crater ring on Mercury 1300 kilometers across. Hills and cratered plains are partially filled with debris. (NASA)

The albedo of Mercury, even less than that of the Moon, implied that the surfaces of the two bodies were similar. Before Mariner photographs proved them correct, many astronomers believed that the surface of Mercury was crater-covered, much like the surface of the Moon. In any event the surface must be extremely irregular, they said. This conclusion is logical when one observes the great increase in brightness between quarter and full as the long shadows of the quarter phase disappear. Mercury is indeed a cratered world, one that has been subject to heavy bombardment and to violent volcanic activity.

Transits

At inferior conjunction (between Earth and Sun), Mercury and Venus usually pass above the Sun, or below it. On occasion, however, a planet passes in front of the Sun and appears as a small black dot moving across the Sun's disk.

Transits of Mercury may occur 13 times during a given century, around May 8 or November 10—these being the times when the Sun is at the line of nodes. Accurate measurements of the motion of Mercury can be made at such times, providing an observational check on conclusions reached by computation. The next transits of Mercury will occur on the dates shown in Figure 17–5, and crossings will occur at the indicated latitudes of the Sun.

FIGURE 17–5   *Transits of the Sun (the large disk) by Mercury that will occur in the remainder of this century.*

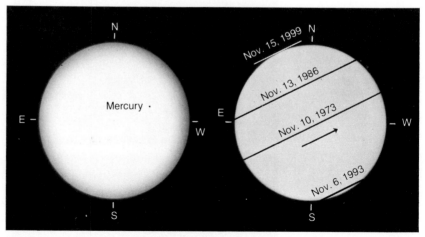

VENUS

The third brightest object in the sky, after the Sun and the full Moon, is Venus. At times its magnitude is −4.4, so bright that it causes shadows on Earth, and can be seen in daytime, if one knows where to look for it. Because the planet is between us and the Sun, Venus cannot be seen at midnight—no more than Mercury can be seen, or the Sun itself. When at eastern elongation, Venus appears in our evening skies; it is an evening star. About five months later the planet has become a morning star, dominating the hours of morning twilight.

Whether it is seen at evening or morning, Venus is impressive. It appears an extremely bright star to laymen unfamiliar with the appearance of planets and, in many cases, people take it to be some eerie extraterrestrial object—neither stellar nor planetary. The ancients knew about Venus; they could not miss it. In the days when all celestial objects were considered stars, Venus, named after the Roman goddess of love and beauty, was thought to be two different stars. The evening star was called Hesperus and the morning star Phosphorus (Figure 17–6).

Motions

Venus moves 35 kilometers per second, taking 225 days to complete one journey around the Sun. Its speed does not vary very much because, of all the planetary orbits, that of Venus is most nearly circular. We recall that the eccentricity of a circle is zero; the eccentricity of the orbit of Venus is 0.007. The difference between the maximum and minimum distances of Venus from the Sun is only about 1.6 million kilometers. The difference in the case of Mercury is almost 24 million kilometers. The sidereal period of the planet— the time required to align with a given star—is 225 days. The synodic period—the interval between two successive alignments of Earth, Venus, and the Sun—is 584 days.

Venus rotates very slowly, taking 243 days to complete one turn. The rotation is retrograde, that is, clockwise as seen from above the north polar region. Because of the slow rotation, there is little flattening of the planet. Venus appears more round than any of the other planets; it is almost a perfect sphere.

The angular distance between the Sun and Venus can be as much as 47°, which means that, unlike Mercury which never exceeds an angular distance of 28°, the planet is often well out of the twilight zone, and so is seen against a black sky.

FIGURE 17–6
Venus is seen at differing positions relative to the Sun. Sometimes it appears in the evening sky, at other times in the morning. Here we see an outer-space perspective.

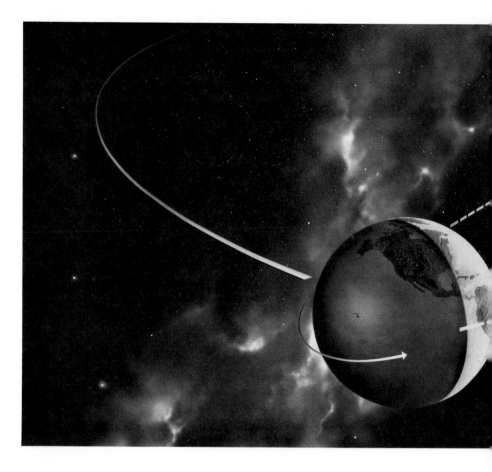

The dates through 1980 when Venus will be at inferior conjunction, eastern and western elongation, are listed below. Greatest brightness, when Venus appears 15 times brighter than Sirius, is reached about a month before and after inferior conjunction.

Eastern elongation	Inferior conjunction	Western elongation
1975, June 18	1975, August 27	1975, November 7
1977, January 24	1977, April 6	1977, June 15
1978, August 29	1978, November 7	1979, January 18
1980, April 5	1980, June 15	1980, August 24

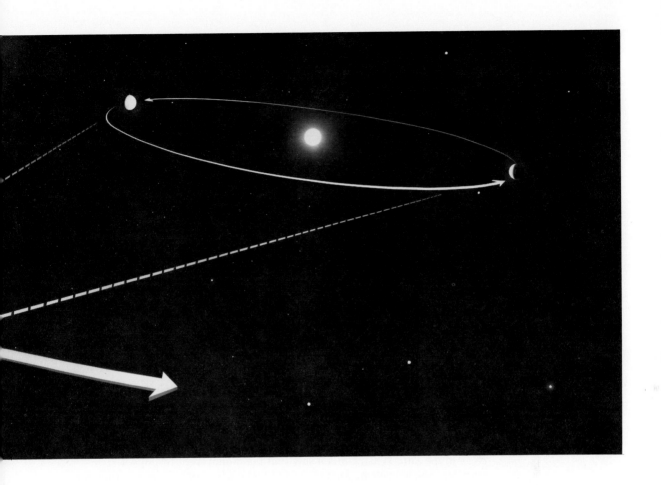

Appearance

To the unaided eye, Venus appears disk-shaped. Telescopically however, one can see the planet go through a complete cycle of phase changes, just as the Moon does. A tremendous stir in the world of science was created in 1610, when Galileo first observed the phases of Venus. The observation provided irrefutable proof of the validity of the Copernican viewpoint. If Venus went around Earth, the phase cycle could not be complete—the full Venus could not be seen (Figure 17–7 and 8).

When Venus is full (at superior conjunction), it is farthest from Earth (our solar distance plus that of Venus), and so appears small—

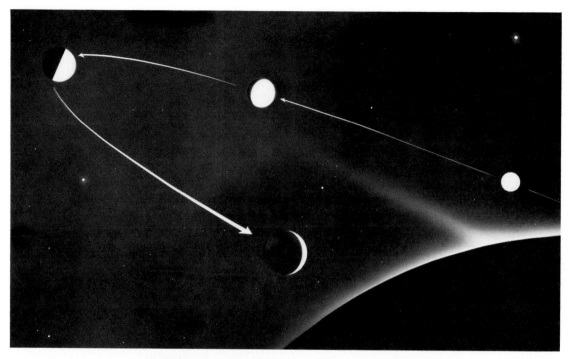

FIGURE 17–7   From Earth's surface we see Venus complete an entire
cycle of phase changes.

FIGURE 17–8   In Galileo's time it was known that Venus was always
close to the Sun. When Galileo saw that the planet went through a
complete phase cycle he knew it had to be in orbit around the Sun (diagram
on right). Were Venus in orbit around Earth, the planet could never appear
full (diagram on left).

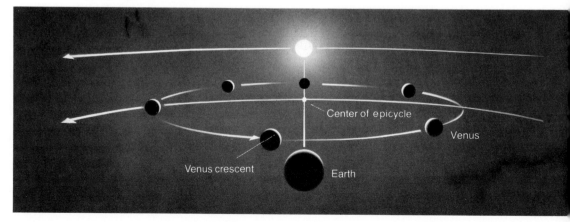

an apparent diameter of only 10″. At the crescent phase, when the distance between Venus and Earth has been reduced from some 257 million kilometers to 43 million, the apparent diameter is well over a minute. Although we see a smaller fraction of the disk, there is a great increase in apparent size, sufficient to produce 2.5 times more light than when the planet is full.

Atmosphere

The surface of Venus has never been seen, for it is hidden beneath an opaque atmosphere. The atmosphere seems laden with clouds that show no differences when seen from Earth, but considerable variation was shown in close-in ultraviolet photographs taken by Mariner 10. The clouds appear so opaque that it is unlikely bright sunlight ever penetrates to the Venusian surface. The skies of the planet are probably continuously overcast. The atmosphere appears to be moving in large convection cells along the equatorial regions. Photographs show high clouds move east to west around the planet in a period of about 5 days. In the polar areas the motion appears to be largely along the surface. Carbon dioxide is the most abundant gas. Hydrogen, perhaps from the solar wind, has also been detected in the atmosphere. Traces of helium and rather abundant amounts of atomic oxygen have also been detected. The amount of oxygen is some ten times greater than the amount detected on Mars.

Venus has no appreciable magnetic field, if any. Therefore, the solar wind penetrates into the atmosphere and so affects its composition. The "wake" of the solar wind is extremely narrow—only one-tenth that of Earth. This is due to the lack of a magnetic field.

Before space probes investigated the planet, one of the more

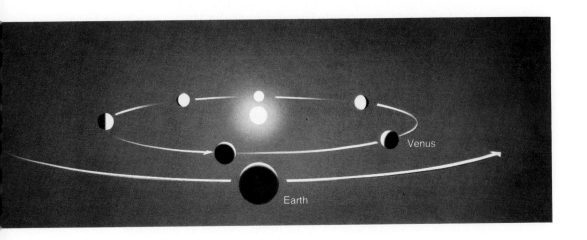

Venus

Earth

FIGURE 17–9
Photomosaic of Venus made by Mariner 10. This view of Venus from 720 000 km taken in ultraviolet light shows atmospheric circulation. Light and dark markings on a series of mosaics show equatorial east-west motions of 100 meters per second, and angular velocity increases with latitude. (NASA)

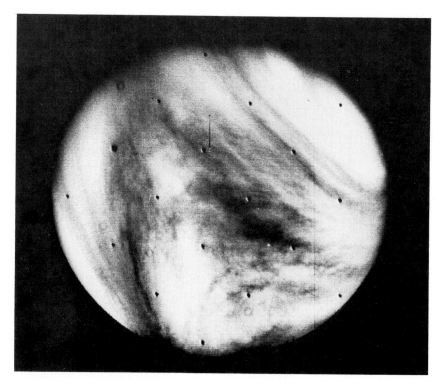

visible effects of the presence of an atmosphere came from observations of the crescent phase, when Venus is close to inferior conjunction. The horns of the crescent (the cusps) are greatly extended, and a twilight glow prevails almost around the disk—effects possible only when an atmosphere is present. Another significant observation is the albedo of the planet. The albedo of the Moon, which we know has no atmosphere, is 0.067; that of Mercury is 0.056; the albedo of Venus is 0.76, even higher than that of Earth. One would expect high reflectivity from an atmosphere, especially one opaque to visible wavelengths of light.

Analyses made by Earth-based equipment indicated that the atmosphere of Venus was largely carbon dioxide. Further data first obtained by the Russian Venera series fixed the carbon dioxide content at 90 to 95 percent—quite a different situation from that on Earth, where the gas makes up only 0.03 percent of the atmosphere. Water vapor was also present, although the amount was extremely small. Other materials such as hydrochloric and hydrofluoric acids have been detected in the atmosphere, and small amounts of nitrogen have been suggested by some of the Venera experiments.

Venus is often called Earth's twin, because the sizes of the two planets are nearly the same. But the similarity ends there, for in other respects Venus is most unlike Earth. Furthermore, it is a planet of mysteries. For example, we have no adequate explanation for the high percentage of carbon dioxide, nor for the small amount of water. Atmospheric pressure at the surface of Venus is one ton per square inch—more than 100 times greater than our pressure at sea level. The clouds of Venus are now thought to be sulfuric acid droplets. Some investigators have noticed slight colorations in Venus's clouds. Variations are quite apparent in Mariner photos of Venus.

Surface

We know that the surface temperature is high, in the order of 500°C, making Venus the hottest of all the solid planets. And it must be extremely dry.

Radio signals from the telescope at Arecibo, Puerto Rico, have been bounced off the surface of Venus, received on Earth, and plotted on a grid. The pattern (map) that results implies a predominantly smooth, rolling surface for the planet, with occasional highlands (Figure 17–10).

Sunset on Venus would be spectacular if we could see it. We notice on Earth that occasionally the Sun becomes redder and more and more flattened as it gets closer to the horizon. This is because sunlight is refracted as it passes through air layers of greater and greater density. On Venus, atmospheric density is so great that the solar image would be flattened and spread all around the horizon. In fact, there would be no discernible horizon. It would be as though we were inside a bowl that curved all around—and that provided few, if any, distinctive features of orientation. (Figure 17–11)

The high temperature of the surface results from the trapping of solar radiation. Based upon the radiation of the Sun and the location of Earth, computation indicates that our temperature should be −30°C. But our actual planetary temperature is close to 16°C. Our atmosphere is dense enough to capture and hold sufficient heat for the balance of inflow and outflow to occur at the higher temperature.

Because Venus is closer to the Sun than we are, it receives a greater amount of energy—1.75 times as much. The opaque clouds keep out a large part of the solar radiation. However, that which gets through does not escape. Radiation absorbed by the surface heats it up, and the longer wavelength radiation from the surface cannot escape from the atmosphere because it is absorbed by carbon dioxide. This insulating process is called the *greenhouse effect*, something one can experience when standing before a sun-drenched window on a cold winter day. Sunlight of relatively short wave-

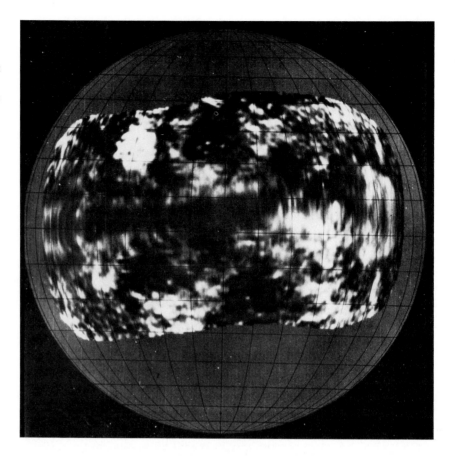

FIGURE 17–10
A radar map of
Venus. Variations in
brightness result from
variations in the
surface. The brightest
areas are the highest.
(Cornell University)

length passes through the glass and into the room. When the energy is re-radiated, the wavelength becomes longer. This longer wavelength cannot penetrate glass, and so the heat is trapped inside the room (or greenhouse). On Venus the abundant carbon dioxide is opaque to infrared radiation; it serves the same function as does the glass in a greenhouse.

Life

A stimulus to the space exploration that flourished in the 1960s was the possibility that astronauts would find life of some kind on the Moon, or at least fossils indicating that life in some form, perhaps most primitive, had existed there at some time in the past. When the Moon was found to be a world that had always been devoid of life, the "life believers" switched their attention from the Moon to Mars, and to Venus. But the possibility of finding life on Venus, or signs of its previous existence, is slim. The planet is extremely hot and dry; the pressure at ground level is 100 times greater than on Earth; the atmosphere is primarily carbon dioxide. It is most unlikely that any organisms recognizable to us could have emerged in such an environment; much less could they, or any signs of them, have survived. If we find any organisms, they may be from unsterilized Russian space probes.

FIGURE 17–11
Artist's visualization of sunset on Venus.

MARS

The only planetary surfaces we have ever observed—with either telescopes mounted on Earth or cameras in spacecraft—are those of Mars (Figures 17–13 to 16) and Mercury. The other planets (the major ones and Venus) are obscured beneath opaque clouds, and Pluto is so small and distant we cannot discern any features upon it.

Mars has been observed since man first appeared on Earth. Because of the blood-red-orange color, the planet was associated by the ancients with war. Mars sometimes reaches magnitude −2.8 and so dominates the night sky. Since it is farther from the Sun than Earth, the planet can be seen throughout the night. Many a sky watcher of ancient days must have gazed at the glowing disk and attached all sorts of mystical significance to it.

Mars has a mass about 11 percent that of Earth, and a radius of 6800 kilometers. Its density is only 3.9; thus its surface gravity is 38 percent of Earth's.

Schiaparelli—The "Canals"

In 1877, the Milanese astronomer Giovanni V. Schiaparelli (1835–1910) was mapping Mars, using an 8-inch refracting telescope.

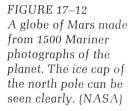

*FIGURE 17–12
A globe of Mars made from 1500 Mariner photographs of the planet. The ice cap of the north pole can be seen clearly. (NASA)*

Among other features that he noted were straight lines which he called channels. The Italian word for channels is *canali*. Whether Schiaparelli originally meant he had seen canals, implying that intelligent creatures existed on the planet, is not clear. Certainly he never denied that life might exist there. This is his answer to a query about life and the canals: "Their singular aspect (that of the *canali*) and their being drawn with absolute geometric precision, as if they were the work of rule or compass, has led some to see in them the work of intelligent beings, inhabitants of the planet. I am very careful not to combat the supposition, which includes nothing impossible."

Lowell—"Life on Mars"

Men have always been intrigued by the possibility that life is not peculiar to Earth. Giordano Bruno, the mystic of the late sixteenth century, taught that all the planets were inhabited. In the early nineteenth century, the *New York Sun* ran the series of articles, described in Chapter 13, about the discovery of creatures living on the Moon and so produced the Great Moon Hoax that excited readers around the world. Even though these newspaper stories were complete fabrications, readers persisted in their belief that there were other creatures on other worlds. The canals of Schiaparelli provided people with a new hope—one that was to be followed energetically by Percival Lowell (1855–1916), who had graduated from college in 1876, one year before Schiaparelli's announcement. Lowell became a writer and lecturer (mainly about life on Mars), and by 1894 he had used part of his fortune to build Lowell Observatory in Flagstaff, Arizona, for the purpose of studying Mars—and the other planets. Lowell was utterly convinced of the existence of life on Mars. In one of his books, he wrote: "In the canals of the planets we are looking at the work of local intelligence now dominant on Mars. Such is what the circumstantial evidence points to unmistakably."

The canals of Mars never showed up in photographs of the planet. However, observers reported they had fleeting glimpses of them, and their sketches reflected the conviction they had seen straight lines. Actually, under the most ideal conditions the very best optics on Earth cannot discern any formation less than 80 kilometers wide.

Conjecture in modern times about the surface of Mars, and whether the planet had canals or life of some sort, continued until the 1960s and 1970s when Mariner probes took close-in photographs and transmitted them to Earth. We still have seen no signs of life, but primitive life is still a possibility. The Viking mission should give us the answer.

Surface

Mars is a crater-covered world. For over a year Mariner 9 studied the planet at close range, taking thousands of photographs, measuring temperature, magnetism, atmospheric composition—and sending back billions of bits of information to be organized and studied by Earth-based scientists.

Astronomers had long known of dust storms on Mars. One was to extend over the entire planet and prevent a view of the surface by Mariner during the first month of its circling Mars. Dust is apparently universal on Mars, not dust as we usually have on Earth, but dust as fine as talcum powder, so fine that winds carry it 50 kilometers above the surface. Because of unequal heating, winds of 160 kilometers per hour develop, and some investigators believe they may attain 450 kilometers per hour. The planet is reddish because the dust is probably a kind of rust, iron oxide, similar to the mineral limonite on Earth.

The dust must have come from rock surface eroded by wind, and perhaps by water that may have been abundant at one time during the evolution of the planet. Presently, there is very little water on Mars (at least in the atmosphere or visible on the surface); perhaps there is enough to cover the planet to a depth of 0.0025 centimeters if it were all condensed out. (The water in Earth's atmosphere would cover Earth to a depth of 2.5 centimeters were it condensed out.)

There may be water beneath the surface, or beneath the polar caps. Occasionally clouds are seen over Martian volcanoes, especially over Nix Olympica, and it is suspected these may be clouds of ice crystals. Similar clouds are seen on occasion at the base of volcanoes, and it is suspected they may be made of water escaping through vents.

About half the total surface appears to be volcanic. Whether the volcanoes are presently active is a question. At this writing, no hot spots have been identified, as would be the case were eruptions in progress. However, there are those who believe that the Martian interior is just beginning to churn and boil, that convection currents within the planet will increase volcanic activity. Perhaps Mars is just now passing through the evolutionary step that Earth experienced long ago in its transition to its present condition.

The Mariner probes answered many questions, but they posed new ones. For example, there are crustal formations that can be

FIGURE 17–13
Nix Olympica, the
great Martian volcano,
rises 24 kilometers
above its surrounding
plain and is 550 kilo-
meters wide at the
base. (NASA)

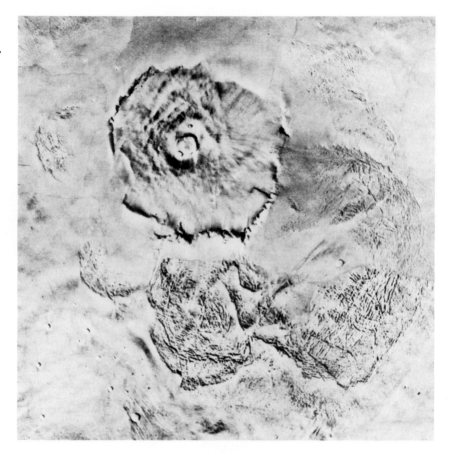

explained by supposing that water flowed through them at one time, gouging out the surface. Was Mars subject to torrential rains at one time? Or, as the planet's orbit changes over periods of thousands of years, are there recurrences of such events? Could the water of these rains be retained in the polar regions—needing only increased temperature to push it through the rain cycle? Again, answers may come from the Viking lander.

The great rift, a canyon 6000 meters deep and in places 250 kilometers across, extends 3700 kilometers along the surface. Previous to the 1970s it was seen as only a dark marking. Unlike the Grand Canyon, the Martian rift was not water-formed. It appears to be the result of the dropping of great blocks at the edge of Martian plates—or at least at weak sections of the crust (Figure 17–15).

Mariner cameras defined clearly objects as small as 100 meters across, and revealed that the surface of Mars is diverse: the polar caps that we already knew about, plains that are smooth, plains that are crater-covered, sections where craters overlap one another, mountainous regions, chaotic terrain, and the volcanic region—Nix Olympica (Snows of Olympus) and Tharsis Ridge. (See Figure 17–13.)

Atmosphere

According to some investigators, the present Martian atmosphere is new, having evolved from the eruptions of volcanoes, primarily Nix

FIGURE 17–14
Mariner 9 photo of Mars showing meandering "river," which some believe is rather convincing evidence that at some time liquid flowed over the surface. (NASA)

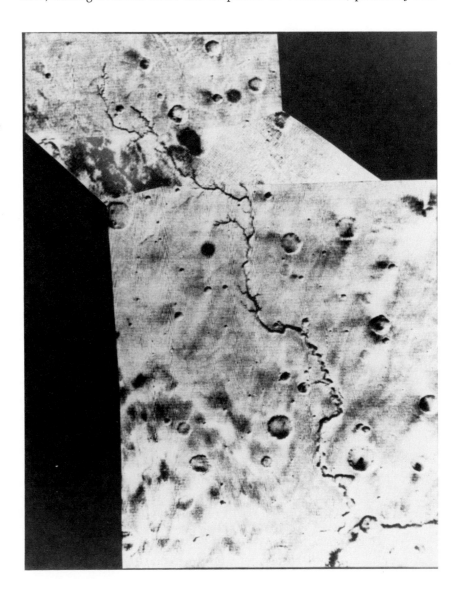

FIGURE 17–15
The great chasm of Mars, stretching several thousand kilometers across an area the size of the United States, was probably produced as result of surface collapse. (NASA)

Olympica. They further surmise that the original atmosphere, formed during the accretion process that produced the planet, was primarily hydrogen and methane. What they are saying is that the Martian atmosphere may be going through evolutionary changes similar to those that Earth experienced.

We know that the pressure and density of the Martian atmosphere is only about $\frac{1}{200}$ that of Earth's atmosphere at the surface, about what they would be at an altitude of some 30 kilometers above the surface of Earth. The main ingredient is carbon dioxide. Attempts to identify other gases, especially nitrogen, have been unsuccessful. Trace amounts of water vapor are carried in the atmosphere. Were it to be condensed into liquid there would not be enough to fill 1 percent of the smallest of the Great Lakes.

However, the atmosphere of Mars is sufficient to pick up dust and carry it from place to place at speeds up to 450 kilometers an hour. The dust may cause erosion. Its redistribution from place to place may explain the changes in brightness that seem to be related to seasonal variations. Before Mariner, some believed that the enlarging darker region that seemed to extend outward from the poles resulted from the growth of mosses or lichens nourished by the water produced by the melting of the polar caps. However, the caps

FIGURE 17–16
Oval tableland near the south pole of Mars shows the light and dark "contour" lines. These may be layered deposits of dust, volcanic ash, and possibly carbon dioxide and water ices. (NASA)

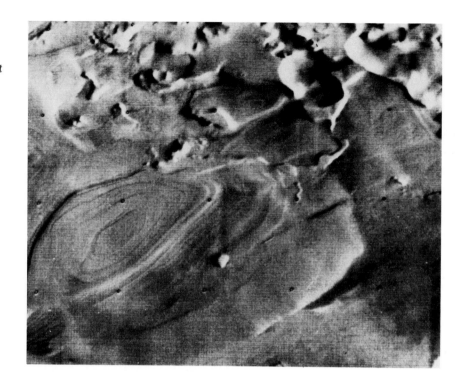

are made of dry ice—solidified carbon dioxide—not water. Some investigators believe that water may underlie the carbon dioxide.

Because the density of the atmosphere is so low, it is not effective as a heat insulator. Also, the gases in the atmosphere do not filter out ultraviolet radiation. Places on the surface reach 25°C during the daytime, but drop to 100° below zero during the Martian night. The ultraviolet radiation would preclude the existence of life such as we know it, at least on the surface, for organic molecules cannot hold together when exposed to intense ultraviolet light.

Phobos and Deimos

In *Gulliver's Travels*, written in 1726, Jonathan Swift describes in considerable detail the fictitious discovery of two small satellites going around Mars. One hundred fifty years later, in 1877 (the same year that Schiaparelli announced the *canali*), Asaph Hall at the U.S. Naval Observatory discovered that there were two satellites associated with Mars, and in many ways they were as Swift had written.

The two satellites, called Phobos (fear) and Deimos (panic)—the companions of the god of war—are extremely small (Phobos in Figure 17–17). Mariner 9 photographed the satellites while waiting

FIGURE 17–17
Pocked surface of
Phobos, photographed
by Mariner 10,
probably resulted
from collisions with
meteoritic masses.
(NASA)

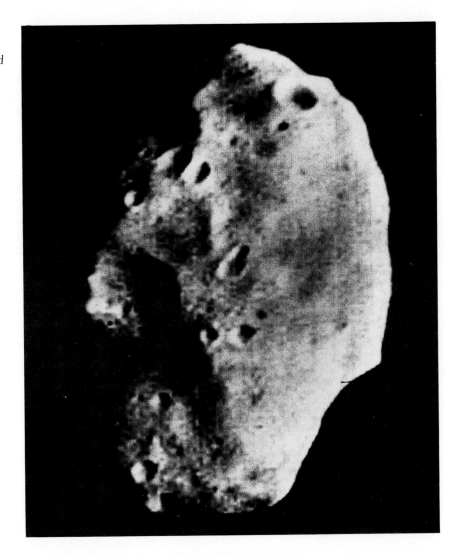

for the Martian sandstorms to quiet down. Deimos, the outer satellite —20 000 kilometers from the surface—is 16 by 9 kilometers, not much larger than Manhattan Island. It completes a circuit of the planet in one and a quarter days.

Phobos, the inner satellite, is only some 9200 kilometers from the surface. It is a bit larger than Deimos, measuring 27 by 19 kilometers. This fast-moving satellite completes a circuit in 7 hours, 39 minutes. Combined with the planet's rotation, the speed of Phobos causes the satellite to rise and set twice each day, and to rise in the west and set in the east. This would be an interesting sight, but not awesome, for the satellite would appear only a fraction as bright as Earth's Moon—more like an artificial satellite as seen from Earth's surface.

Because of its low escape velocity, Phobos has been suggested as a likely landing place for spacemen to survey Mars. It would make an ideal observing platform, and one easy to take off from, for the launching rocket would need very little fuel.

JUPITER

The planet Jupiter is becoming the most interesting object in the solar system, except for Earth. Recent observations from Earth, and first-hand studies by the Pioneer spacecrafts 10 and 11, have revealed that Jupiter and its satellites are in several respects quite different from what had been thought.

Appearance

Jupiter is the second brightest planet, and the fourth brightest object in the sky. Over 142 000 kilometers in diameter, Jupiter is also the fastest rotator of all the planets, with a day only 9^h55^m long. This means that a point along the equator at the cloud tops is moving 35 400 kilometers per hour.

Early observers noticed the polar regions of Jupiter are large and dusky. The rest of the atmosphere is divided into light and dark bands of clouds. The lighter bands are called *zones*, and have a subtle range of colors from yellow to bronze. The dark bands are known as *belts*, and are predominantly blue-gray. Superimposed on this general appearance are a few lighter and darker spots, and one huge region known as the Great Red Spot. It was first noticed in the seventeenth century, and has since disappeared and reappeared several times.

Radiation from Jupiter

Earth-based studies showed some time ago that Jupiter is a strong source of both thermal and non-thermal radiation, principally in the infrared and radio regions of the electromagnetic spectrum. Jupiter radiates 2½ times more thermal energy than it receives from the Sun. This is probably a result of left-over primordial heat. Jupiter is so large that it has not yet cooled off after 5 billion years from the formation of the Solar System.

In the radio region of the spectrum, Jupiter gives off three distinct types of radiation: (1) *centimeter waves*, from motions of hot molecules in the atmosphere; (2) *decimeter waves*, which are synchrotron radiation from electrons in Jupiter's strong magnetic field; and (3) *decametric waves*, believed to be generated by tremendous lightning storms in the ionosphere. The bursts of decametric waves, first discovered in the 1950s by Dr. B. F. Burke and Dr. K. L. Franklin, have an energy equivalent to several hydrogen bombs. Strangely, the radio waves are modulated by the satellite Io in a manner not yet understood. Except for the Sun, Jupiter is the strongest source of radio waves received on Earth.

Atmosphere

The composition of Jupiter's atmosphere is about 80 percent hydrogen, 20 percent helium, and about 1 percent everything else. In other words, it is about the same as the composition of the Sun. This is hardly surprising, since Jupiter is the next most massive object in the Solar System after the Sun.

The atmosphere is thought to be composed of at least four layers: (1) a top layer of ammonia ice, (2) a lower layer of ammonia compounds, (3) a layer of water ice, and (4) a layer of water droplets. The water has not been observed but is predicted from studies of the atmosphere. In all, the atmosphere makes up about 1 percent of the mass of the planet.

In the upper regions of the atmosphere, hydrogen, along with the gases ammonia and methane, may be combining to produce complex hydrocarbon molecules. Recently, ethane and acetylene have been discovered by Earth-based astronomers.

FIGURE 17–18
Jupiter in blue light with South at the top. Large, dark spot and bands show clearly as also does Ganymede (upper right) and its shadow (left of center top). (Hale Observatories)

The belts and zones are the result of the global weather structure of Jupiter. The light zones are rising and cooling areas; the dark belts are descending. The rapid rotation of the planet spreads out these convection areas into the planet-circling bands we observe. Localized storms such as the Great Red Spot resemble hurricanes on Earth except that some are regions of higher pressure (hurricanes are low-pressure areas on Earth). The Red Spot is over 40 000 kilometers across.

Above the visible atmosphere is Jupiter's ionosphere, composed of gases with temperatures over 1000°C. It is heated by high energy particles and by energy pumped up from below.

Below the Clouds

The spacecraft Pioneer 10, and later Pioneer 11, revealed that Jupiter is not solid, but liquid. Studies of the gravity of the planet and also theoretical models indicate that 1000 kilometers below the visible cloud tops the temperature is 2000°C, and that 3000 kilometers down the temperature is 5500°C and the pressure is 90 000 times that on Earth. Some 25 000 kilometers below the cloud tops the pressure reaches 3 million atmospheres. The central temperature is on the order of 30 000°C, five times the surface temperature of the Sun! Jupiter possibly has a small, rocky core, but this would be a very small part of its mass.

FIGURE 17–19
Pioneer 10 photograph of Jupiter, showing the shadow of Io, taken from distance of 2.5 million kilometers. (NASA)

Satellites

Jupiter has the largest retinue of satellites of any planet, 13 known so far. The brightest ones, Io, Europa, Ganymede, and Callisto, were discovered by Galileo in 1610, and called the Galilean satellites after him. Callisto and Ganymede are both about the size of the planet Mercury, and Io and Europa are slightly smaller. Pioneer 10 showed that Ganymede has several mare areas, like our Moon, and a bright north polar region. Io is the smallest body in the Solar System known to have an atmosphere. It seems to be surrounded by sodium, hydrogen, and nitrogen. Why this is so is still a mystery.

The outer satellites revolve about Jupiter in a retrograde direction. They are probably captured asteroids, and it would not be surprising if Jupiter had several dozen smaller satellites stolen from the asteroid belt.

Magnetic Field

Jupiter's magnetic field is ten times stronger than that of Earth at the surface of our planet. It extends out over 3 million kilometers from the giant planet. The magnetic pole is tilted 10° from the rotation axis.

The magnetic field acts like a slingshot and ejects high-energy particles off into space. They were first identified by Pioneer 10, although they had previously been detected from Earth and not identified with Jupiter. The method by which they are ejected has some similarities to the bursts of energy from pulsars, according to Dr. James Van Allen, the discoverer of the radiation fields around the Earth. Perhaps these energy bursts from Jupiter will shed some light on pulsar mechanisms.

Life on Jupiter

Because of the "primordial soup" of Jupiter's atmosphere, it is possible that some of the complex hydrocarbons may combine to form some simple form of life. This life would, however, have to obtain energy from the chemical components of the atmosphere, not from photosynthesis, since the sunlight received by Jupiter is only $\frac{1}{27}$ of that received on Earth. If such life exists, it would have to have some means of stabilizing its altitude in the atmosphere to avoid being carried down into the high temperature regions or upward into the destructive flux of solar ultraviolet radiation.

At the present time, we cannot say whether life exists in the atmosphere of Jupiter. Such a discovery would probably have to come from on-the-spot investigation.

SATURN

Saturn, with a magnitude of −0.4, is the outermost planet readily visible to the unaided eye (Figure 17–20). Until 1781 when Uranus was discovered, Saturn was considered to mark the boundary of the

FIGURE 17–20
The rings of Saturn are presented to us so that we can see the northern and southern sections. Edge-on, the rings disappear. Because of inclination of the rings, we see them from above in the illustration below, from beneath in the upper illustration on the next page, and edge-on in the remaining illustration.

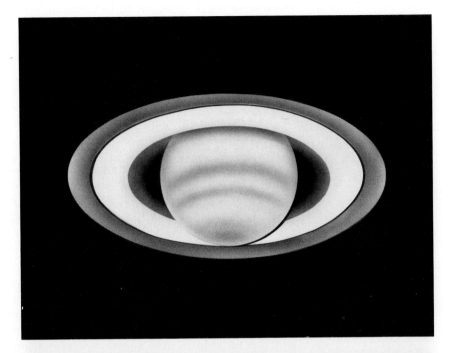

solar system. When Galileo first observed the planet, he noticed there were what appeared to be blobs on either side of it. He wrote, "Saturn has ears." Not until half a century later, in 1655, were the "ears" to be explained when the Dutch astronomer Christiaan Huygens described the rings of Saturn.

When the rings are almost edge-on to us, we see them as a thin dark line across the surface of Saturn. When exactly edge-on the rings are invisible. However, when the southern face is toward us, or the northern face, the rings become one of the most impressive sights of the solar system. There appear to be three major rings. The outermost, or A ring, is some 270 000 kilometers in diameter. The central B ring is brightest of all. And the innermost visible ring, the C or "crepe" ring, is again rather faint and nebulous. A fourth, extremely faint ring, located within the crepe ring and reaching almost to the planet, has been identified by Pierre Guerin, a French astronomer, and is visible only on photographs.

All the rings are made of discrete particles, probably about 1 centimeter in size. This is concluded from the measured heating and cooling rates of the rings. Small particles cool off faster. The rings are within *Roche's limit*. In 1850 the French astronomer Edouard Roche computed that if two fluid bodies having the same density should be within 2.44 times the radius of the larger, the smaller mass will not be able to stay together. Gravitational force of the larger mass will produce such extensive tides on the satellite body that it will be shattered. Perhaps the rings of Saturn may at one time have been satellites of the planet—or a single satellite that experienced the explosive forces that reduced it (them) to the small particles we now observe.

The particulate structure of the rings also explains the observed differences in revolution periods of the inner and outer portions of the rings. The inner portions of the rings revolve more rapidly than the outer regions, in agreement with Kepler's third law of motion. If the rings were solid, the periods would be the same throughout.

Although the rings have a tremendous diameter (some 270 000 kilometers), they are extremely thin, not exceeding 2 kilometers, and probably only about 50 meters thick.

Atmosphere

Like Jupiter, Saturn is covered with an extensive atmosphere of hydrogen and helium. Traces of methane have also been observed. About a billion and a half kilometers from the Sun, Saturn is a cold planet, so cold that any ammonia would be frozen into crystals, making methane the dominant gaseous compound. The atmosphere may comprise more than half the total diameter of the planet, and any solid core, if one exists, would be only some 38 000 kilometers across. The overall density of the planet (0.69) is the lowest in the solar system. This is less than water, which means that, curiously, the planet would float on water were there an ocean large enough to hold it.

Its low density means the planet is largely gaseous; this coupled with the rapid rotation (just slightly greater than 10 hours) causes considerable flattening, and the polar diameter is 12 800 kilometers less than the equatorial diameter.

Satellites

Each single discrete particle composing the rings of Saturn might be considered a satellite. However, in addition to these particles, ten satellites have been discovered—the first by G. Cassini in 1684 and the last by A. Dollfus in 1966. Their sizes range from about 250 kilometers in diameter (Phoebe) to 4600 kilometers (Titan), about the size of Mercury. Distances from the planet range from 160 000 kilometers (Janus, the last to be discovered) to 12.9 million kilometers (Phoebe). All the satellites move directly around the planet, except the outermost (Phoebe), which moves retrograde.

Investigations reveal that Titan possesses an atmosphere. Methane and hydrogen have been identified, but probably other substances also exist there. The surface pressure is higher than the pressure at Earth's surface.

The American astronomer R. E. Murphy has found that some of the inner satellites have an albedo greater than 0.6, which is very high. Iapetus has one side with an albedo of 0.04 and the other side an albedo of 0.28. This difference is probably due to a sheet of ice on the surface. Because of their small mass, the satellites cannot hold atmospheres.

URANUS

In ancient times, the number 7 had mystical significance. It was appropriate that there should be seven special objects in the sky— the Sun, the Moon, and the five planets, Mercury, Venus, Mars, Jupiter, and Saturn.

In the early seventeenth century when Galileo saw the satellites of Jupiter, questions about the agreement of the celestial world with man's beliefs (and superstitions) were raised. In 1781 when Herschel discovered another planet, considerable consternation was the result. Now, when Earth was included, there were seven planets in the solar system. Equally amazing, the solar system was twice as large as believed previously—Saturn is 9.5 a.u. from the Sun, whereas the distance of Uranus is 19.18 a.u. (Figure 17–21).

Discovery

William Herschel, a professional organist and composer, became intensely interested in astronomy and found he was spending more and more time at it, especially after he moved to England from Germany. He was thoroughly familiar with the sky, having devoted himself to making star counts and charting star positions. One evening he noticed a disklike object among the stars. It was first reported to the Astronomer Royal as a comet. However, repeated observations by Herschel and others confirmed that the object was a planet. Apparently the object had been seen at least 20 times before Herschel saw it, but it was always thought to be a star, partly because it did not appear to move. Actually, its motion during a short time is slight, since it has a period of 84 years.

In loyalty to King George III, Herschel called the new planet "George's star." Fortunately the name did not stick. The German astronomer Elert Bode suggested the name Uranus, after the god that represented Heaven and was the husband of Earth.

FIGURE 17–21
Uranus, showing three of its satellites as photographed with the 120-inch telescope. (Lick Observatory)

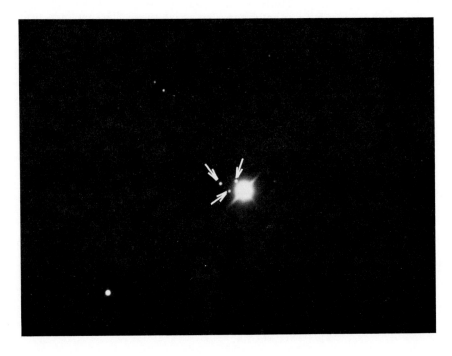

Herschel's discovery has sometimes been called accidental. It certainly is true that he was not looking for a planet when he sighted Uranus; however, when he saw the planet he sensed the significance of the observation. About the discovery Herschel wrote: "It has generally been supposed that it was a lucky accident that brought this star into my view; this is an evident mistake. In the regular manner I examined every star of the heavens, and it was that night its turn to be discovered. . . . I perceived the visible planetary disk as soon as I looked at it."

Motions

Uranus moves about 6.7 kilometers per second in its motion around the Sun, taking some 84 years to complete an orbit. Earth moves 29.6 kilometers per second. Like Jupiter and Saturn, Uranus spins rapidly on its axis, completing a rotation in just under 11 hours. At the Uranian equator, velocity is some 13 600 kilometers per hour.

Some observers report they have seen cloud bands in the atmosphere of the planet, similar to those seen on Jupiter and Saturn. Methane is the most apparent substance in the atmosphere, whereas molecular hydrogen is the most abundant. The temperature of the planet is very low, perhaps −200°C.

Tilt of the Axis

The tilt of the axis of rotation of Uranus to the plane of its orbit is unique in the solar system (Figure 17–22). Earth's axis is tilted 23½°; the axis of Uranus is tilted at almost a right angle. Curious effects result. For example, at times the north pole of the planet is pointed almost directly at the Sun. At this distance, the Sun has become quite ineffective as a light and heat source, although it is almost directly overhead at the pole. Forty-two years later, the north pole is turned away from the Sun continually (Figure 17–23).

An astronomy buff who knows where and how to look can see Uranus with the unaided eye. Chances are that the average observer has not, however, because its magnitude is 5.7. Generally, we say the limit of seeing without optical aid is magnitude 6.0; however, this assumes excellent vision, skill in observing, and exceptionally good sky conditions. The undesirable sky conditions that are the rule reduce the limit considerably.

Uranus has five satellites ranging from the smallest—Miranda (560 km across) to Titania (1760 km across). The two outer satellites, also the largest, were discovered by William Herschel in 1787. The innermost satellite, Miranda, which is some 123 000 kilometers from the planet's surface, was discovered by G. Kuiper in 1948.

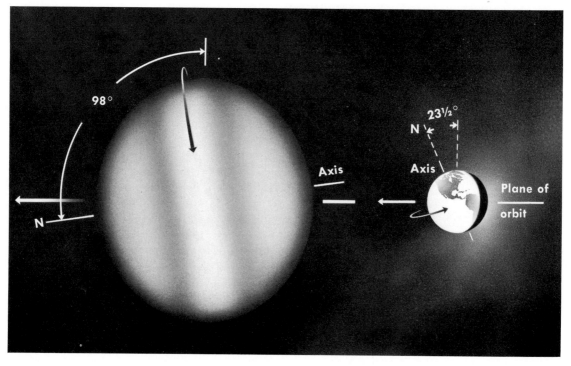

FIGURE 17–22   *The axis of Uranus is tilted at almost a right angle to the plane of the ecliptic.*

NEPTUNE

Neptune was discovered before anyone had seen it (Figure 17–24). Astronomers knew it existed because their calculations indicated that a considerable mass was affecting the orbits of the outer planets, especially the orbit of Uranus.

After three observations of a planet's position are available, an orbit can be constructed. Once this is done, predictions of future locations of the planet can be made.

For about 20 years after it was discovered, Uranus behaved as predicted. But by 1840, the planet was not at the location it should have occupied. The variation was not great, but no variation is allowable; there must be an explanation for any exception. Nearby planets, Jupiter and Saturn, affect Uranus, but even when the effects of these two planets were considered, the variation could not be explained completely.

FIGURE 17–23   Because of the tilt of its axis, Uranus experiences long periods of daylight and darkness. (Inner orbit—Earth)

FIGURE 17–24
Neptune and Triton
(Nereid in far upper
right) as photographed
with the 120-inch
telescope. (Lick
Observatory)

Adams and Leverrier—The Search for Neptune

Some people thought the laws of gravitation did not operate at great distances. Others believed there was another planet out beyond Uranus that was affecting the orbit. John Couch Adams, an English student recently graduated from college, decided in 1843 to search for the planet. By 1845 he had worked out a solution: if a planet of certain mass were located at a certain position, the observed perturbation of Uranus could be explained.

Adams requested Sir George Airy, Astronomer Royal, to search a given area of the sky for the elusive planet. Airy was not convinced. Adams was a young man, inexperienced in astronomy; the area of the sky he wanted explored was relatively uncharted, and there were probably other explanations for the behavior of Uranus. For almost eight months nothing happened.

Meanwhile, Urbain J. Leverrier, a French mathematician, attacked the same problem. In 1846 he published a paper which decribed the sky area where the unknown planet should be found. Sir George Airy read that paper and recalled that this was the same region Adams had asked him to explore. Hastily the region was searched, but somehow—perhaps because of faulty mapping—the planet was not identified.

Leverrier had written to Johann G. Galle, astronomer at the Berlin Observatory, asking him to search the sky. That same night the planet, later to be called Neptune, was observed.

At the time there was considerable controversy over whether Adams or Leverrier should be credited with the discovery. We generally favor both men, for both of them arrived at the same conclusion at essentially the same time.

The story of Neptune was a triumph for mathematical reasoning. It was the first time that a planet was known to exist before anyone had seen it.

Appearance

Neptune is 30 a.u. from the Sun and thus extremely cold, less than 60°C above absolute zero. Ammonia would be frozen solid, making methane and molecular hydrogen dominant in the atmosphere. Its magnitude is +7.6, making it far beyond unaided-eye visibility. Telescopically the planet is somewhat greenish-blue because its atmosphere absorbs the longer wavelengths. Like the other major planets, Neptune spins rapidly on its axis, completing a rotation in 16 hours. It takes 164.8 years to complete a revolution around the Sun.

Two satellites revolve around Neptune. Triton, the larger, is larger than the Moon and also quite close to the planet—some 350 000 kilometers away. It speeds around the planet in a bit under 6 days, at a velocity greater than 16 000 kilometers per hour.

The smaller satellite, Nereid, is probably less than 550 kilometers across and at a distance of 5.6 million kilometers from the planet.

PLUTO

The discovery of Neptune helped astronomers explain observed perturbations in the orbit of Uranus. But Neptune's effect was not sufficiently great to provide a complete explanation. W. H. Pickering and Percival Lowell believed there was another planet, one beyond the orbit of Neptune, that was affecting Uranus. At Lowell Observatory in Arizona, they searched for the planet until the death of Lowell in 1916.

Discovery

In 1929, Lowell's brother gave the observatory a telescope that could photograph a large section of the sky at one time. Also, a new tool called the blink microscope, or the *blink comparator*, had been invented. With this instrument two plates of the same sky region can be viewed rapidly one after the other by the operator. Stars on the two plates do not change position. However, a planet which has changed position in the interval between the two exposures appears to jump back and forth when seen on first one plate, then the other. At that time Clyde Tombaugh resumed the search, using the new telescope camera and the blink microscope. On February 18, 1930, he was studying photographs made on January 23 and 29 when an image jumped from side to side. It turned out to be planet number 9, and was so announced on March 13, 1930. It was named Pluto, and, quite appropriately, its symbol is made of the letters P and L, the initials of Percival Lowell.

Orbit

The orbit of Pluto turned out to be just about as predicted. However, the discovery itself may be one of the outstanding "accidents" in astronomy, for calculations reveal that the effects on Uranus that Lowell attributed to the new planet could not have been caused by Pluto; its mass is too small. The variations in the orbit of Uranus, which amounted to only 2 seconds of arc, probably resulted from errors in the computation of the effects of other planets on Uranus.

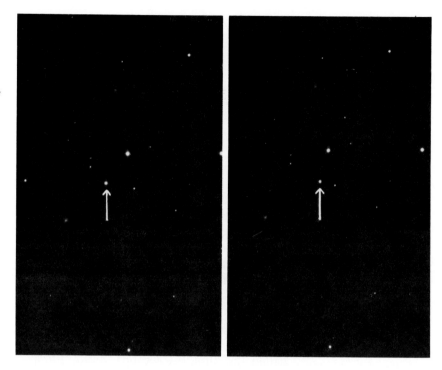

Sunlight takes 5½ hours to reach Pluto. From that distance the Sun appears as a bright star, the light received being equivalent to that falling upon Earth from 250 full Moons. The distance to Pluto is 39.44 a.u., ranging between 7.22 billion kilometers at apogee and 4.3 billion at perigee. When at perigee, as it will be in 1987, Pluto is closer to the Sun than Neptune. Because of the high eccentricity of the orbit (0.249) and the small mass and size, Pluto is considered by some investigators to have been a satellite of Neptune. While that may have been true at some stage of the evolution of Pluto, it now behaves as a full-fledged planet.

The orbital paths of the planets are in essentially the same plane—inclined by only a few degrees. Pluto, the exception, moves in a path inclined 17.2° to the ecliptic. This inclination makes it impossible for a collision to occur when Pluto passes within the orbit of Neptune. At the closest, the two planets will be half a billion kilometers apart (Figure 17–26).

Features

Because of its great distance, Pluto is extremely dim; the faintest visible stars are thousands of times brighter. Its magnitude is +14, which means that at least a 10-inch telescope is needed to detect its presence. Its size and mass have been calculated by various investigators, and there is a considerable range in results. The bulk of data seems to point toward a diameter between that of Mercury and Mars, and a mass of perhaps 0.11, about the same as the mass of Mars.

An atmosphere is not likely. The planet is probably an intensely cold (−250°C), dead, solid world, perhaps even ice-covered, as some suspect.

Careful measurements of the brightness of Pluto reveal a regular cycle of changes that occur at intervals of 6.387 days. The changes are probably the result of different faces of the planet being turned toward Earth; the changes in reflecting ability of the faces (due to variations in surface features) account for the observed brightness variations. This is the best observation presently available that enables one to draw conclusions about the rotation period of the planet.

It is questionable whether Earth would ever be observed from the bleak, solitary platform of Pluto. Our planet would never be more than 1° from the Sun; so it would be quite impossible for instruments to separate it from the solar glare.

After 1930, Clyde Tombaugh continued his sky searches for several years, seeking other planets that might help explain orbital perturbations of the outer planets. None were found, however. Astronomers generally do not feel it necessary to continue the search, for they feel there are explanations within the solar system for the perturbations that they observe.

REVIEW

1   There are nine planets in the solar system: the four major planets, Jupiter, Saturn, Uranus, and Neptune, and the five more dense terrestrial planets, Mercury, Venus, Earth, Mars, and Pluto.

2   Mercury, the smallest planet and nearest to the Sun, is difficult to observe; however, Mariner 10 showed it to be a crater-covered world surrounded by a tenuous atmosphere of rare gases.

3   Venus, studied intensely by Earth-based and Mariner-based instruments, is a hot, carbon-dioxide-covered world that rotates retrograde in 243 days.

4   Mariner studies of Mars reveal it is a crater-covered world that may have had water at one time; however, no signs of

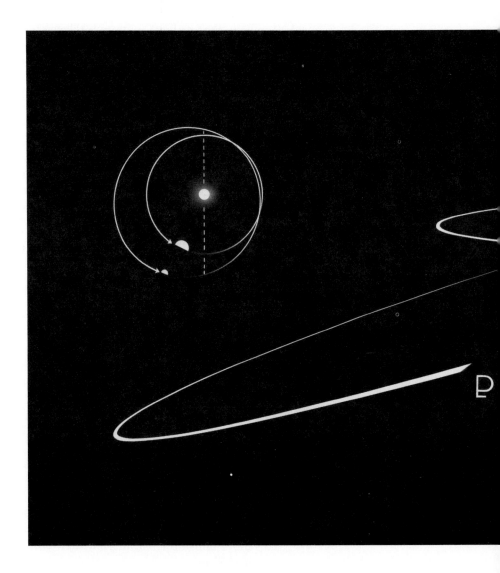

FIGURE 17–26
When at perigee,
Pluto is closer to the
Sun than is Neptune.

life have been revealed.

5 Jupiter, the largest planet, contains considerable free hydrogen and hydrogen compounds, has an extensive radiation
field, gives off more energy than it receives, and may be a
planet upon which life is emerging.

6 Saturn's rings are a distinctive feature of the planet. It has
ten satellites and colored cloud bands.

7   Uranus, the first planet to be discovered, and found by William Herschel in 1781, doubled the size of the solar system.

8   Neptune's presence was predicted by Adams and Leverrier before its discovery in 1846.

9   Pluto, discovered by Clyde Tombaugh in 1930, is presently the outermost planet of the solar system, but it will be inside the orbit of Neptune in 1987.

QUESTIONS

1 The rotation period of Mercury (59 days) is two-thirds of its period of revolution (88 days). How would these circumstances lead observers to believe that the periods of rotation and revolution were the same?

2 The synodic period of Mars is 780 days, but its sidereal period is 687 days. Explain why this can be so.

3 The mass of Saturn is about 95 times greater than the mass of Earth, yet its surface gravity is only 1.13 that of Earth's. Explain why this is so.

4 The perihelion distance of a planet is the product of its major axis and 1 less its eccentricity; and the aphelion distance is the product of its major axis and 1 plus its eccentricity. Find the aphelion and perihelion distances of Pluto with a major axis of 39.52 a.u. and an eccentricity of 0.25. Find the aphelion and perihelion distances of other planets. (Use the data in the table on pages 434–435.)

5 Define synodic period and sidereal period. Diagram the synodic and sidereal periods of the Moon.

6 When Galileo observed the phases of Venus, he was certain that Venus went around the Sun. Why could he draw this conclusion? Diagram if necessary.

7 Occasionally when Venus is between Earth and the Sun (inferior conjunction) it can be observed. Explain why this can be so. How often does it occur?

8 When Mercury and Venus are at eastern elongation, they are visible in the west; and when at western elongation, they are visible in the east. This sounds contradictory. Explain, using a diagram if needed.

9 The *World Almanac* provides tables of the planets which tell when and where they are visible. Check the data, and try to locate each of the unaided-eye planets.

10 In terms of its phases, what is the most opportune time for observing Venus? Why is this so?

11 How does the "greenhouse effect" alter the temperature of Venus? of Earth?

12 What are some of the conditions on Venus that astronomers find elude adequate explanations?

13 Mariner 9 produced revealing photographs of the Martian surface. What do these photographs imply about (a) the evolution of the planet and (b) its present condition.

14 Why is Jupiter sometimes referred to as the planet of emerging life?

15 With a low-power telescope, observe the four Galilean satellites of Jupiter and plot their changing positions. Check your data against

the *Observer's Handbook* (publication of the Astronomical Society of Canada, 252 College Street, Toronto 130, Canada).

16 Explain the nature of Saturn's rings and the manner in which they might have come into existence.

17 What are the various radiations in the radio range received from Jupiter, and what are explanations for them?

18 The story of the planets is replete with famous names in astronomy. You might find it of interest to recall contributions of Ptolemy, Copernicus, Brahe, Kepler, Newton, Galileo, Airy, Leverrier, Adams, Galle, Flamsteed, Halley, Lowell, Tombaugh, and Herschel, to mention a few.

READINGS

Asimov, Isaac. *Jupiter, the Largest Planet*. New York: Lothrop, Lee & Shepard Company, 1973.

Kuiper, Gerard, ed. *Planets and Satellites*. Chicago: University of Chicago Press, 1961.

Sky and Telescope. "Venus Observed by Mariner." *Sky and Telescope,* April 1974.

Watts, R. N. "Pioneer Observes Jupiter." *Sky and Telescope,* February 1974.

SPACE DEBRIS

There are probably occasional solitary stars between galaxies; atoms and molecules appear in space between stars; and in space between the planets there are tons of dust and gases. In the universe, "empty" space does not exist. Although the density of material in interstellar and intergalactic space is extremely low, its volume is so great that the total amount of material is impressive.

On the first night of the nineteenth century the first asteroid was discovered, confirming the beliefs of Johann Titius and Elert Bode that the space between Mars and Jupiter was not empty.

Subsequently, the orbits of some 1500 asteroids have been

The asteroids move in the vast space between Mars and Jupiter. The orbits of some 1500 of them have been determined. While there are probably thousands of them, altogether they would make a sphere of only some 800 kilometers.

determined. Undoubtedly there are thousands more of them. They may be smaller concentrations of material that consolidated some 5 billion years ago but that never added up to an amount great enough to be regarded as a planet. In this chapter, we will deal with many such smaller concentrations of matter, including comets, meteoroids, and tektites as well as asteroids.

ZODIACAL LIGHT

Every day several thousand metric tons (perhaps 20 000 or more) of cosmic dust—micrometeorites—fall upon Earth. While the amount is tremendous, it does not significantly alter the much greater mass of Earth, which is 6×10^{24} kilograms. We are not aware of this downfall; most of it enters the sea, and the rest is spread sparsely across the countryside. However, under good conditions, the "dust" that apparently surrounds Earth can be seen as a soft glow in the skies of twilight. It is called the *zodiacal light* because it appears in the zodiacal belt, the belt around the sky through which the Sun appears to move. On rare occasions it has been seen, and photographed, during total solar eclipses. This was first done by Charles H. Smiley 4500 meters up in the Peruvian Andes during the eclipse of 1937. In the tropics, and occasionally at other locations, the light can be seen as a soft cone-shaped glow along the horizon, originating at the location where the Sun was last seen. Similarly, under good seeing conditions, the glow can be seen along the eastern horizon immediately preceding sunrise (Figure 18-1).

In the tropics, the belt of the zodiac is vertical to the horizon, or nearly so, throughout the year. At higher latitudes, the belt is often almost parallel to the horizon. When it appears, the zodiacal light is discernible for only about an hour, and after the Sun is some 18 degrees below the horizon. Therefore, in upper latitudes there is not sufficient contrast for the light to be visible. Generally, the zodiacal light is a tropical phenomenon.

False Dawn

The brightness of the zodiacal light is about equal to the dim parts of the Milky Way. Therefore, it can be seen only when the dimmer parts of the Milky Way are above the horizon. This would be in springtime. Earliest reference to the light pertains to its morning appearance. In many countries religious festivals were timed to begin at sunrise. Apparently ancient priests often mistook the zodiacal light for the glow of sunrise (which occurs about one hour later)

FIGURE 18–1
The zodiacal light appears as a cone of soft light. It extends into the sky from the location of sunset (and sunrise) as shown in the inset.

and so erred in timing the ceremonies. For this reason, the zodiacal light of morning was called *false dawn*. It is still referred to in this way in some parts of the world.

In the seventeenth century, Giovanni Domenico Cassini (1625–1712), the Italian who later became a French citizen, made some conclusions about the zodiacal light. One of his assistants, Niccolo Fatio, made several observations of the light and reported his sightings to Cassini. It occurred to Cassini that the light might be caused by sunlight reflecting from small particles moving in an orbit around the Sun. This explanation has held up through the centuries and is still widely accepted (Figure 18-2).

If one were to see the zodiacal light, it would appear as a soft haze, cone-shaped, and tapering quite steeply to the tip. Experienced observers have occasionally noted that the evening cone and the morning cone appear to be joined together by a nebulous band. If this is so, the particles that produce the light must extend far beyond Earth.

FIGURE 18–2   *The zodiacal light may be produced by sunlight reflecting from microparticles that are in orbit about Earth. The dashed line is the observer's horizon.*

Gegenschein

The presence of the gegenschein is another observation supporting the belief that the particles extend beyond Earth. The gegenschein is a faint glow sometimes seen directly opposite the Sun's location—the antisolar point. It was first reported by the German explorer Alexander von Humboldt (1769–1859), who said he had seen the zodiacal light and also the counterglow of the Sun. Humboldt used the German word *Gegenschein* for "counterglow," and the name has persisted. Probably a band of tiny particles extends a million kilometers or so beyond Earth. Each of the particles directly opposite the Sun would catch light and reflect it back to Earth, much as the full Moon does. Particles at other locations would also reflect light, but scattering would make it so diffuse it would not be seen by an Earth-bound observer.

The minuscule particles that produce the zodiacal light, and very likely the gegenschein too, are arranged in a doughnut-shaped band that tapers off as distance increases. Close to the sun there would be no particles because the temperature is high enough to vaporize them.

The apparent size of the particles poses a mystery as to their origin. Small grains are not heated evenly by the Sun; the Sun side becomes hot while the side away from the Sun remains cool. The temperature spread produces a shift in the center of balance of a particle, causing a slight drop in velocity which results in a motion slightly inward toward the Sun. Over and over the process is repeated, causing the particles to follow a spiral rather than elliptical path.

It has been estimated that large particles—a centimeter across —would spiral into the Sun in about 20 million years. Smaller particles would last only about a million years. At this rate, after only a few million years, space should be swept clean of small particles, and only the large ones should remain. But the lights we are talking about are produced by very small fragments. Could they have been produced only a few million years ago? If so, where did they come from?

Origin

Comets are possible sources. The head of a comet contains particles of many sizes imbedded in frozen gases. Should a comet approach a planet, such as Jupiter, the comet head is pulled apart to some degree. The gases evaporate, and the disconnected solid particles continue in orbit around the Sun. They may diffuse widely and replenish the cloud of particles responsible for the zodiacal light.

Some contend that, while comets may be the source of some of the interplanetary dust, asteroids may also contribute. Occasionally there must be collisions between asteroids which produce small particles and dust. Most of the debris would remain in the same orbit that the asteroids were in originally; but it is reasonable to suppose there might be diffusion into other regions, and organization into the particle bands believed to exist.

COMETS

It has been said that comets are as near to nothing as anything can be and still be something. The description refers to the tail of a comet, which is so nebulous that one could pass through without being aware of its presence.

Until the latter part of the seventeenth century, comets were thought to originate spontaneously. However, in 1680 Isaac Newton determined that a comet which appeared that year moved in an elliptical orbit, which meant it had a regular period. Edmund Halley, who was to become Astronomer Royal, made the next step. His analysis

FIGURE 18–3   The orbit of Halley's comet is tilted some 17 degrees to the plane of the solar system. The ascending node, the point where it rises above the plane, lies just inside the orbit of Mars. In 75 years the comet completes an orbit.

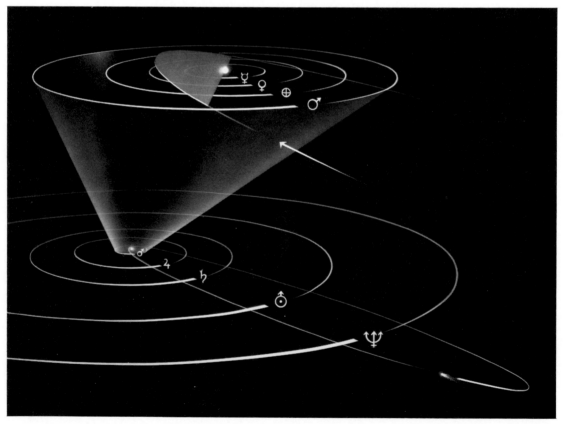

of information available concerning the comets that appeared in 1531, 1607, and 1682 implied that all three were related; indeed, they were reappearances of the same comet. He predicted that the comet would make another appearance 75 years later. Other mathematicians studied Halley's figures and agreed with his conclusions. However, they maintained that the gravitation of Jupiter and Saturn would slow the comet by 618 days. They concluded that the next appearance would be April 1759 (Figure 18-3).

One month earlier than predicted—March 12, 1759—the comet was seen. Unfortunately, Halley did not live to see confirmation of his belief. Now there was no question: comets were a part of the solar system; they moved in elliptical orbits around the Sun. A few comets, some of them listed in Table 18-1, have been seen over and over again.

Table 18-1   *Periodic comets observed 10 or more times*

Comet	Period (years)	Closest approach (a. u.)	Number of observations
Encke	3.30	0.338	45
Tempel II	5.30	1.391	12
Pons-Winnecke	6.12	1.159	15
D'Arrest	6.70	1.378	10
Faye	7.41	1.652	14
Halley	76.03	0.587	29

Periodic comets move in elliptical orbits, while those that appear only once seem to move in parabolic or hyperbolic paths. When the comet is at perihelion, it is difficult, if not impossible, to distinguish a parabola from a very long ellipse.

Naming

Comets are named after their discoverers. Quite often different observers discover them simultaneously. In such cases, the comet bears two names, or even three. For example, we have Comet Ikeya 1963a —the "a" meaning it was the first comet discovered in 1963. After the discovery year is completed, comets are numbered in the order in which they pass through perihelion, the closest approach to the Sun. For example, Ikeya-Seki 1967n (the fourteenth comet discovered in 1967; it was found by two Japanese, Kaoru Ikeya and Tsutomu Seki) became Ikeya-Seki VII, the seventh comet to go through perihelion.

It used to be that most comets were discovered by amateurs.

since careful time-consuming scanning of the sky was required. When professional astronomers have access to a telescope, they are concerned with some particular research problem and so cannot engage in sky scanning. However, now wide field telescopes photograph large segments of the sky. When the photographs are scanned, which can be done at any time, comets show up clearly. In this way, seven or eight new comets are discovered on the average every year. Some 2000 comets have been observed and photographed. But there are probably millions of them in the solar system; some astronomers say there may be 100 billion.

Most comets are not exciting. They appear as fuzzy stars. Only occasionally does one get close enough to the Sun to develop a tail (Figures 18–4 and 18–5).

FIGURE 18–4   *As a comet approaches the Sun, gases stream from the head. The gases always point away from the Sun, pushed by the solar wind. The gases are so nebulous that Earth may pass through them, as happened in the 1910 visit of Halley's comet, without any noticeable effect.*

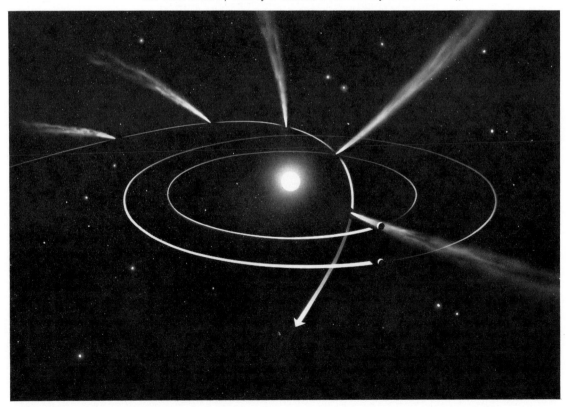

FIGURE 18–5
The apparent length of the gas streamers (often referred to as the "tail," in spite of the fact that they precede the comet head as much as they follow it) depends upon location and direction of motion. The tail at left appears larger than the one at right (as shown by the double-headed arrows), but the opposite is true.

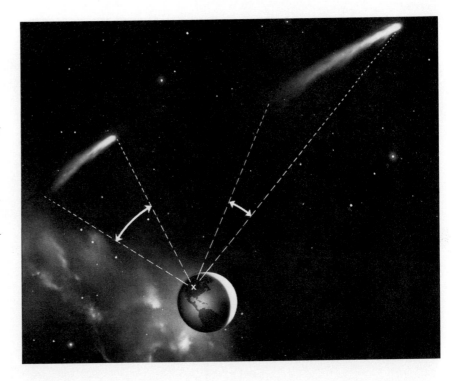

Composition

At one time comets were called "flying gravel banks." However, some years ago Fred Whipple of Harvard Observatory suggested that comets might better be called "dirty snowballs." The head of the comet is probably made of frozen hydrogen, helium, carbon, nitrogen, oxygen, and simple compounds of these substances. Imbedded in the snow and ice would be larger particles, and dust, substances that do not vaporize readily. As the comet gets nearer the Sun, heat causes some of the frozen gases to vaporize. Solids held in the "snowball" would be set free and might eventually produce spectacular meteor showers. Ultraviolet radiation would break down complex molecules such as ammonia (CH_4) and methane (NH_3) into simpler fragments—C_2OH, C_3, CH_2, CN, NH. The latter have been identified spectroscopically, whereas the more complex molecules have not. However, observations in 1974 of Comet Kohoutek at radio wavelengths revealed the presence of methyl cyanide (CH_3CN) and hydrogen cyanide (HCN).

Occasionally bright gases are seen jetting from comet heads in directions opposite those of the tail. If comets are indeed "dirty snowballs," the jets can be explained (Figure 18–6).

FIGURE 18–6   Comet jets (antitails) may result because of rapid
expansion and ejection of gases, or they may be due simply to perspective.
Gases detached from the mass lie behind the comet in the plane of the
orbit. These gases, reflecting sunlight, appear to be connected to the comet
itself.

Porous layers of solids may form a crust around the icy material. When the comet nears the Sun, radiation may penetate the
crust here and there, vaporize a local region of ices, and the gases
that are released may jet explosively through the opening. The crust,
if it exists, may also prolong the life of a comet by reducing the rate
of evaporation. It would seem that solar gravitation would destroy
a comet in a relatively few years; yet, as we know, some comets last
for thousands of years.

The snowball theory is an attempt to describe the nucleus of a
comet: a part that we do not usually observe. The main parts of a
comet are the *tail*, the *nucleus*, which is made of the frozen materials
and dust mentioned above, and the *coma*, which is hazy glowing
material surrounding the nucleus.

The nucleus of a small comet may be only a few kilometers or so across, but the coma may extend into space for thousands of kilometers. The nucleus of Halley's comet is believed to be about 30 kilometers in diameter, but its coma is several thousand kilometers across. During its last approach in 1909–1910, the tail extended 250 million kilometers.

The Source of Comets

All comets belong to our solar system, and periodic comets move in elliptical orbits around the Sun. It appears that comets which we observe may originate from a region 50 000 to 150 000 a.u. from the Sun. There may be a vast cloud of comets out there—halfway to the nearest stars—that are moving in essentially circular orbits. Occasionally a comet is deflected and moves toward the Sun. Should it continue to move toward the Sun it might become visible. Perhaps only one in a hundred thousand would ever go through such a history. There may be 100 billion comets in the comet cloud. Yet comets are so nebulous that all of them together may add up to a mass only one-tenth that of our own planet.

The carbon, nitrogen, oxygen, iron, and other materials in comets may be material left over from the original great cloud from which the Sun and planets were formed billions of years ago. There appears to be a virtually endless supply of them, for each year new comets are discovered and each year comets go out of existence, victims of the shattering effects of the Sun.

METEORS AND METEORITES

As a comet disintegrates, the solid particles are spewed into space. But since each particle is affected by the Sun's gravitation, the particles remain in orbit around the Sun. In some cases the particles are spread evenly throughout the path of the comet; and in others they are clumped together into clusters.

When Earth passes through the path of a comet, the particles are intercepted and meteor showers are produced. Should Earth intercept comet debris that is clumped, the resultant shower is extremely heavy—as many as a thousand "shooting stars" per minute are seen. This is what happened with the Leonid meteor shower of 1833, and again in 1966.

Meteors, streaks of light caused by the passage of particles through our atmosphere, occur every night of the year. With good seeing conditions, a team surveying the whole sky should be able to see five or six of these sporadic, unpredictable shooting stars in an hour. However, on certain nights many more are visible. These are the recurring meteor showers, some of which are listed in Table 18-2, together with the comet with which the shower is associated.

We notice that the showers often take their names from constellations (Perseids, Draconids, Leonids) or from particular stars (Eta Aquarids, Alpha Capricornids). (The suffix *id* is from the Greek and means "daughters of.") This is because the meteors seem to emanate from that particular region of the sky—the radiant of the shower.

Table 18-2   *Major meteor showers*

Shower	Date of maximum	Duration (days)* of peak activity	Hourly** rate	Remarks
Quadrantids	January 3	0.5	40	Permanent annual.
Lyrids	April 21	2	15	Permanent annual. Associated with Comet 1861 I.
Eta Aquarids	May 4–6	18	20	Permanent annual. Possible association with Halley's comet.
Arietids	June 7	20	60	Permanent annual; daytime.
Zeta Perseids	June 9	15	40	Permanent annual; daytime.
Beta Taurids	June 29	10	20	Permanent annual; daytime. Associated with Encke's comet.
Delta Aquarids	July 29	20	20	Permanent annual.
Perseids	August 12	5	50	Permanent annual, associated with Comet 1862 III.
Giacobinids	October 10	0.10	100	Periodic, associated with Comet Giacobini-Zinner (1946 V).
Orionids	October 21	8	25	Permanent annual. Possibly associated with Halley's comet.
Taurids	November 3–10	30	10	Permanent annual. Associated with Encke's comet.
Leonids	November 16	4	15	Permanent annual. Also periodic (33 years) with greater strength. Associated with Comet Tempel-Tuttle (1866 I).
Phoenicids	December 5	0.5	50	Newly discovered (1956), possibly periodic.
Geminids	December 13	6	50	Permanent annual.
Ursids	December 22	2	15	Discovered in 1945. Permanent annual. Associated with Comet Tuttle (1939 X).

* In which the number of meteors will be ¼ or more of the rate at maximum.
** In a single shower, on a Moonless night, looking in the direction of the zenith.

Actually *meteoroids*, the small particles that produce meteors, stream through space parallel to one another. They appear to stream from a point only because of illusory effects, as shown in Figure 18–7. The situation is similar to the manner in which railroad tracks seem to converge as distance increases.

The larger part of meteoric (or cometary) debris is microscopic. It falls to Earth without producing any noticeable light, or it may fall during daylight hours, and so we are not aware of it. It is difficult to determine the total amount that falls upon Earth during 24 hours, but, as mentioned earlier, 20 000 metric tons is not unreasonable.

The number of shooting stars that may be anticipated during a given shower is indicated in the table. In actual practice, the number may vary considerably from the prediction. During its journey, a

FIGURE 18–7 *Meteors appear to radiate (solid arrows on a plane at right angles to the viewer) from a point in the sky called the radiant (heavy dashed line). (To illustrate the phenomenon, we show the events occurring far from Earth. Actually, meteors appear in Earth's atmosphere.)*

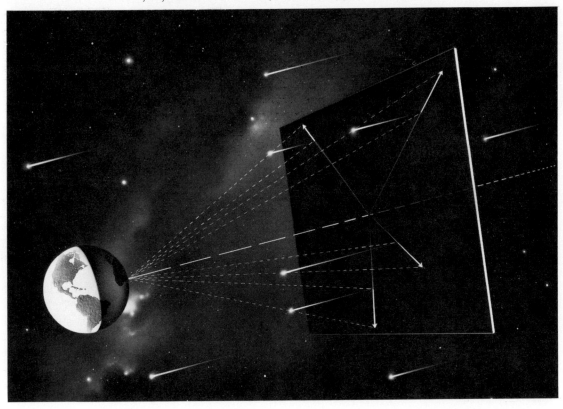

meteoroid swarm may be deflected from its orbit; it may be speeded up or slowed down; it may even disappear completely.

In 1966 the Leonid shower, which occurs annually, was expected to be especially spectacular, just as sensational as it had been in 1833. This is because every 133 years, or thereabouts, Earth passes through that portion of the path of Temple's comet where the debris is heavily concentrated. In anticipation of the event, during the hours before sunrise newsmen and astronomers flew figure-eights high above the eastern coast of the United States. But only a handful of meteors were seen. With considerable disappointment, the flight was terminated just before daybreak. At just about that hour, it was 5:00 A.M. in Denver, Colorado. There, against a black sky, observers saw the most awesome display of the century. At its peak they could see 1000 shooting stars in one minute—the sky was literally filled with light streaks.

Time for Viewing

The best way to hunt for meteors is with a team of observers, so that the whole sky may be observed simultaneously. Also, the best time to view them is after midnight. The velocity of meteoroids through space is 42 kilometers per second. Earth's velocity through space is 29 kilometers per second. In the hours before midnight, an observer is moving away from oncoming meteoroids, so their effective velocity is 12.5 kilometers per second—the difference between Earth's velocity and that of the particles. After midnight, it is possible for the observers to move toward the particles, so the velocity may be much greater—some 70 kilometers per second. The higher velocity meteoroids produce brighter "shooting stars." Also, more of them are intercepted by Earth just before sunrise. Earth catches up to them.

Meteorites

When a meteoroid is large enough to penetrate our atmosphere, and all, or part, of it lands upon the surface, it is called a *meteorite*. Similarly, objects that fall upon other planets or satellites are also called meteorites.

Until the beginning of the nineteenth century scientists believed it impossible that "stones could fall from the sky." In 1794 a German physicist, Ernst Chlandri, reported that he believed stones and metals he had seen had come from outside Earth. Later, on April 26, 1803, it was reported that stones had been seen falling from the sky in the city of L'Aigle, France. The French Academy directed Jean Baptiste Biot to investigate the report. After questioning witnesses and searching out some 2000 pieces of stone, the largest of which weighed 44 kilograms, Biot was convinced that the stones really had fallen from the sky. Thus began the serious study of meteorites, which, until Moon rocks were brought to Earth, were the only extraterrestrial materials man had ever handled. No one knows how many fall upon Earth every day; however, the number must be extremely high. Most are small, but on rare occasions they are very large. Falls in excess of 9 metric tons are listed in Table 18–3.

Table 18-3   *The largest meteorite finds,*
each over 10 tons (1 ton = 2000 lb)

Hoba West	Grootfontein, Bechuanaland, S. W. Africa (estimated 60 tons), one mass.
Ahnighito	Cape York, Greenland (34 tons). Found along with the "Woman" (3 tons) and the "Dog" (900 pounds). (On display at the American Museum–Hayden Planetarium.) A fourth meteorite (Savik) weighing 3.5 tons found in the same region. On display in Copenhagen.
Bacubirito	Senaloa, Mexico (estimated at 30 tons), one mass.
Santa Catharina	Brazil (estimated at 25 tons), several masses.
Chupaderos	Chihuahua, Mexico (23 tons), two masses.
Xiquipilco	Ixtlahuaca, Mexico (estimated at 20 to 25 tons), many masses.
Meteor Crater (Canyon Diablo)	Arizona (estimated at 20 to 2000 tons), many masses.
Bethany	Great Namaqualand, South Africa (16.5 tons), 51 masses.
Willamette	Willamette Valley, Oregon (14 tons), one mass on exhibit at the American Museum–Hayden Planetarium.
El Morito	Chihuahua, Mexico (11 tons), one mass.

Meteorites are not associated with comets. They appear to originate in the asteroid belt, the region between the orbits of Mars and Jupiter. Among the famous meteorites is Ahnighito, which is on display at the American Museum–Hayden Planetarium in New York City. Because of its weight, slightly over 30 metric tons, the task of putting it aboard ship, unloading it, and moving it to the museum was formidable. However, Admiral Peary finally succeeded. After several years and many aborted attempts, he moved it from Cape York, Greenland, to the navy yard in New York City. After seven years it was finally drawn by horses to the Museum of Natural History (Figure 18–8).

FIGURE 18–8
Ahnighito (32 438 kg),
on display at the
American Museum-
Hayden Planetarium,
is the largest
meteorite ever
recovered.

Canyon Diablo. Perhaps the most famous meteorite crater is the Canyon Diablo crater in Arizona. The formation is commonly called the Winslow Crater, or the Barringer Crater; but Canyon Diablo is the name of the nearest post office, and traditionally that is the source of the name given to a new-found meteorite.

It is believed that some 20 000 years ago a meteoritic mass about 15 meters across and weighing at least 50 000 metric tons (estimates go as high as 2 million tons) crashed into Earth, generating enough heat to produce a mammoth explosion. The meteorite itself was shattered, some of it being buried beneath the floor of the crater —a huge hole over a kilometer across and almost 200 meters deep.

A mining engineer by the name of Barringer figured that there must be a tremendous amount of rich iron ore under the crater. He sank shafts and did strike iron, but the deposit was not rich enough for Barringer to continue, so the project was abandoned. However, fragments of nickel and iron totaling some 1800 metric tons have been found in the crater and in the region surrounding it. The largest single fragment weighs 635 kilograms (Figure 18–9).

FIGURE 18–9
The Canyon Diablo
crater near Winslow,
Arizona, is about 1.5
kilometers across and
180 meters deep.

Tunguska. Another famous meteorite event occurred in Siberia in 1908. Because of the effects caused by the fall, many investigators believe that the object was not a meteorite, but rather the head of a small comet.

Witnesses saw an extremely large, bright fireball that streaked across the sky, and seconds later they heard a thunderous explosion. According to reports, a witness close to the center of the explosions was deafened by the loudness. In a few seconds, a herd of 1500 reindeer disappeared, except for a few charred carcasses that were not vaporized. Earth vibrations were recorded 4800 kilometers away, and the sounds were heard for 1400 kilometers.

When the area was investigated, no meteorites were found. However, there were at least 100 craters of various sizes, up to 100 feet (about 160 kilometers) across. In an area 6 kilometers across, the trees had been laid flat on the ground, their tops pointing away from the center. If the object were a comet, the tremendous heat and pressure wave could be accounted for. Also, a comet would leave little if any trace, for the fragments composing it would be small, and much of it would have been ice that would have melted and evaporated. The Tunguska fall of 1908 continues to intrigue investigators, for a full and adequate explanation is still forthcoming. Some have proposed that the object was not a comet, or a meteorite, but a black hole, one that went right through Earth and emerged in the North Atlantic.

Classification. Meteorites can be classified into three main groups, depending upon their composition. The first are the irons, or *siderites.* Although iron is the principal ingredient (90 to 95 percent), other materials are also found, especially nickel.

The second group are the *siderolites,* a mixture of metal and stone. They are found only occasionally, perhaps because the stony material breaks down rather rapidly, leaving the more resistant metal.

The third group are the stony meteorites, or *aerolites.* The material in them is mostly silicates; however, iron and nickel may also appear, though the amount is only 10 or 15 percent of the total. These are the most common of all falls, though they are rare in collections. Meteorites seen to fall and so located immediately most often prove to be stony. However, unless recovered at once, stony meteorites may never be identified, for they look much like terrestrial rocks; they also disintegrate fairly rapidly through natural processes of erosion.

Only an expert can positively identify a meteorite. In the metals, a sure clue is the presence of *Widmanstaetten figures,* or lines, named after A. B. Widmanstaetten, a Viennese scientist who first described them in 1808. When a face of an iron meteorite is cut, polished, and etched, patterns such as those shown in Figure 18–10 appear. Apparently very slow cooling, a degree in a million years, causes the metal crystals to line up in this fashion.

All recovered meteorites, regardless of type, are associated with sporadic falls, not with meteor showers. However, spectroscopic studies of the light observed during showers reveals that the meteoroids producing them are made of essentially the same materials as those that fall on Earth.

Origins. Perhaps meteoroids originate among the *asteroids*; in fact, they may be asteroids. They are among the oldest things we have ever dated—going back some 4.6 billion years, to the beginnings of the solar system itself. Asteroids may have been quite large originally, and extremely hot, so hot that the stony materials and metals melted and separated because of differences in density. During the next billion years or so, the asteroid may have cooled slowly, producing the Widmanstaetten figures we now observe. A cooling rate of a degree in a million years would be reasonable if the asteroid were a few hundred kilometers in diameter. At some stage, collisions may have occurred between asteroids, or between an asteroid and a comet. The pressure at impact may have been great enough to

produce the diamonds that are occasionally found in meteorites. Also, the collision which produced shattering would result in the fragments' following new orbits. Attraction by other asteroids and planets would also modify orbits. In any event, the orbits were changed so that Earth intercepted them. We expect that the impacts by meteorites during the evolution of Earth were much more numerous than present evidence indicates. Impact craters would have long since disappeared—filled by glacier action, eroded away, and perhaps buried beneath crustal plates. A considerable portion of Earth's mass must be made of meteoritic materials that impacted on our planet, just as they did on the Moon, on Mercury, and on Mars— a fact we know well because of the presence of numerous craters that have persisted at those locations because of the absence of agents of erosion.

ASTEROIDS

In the eighteenth century people believed there were planets in the solar system that had not yet been discovered. They were not concerned with what might lie beyond Saturn—which was, so they thought, the boundary of the solar system. But their knowledge of the arrangement of the planets out to Saturn implied that there were unfilled gaps. Very likely these gaps held planets, could they but find them.

In 1741 the idea was reported in a book written by Christian Freiherr von Wolff. About a decade later, Johann David Titius (1729–1796) of Wittenberg published an analysis of the distances of the planets from the Sun and their relationship to one another. He pointed out that the distances of the planets from the Sun were "in proportion to their increase in size. . . . But now notice that from Mars to Jupiter there is a deviation of this exact proportion . . ." Titius suggested that the space between Mars and Jupiter might contain outer satellites of Mars or, perhaps more likely, of the planet Jupiter.

One of those who read the works of Titius was Johann Elert Bode (1747–1826), who was director of the Berlin Observatory. Bode was impressed by the work of Titius and at a meeting held in 1796 encouraged his colleagues to search the area between Mars and Jupiter systematically. He felt there was a planet to be found—not merely satellites of planets. However, the first object discovered in that region is credited to none of the delegates who attended that meeting, but to Guiseppi Piazzi (1746–1826), who was director of a small observatory on the island of Sicily. Piazzi had been correcting a catalog of star positions, and on the night of his discovery (January 1, 1801), he happened to be observing the Taurus region. He saw a "star" that was not listed in the catalog. The next night he saw it again, but in a slightly different location, so the object could not be a star. To Piazzi it appeared to be a comet, but when Bode heard about the discovery he believed it to be a planet.

It so happened that Piazzi became ill before his observations were completed. The next time he was able to observe, he could not find the object. Since no orbit had been worked out, no other astronomer could find the object either. Fortunately, Karl Friedrich Gauss (1777–1855) was able to compute the orbit of the object from the slim data Piazzi had assembled. One year after the discovery, and after the object had moved into a favorable viewing position, the object was sighted. It was given the name Ceres, the ancient Roman goddess of agriculture.

In the next few years three more asteroids were discovered, two by an amateur astronomer named Wilhelm Olbers, and one by Karl Harding. Information about these asteroids is given in Table 18–4.

Table 18–4 *The four largest asteroids*

Name	Discovered	Diameter (kilometers)	Brightness (magnitude)
Ceres	Piazzi	765	7.40
Pallas	Olbers	487	8.00
Juno	Harding	189	8.70
Vesta	Olbers	377	6.50

Number

No discoveries were made in the next several decades. But discoveries came rapidly toward the end of the century when photography became an important tool of the astronomer. Asteroids, even dim ones, show up as light streaks on star-field photographs. Presently, about 1600 asteroids have been studied, their orbits have been worked out, and their appearance at a given location can be predicted. There are probably a good many more. Some believe that 40 000 to 50 000 are within range of a 3-meter telescope, which can photograph objects down to magnitude 21. Asteroids which come closest to Earth are listed in Table 18–5.

Table 18–5 *Asteroids which come nearest Earth*

Asteroid	Discoverer (year)	Closest approach (million kilometers)	Period (years)	Diameter (kilometers)
Eros	Charlois and Witt, 1898	22.20	1.76	32.0
Amor 7	Delporte, 1932	16.60	2.67	2.4
Icarus	Baade, 1949	6.40	1.10	0.8
Apollo	Reinmuth, 1932	4.00	1.81	1.6
Adonis	Delporte, 1936	1.90	2.76	1.6
Hermes	Reinmuth, 1937	0.76	2.00	1.2

It has been suggested that if all the asteroids were put together in a single ball, the diameter of the sphere would be about 800 kilometers, hardly enough to make a planet, as suggested by Bode, or even a respectable satellite, as suggested by Johann Titius (Figure 18–11).

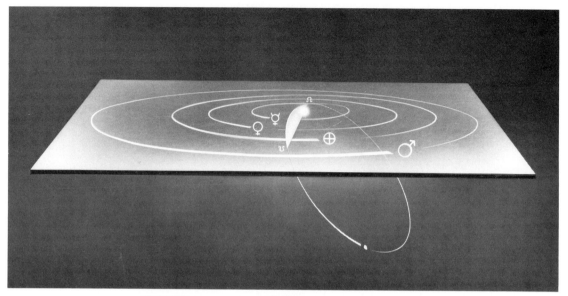

FIGURE 18–11 *At perihelion, Icarus is closer to the Sun than Mercury; at aphelion, it is between Earth and Mars. Every 19 years, Earth and Icarus are only 6 400 000 kilometers apart.*

Sources

Olbers suggested that the asteroids are remnants of a planet that ruptured some time long ago. Because of the small total mass, and because of difficulty in establishing how early the shattering would have occurred, the theory is not generally supported. Rather, it is believed that asteroids and planets were formed at essentially the same time, in substantially the same manner, and from the same basic materials. Smaller asteroids may be remnants of collisions between larger bodies. A basis of the latter argument is the fact that asteroids seem to fall into similar groups, as would be expected if they were collision debris.

The Trojans

Clusters of asteroids, called the *Trojans*, move in Jupiter's orbit. In 1772, the French mathematician Joseph Louis Lagrange (1736–1813) predicted there were two points in Jupiter's orbit where asteroids should be found—one-sixth of an orbit ahead of the planet and one-sixth behind. A hundred years after his death, Lagrange was proved correct when asteroids were discovered at these locations.

So far, fourteen asteroids have been located. Probably there

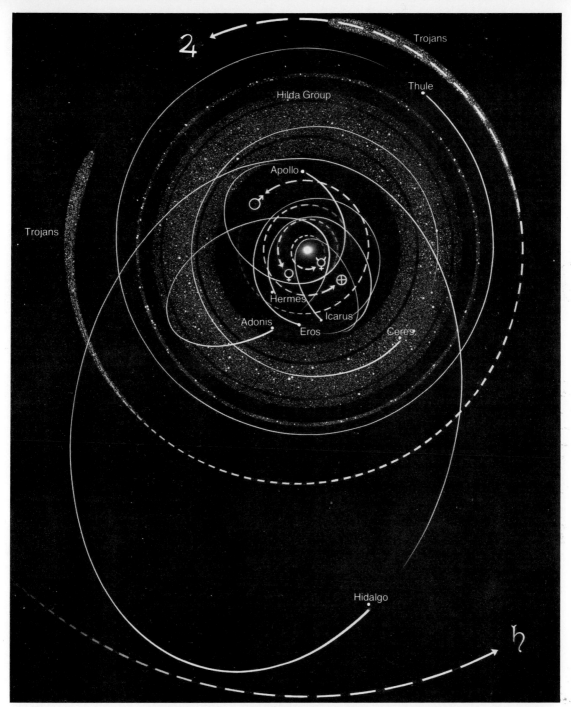

FIGURE 18–12   *Dashed lines are orbits of planets, and solid lines are the orbits of asteroids. Icarus makes the closest approach to the Sun; Hidalgo has the most distant apogee. The asteroids move in belts separated by gaps. The Trojan asteroids move in Jupiter's orbit—60 degrees ahead and 60 degrees behind the planet.*

are many more, but they escape observation because at the distance of Jupiter such small objects are extremely elusive targets. The asteroids east of Jupiter are named after Greek heroes of the *Iliad*—Ulysses, Nestor, Agamemnon—while those west of Jupiter are named after Trojan heroes—Priam, Aenas, Troilus.

The essential period of the Trojans is 12 years, the period of Jupiter. However, the asteroids appear also to move in complex manners among themselves, although generally preceding Jupiter, and following it, by one-sixth orbit.

TEKTITES

Many researchers believe that tektites are terrestrial in origin, and so should not be included in this discussion of debris of the solar system. We are inclined toward the theory that tektites originate on the lunar surface and approach Earth in long, sweeping trajectories, and so qualify for inclusion.

FIGURE 18–13   *Tektites have been found at the indicated locations. Open circles are sites of meteorite impacts that may have produced tektites. The dot in the Atlantic indicates a single specimen found on Martha's Vineyard, off the coast of Massachusetts.*

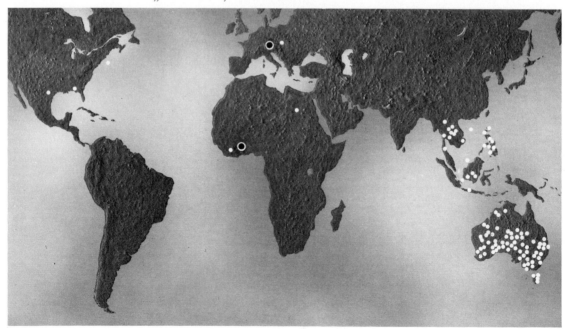

Tektites are found at various places on Earth—Africa, Indochina, Czechoslovakia, and Georgia and Texas in the United States. They are small bits of glasslike rock, most of them only a millimeter or so across and weighing only a few grams. The word comes from the Greek *tektos*, which means "molten." Tektites occur as rods, drops, spheres, buttons, and dumbbells, reminding one of dark glass that has been melted and then solidified. The richest source is the Philippine Islands, where some half-a-million are believed to have been recovered. Some of them are very large, having weights between 200 and 700 grams (Figure 18–13).

Tektites are usually named after the region where they are found. In most cases the connection is obvious. Where it is not, the name may be derived from a geologic formation in the region (the Moldau River, for example), or from some other distinctive feature of the region, such as an Indian tribe (the Bediasites from Texas). Recognized finds, their locations, and estimated numbers are given in Table 18–6.

Table 18–6   *Tektites*

Name	Number	Region where found
Australites	40 000	Australia
Bediasites	2 000	Texas
Billitonites	7 500	Isle of Billiton (Java Sea)
Indochinites	40 000	Thailand
Ivory Coast	200	Africa
Javaites	7 000	Borneo
Moldavites	55 000	Czechoslovakia
Philippinites	500 000	Philippine Islands

When we study a family of tektites, we find that they are all similar and of substantially the same age. All tektites, from whatever group, are very young, the oldest being about 45 million years old, the youngest only a few thousand years.

Origin

The origin of tektites remains a mystery. One school contends that tektites were formed when meteorites, or the heads of comets, crashed into Earth. If this were so, remains of craters made by the collisions should be discernible, for there has not been enough time for the craters to have been eroded away completely. In some cases, impact areas have been identified. But in other cases—that of

Australites, for example—no crater of the proper age has been found. Suggestions that the fall may have occurred in the open sea are discounted because a great tidal wave, of which there would still be signs, would have been produced.

Believers in the impact theory point out, however, that the crater might be far removed from the tektites. The collision might have melted terrestrial rock and thrown it high above Earth where it went into an Earth-circling orbit. When the orbit decayed, the tektites fell to Earth, sometimes in clusters, and sometimes as sporadic "single" falls. Long passages through the atmosphere would produce heat which would melt the rock fragments and cause them to assume the shapes associated with tektites (Figure 18–14).

FIGURE 18–14
Tektites may have been produced by meteorites in collision with Earth. Fountains of fragments may have been sprayed over the area. Other fragments would have gone into Earth-circling orbits, to land at some distant location.

There are investigators who believe tektites were produced by volcanic action or by lightning striking Earth. Or tektites may be pieces of the Moon, perhaps thrown into space by violent lunar volcanoes or by meteorites colliding with the Moon.

Lightning fuses sand occasionally, but the shapes that result do not match those of tektites. It is true, also, that glass is produced in volcanoes, but volcanic glasses and tektites do not match chemically, and volcanoes are not features common to the areas where tektites have been found.

Lunar collision. An interesting study of tektites relates them to collisions between the Moon and cosmic fragments. The conclusion is that high-speed meteorites crash into the Moon and blast debris into space, the velocity of the fragments being great enough for them to go into Sun-circling orbits. During millions of years Earth would intercept large amounts of this debris (Figure 18–15).

The fragments would go into Earth-circling orbits. As the orbits decayed, gravitational forces and other stresses would cause large fragments to break into smaller pieces. While approaching Earth, the smaller fragments would become molten. Ablation, or melting, of the leading edges could produce the odd shapes associated with tektites.

Bombardment, either of the surface of the Moon, or Earth, is generally accepted as an explanation for the formation of tektites. It was hoped that analyses of lunar rocks would provide answers to the dilemma. For the most part, the chemical structures of the two do not agree. However, microtektites found in ocean silts do have compositions similar to those of lunar rocks. Also, selected lunar rocks prove to be very much the same as tektites in percent distribution of components. For example, consider lunar rock number 12013 and one of the Java tektites, as shown in Table 18–7.

Table 18–7 *Composition of lunar rock and tektite compared*

Compound	Lunar rock 12013	Java tektite
Silicon dioxide	61.0 percent	63.5 percent
Titanium oxide	1.2	0.8
Aluminum oxide	12.0	12.6
Iron oxide	10.0	8.5
Magnesium oxide	6.0	6.8
Calcium oxide	6.3	3.8
Sodium oxide	0.69	0.7
Potassium oxide	2.0	1.5

FIGURE 18–15   *Tektites may be lunar fragments, thrown into Earth-circling orbits when gigantic meteorites crashed into the Moon. Fragments would be strewn along Earth as shown.*

The similarities between the two are impressive—so impressive that one could not say they did not have a similar origin.

Maybe tektites really are pieces of the Moon. The possibility is certainly a strong one. In spite of considerable study of tektites, however, we still lack positive explanations for their origin, for why they are found in only limited areas, and for why they occur in clusters generally. Perhaps, on the other hand, they should be discussed in a chapter devoted to Earth, and not in one that considers space debris.

REVIEW

1 Interplanetary space is far from empty since it contains gases, dust particles, meteoroids, asteroids, and comets.

2 There may be millions of comets in the solar system, many moving in elliptical orbits and making periodic passes through perihelion—the location nearest the Sun.

3 Comets may originate in deep outer space, in a region extending halfway to the nearest stars.

4 Each year thousands of metric tons of meteoritic material, which may be debris left over from the days of formation of the planets, fall upon Earth.

5 Showers of meteors are usually associated with a particular comet, while meteors which occur sporadically appear to be unrelated.

6 Meteorites (meteoroids that fall upon Earth) occur as iron, stones, or stony irons.

7 Meteors are sometimes extremely bright and are called fireballs; sometimes they are both bright and crackling and are called bolides.

8 Some 1500 asteroids have been discovered in the region between Mars and Jupiter since 1801 when the first one was observed by the Italian astronomer Piazzi.

9 Tektites, curiously shaped glasslike formations, may originate on the Moon as the result of collisions with large meteorites.

QUESTIONS

1 Explain the appearance of the zodiacal light and the reason for its occurrence. Why is it primarily a tropical phenomenon?

2 Although rotation of asteroids cannot be observed telescopically, the rotation periods of a few have been determined. How is this possible?

3 What are the differences between meteors, meteoroids, and meteorites?

4 How can one determine whether a given specimen is a meteorite?

5 What is the best time for observing meteors? Why?

6 Some meteor showers are seen at about the same time every year. Others are seen less frequently. Why is this so?

7 How is light produced in a "shooting star" and in the trail of the phenomenon?

8 Explain the radiant of a meteor shower. What does it prove?

9 Why do small particles in circular orbits eventually spiral in toward the Sun when sunlight falls upon them?

10 What are reasons for believing that meteorites may be small asteroids?

11 Each year new comets are discovered. This means there must be a source for them. What are explanations that have been offered?

12 How do we know that the mass of a comet is much smaller than the mass of Earth?

13 Explain the reasons why the tail of a comet always points away from the Sun.

14 Find the period of comets that have aphelion distances of 2000 and 10 000 a.u. and that graze the Sun at perihelion.

15 How would a comet appear to an observer on Mars and on Mercury?

READINGS

Brown, Peter. *Comets, Meteorites and Men.* New York: Taplinger Publishing Co., Inc., 1974.

Kuiper, Gerard, and Barbara Middlehurst, eds. *Moon, Meteorites and Comets.* Chicago: Chicago University Press, 1963.

Lyttleton, Raymond A. *Comets and Their Origin.* Cambridge: University of Cambridge Press, 1953.

Watson, Fletcher. *Between the Planets.* Cambridge, Mass.: Harvard University Press, 1962.

Whipple, Fred L. "The Nature of Comets." *Scientific American*, February 1974.

Wood, J. A. *Meteorites and the Origin of Planets.* New York: McGraw-Hill Book Company, 1968.

THE SEARCH
FOR LIFE-
IS ANYBODY THERE?

CHAPTER 19   There have been significant milestones in man's journey toward
understanding who and what he is, the nature of his world, his
place in the universe. Some examples are the heliocentric solar
system of Copernicus, and of Aristarchus before him, the
dimension and structure of the Galaxy, the dimension of the
universe, and the impermanence of the stars. But certainly the
most mind-stretching discovery that man could ever make would
be the existence of intelligent creatures at some location out
among the stars. Upheavals of immense proportion would rock the
foundations of many of men's religious beliefs. Tenets of schools
of philosophy would have to be modified, and political systems
reshuffled. This would be a changed world indeed.

*Barnard's S-nebula in
the Ophiuchus region
of the Milky Way
reminds one of a
question mark: Are
other civilizations out
there among the
multitude of stars?
(Hale Observatories)*

In the entire universe, Earth is the only place where life of any kind is known to exist. But there is strong feeling that events out of which life evolved on Earth must have occurred at other locations. That these processes could have happened only once—a single accident in billions of years—is not credible. All other phenomena—the creation and destruction of stars, collisions of asteroids, meteorites crashing into planets and satellites, the disintegration of comets— happen over and over again, millions, even billions, of times. There is no logic to believing in the uniqueness of life. This is especially true when one considers the dimensions of space, and the number of environments there must be which are conducive to the appearance of life in much the same manner that we believe life emerged on our planet. At the same time, we must not dismiss counterarguments which state it is unlikely that man could appear anywhere else in the universe in the same shape and form that prevail on Earth. This approach has been advanced by many scientists, especially George Gaylord Simpson, the evolutionary biologist now associated with Harvard University. Man is the result of billions of years of changing environmental factors; a single cosmic particle of a certain energy level might have struck a cell in an early organism that resulted in some peculiar characteristic that persisted and is still present in modern man. To expect that such an event would happen again here on Earth, or anywhere else, in precisely the same way and to an identical organism is questionable. Simpson points out that man is nonrepeatable. If intelligent creatures inhabit other worlds, one should not expect them to look anything like Earth men.

POSSIBLE LOCATIONS

As discussed earlier, it is generally accepted that first-generation stars originated out of primordial clouds of hydrogen and helium. Eddies developed in the clouds, producing concentrations of particles. The resultant increase in mass produced a gravitational field that attracted additional particles. A process of accretion was inaugurated. After millions of years the mass had reached proportions great enough for the gravitational force to generate heat. The temperature went up and up, ultimately reaching several millions of degrees, high enough to initiate nuclear reactions. The disorganized gas cloud had become a self-sustaining generator of energy; it had become a "hydrogen-burning" star.

That new star would go through a life cycle during which hydrogen nuclei would be the basic material out of which other elements would be created by nuclear fusion. Assuming adequate mass, after millions of years, it would become a supernova in which temperatures would go so high that heavier elements would be created. Vast amounts of material would be thrown into space to become the material out of which other stars would evolve in similar fashion. These would be stars made of "used" material, such as our Sun, a second-generation star.

In all this star-making process the gas clouds might become two or more stars rather than a single star; they might become a single star and formations of solid material, no one of them massive enough to have become a star. Or the cloud might become resolved into two or more stars, plus solid, cold masses held in the system by the strong gravitation of the stars.

In our galaxy the process has transpired many times; presently there are perhaps 100 billion stars. Some cosmologists believe that among them it would be unusual to find a star that is truly alone. The more common condition, they believe, is for a star to be part of a star system—two or more stars that revolve about one another—or for a star to be accompanied by solid masses (planets, if you will), debris left over from the process of star formation. This would be material that coalesced out of the same cosmic cloud that gave birth to the star, but that was not pulled out of its star-circling orbit. And there is no reason why there should be only a single mass; more likely there would be many of them, varying in size, in velocity in orbit, and in distance from the central star.

STARS AND PLANETS

No one can say how many stars there are. In our own galaxy, one of medium size, 100 billion is a conservative figure. But there may be 10 billion galaxies in range of the largest telescopes—some of which contain many more stars than does our Milky Way Galaxy. Possibly there are planets associated with each one of these billions upon billions of stars. Very likely only a fraction would be planets such as our own; but even if the fraction were one-millionth of 1 percent, this would work out to at least 1000 such planets in our galaxy alone.

This is an ultraconservative figure. Stephen Dole estimated that in our own galaxy there are 600 million planets which are capable of supporting life as we know it. He calculated that one of these planets is within 27 light-years of Earth, two only 34 light-years away, five at a distance of 47 light-years, ten within 59 light-years, and fifty or so at a distance of 100 light-years. Shklovskii and Sagan in their book *Intelligent Life in the Universe* put the number of civilizations in our galaxy technically superior to our own at about one million.

PLANETARY REQUIREMENTS

Stars vary a great deal in surface temperature, brightness, color, size, and length of life, as is made apparent in the H-R diagram. Although it is not unreasonable to suppose that planets may be as common as stars (if not more so), only a relatively small number of stars could support planets that would have those conditions which are believed essential to the emergence of life, and to its evolution.

The development of life forms from physical elements to biological entities, and from simple molecules to organic systems, requires 2 billion years, conservatively. This means that during that time physical and chemical environmental conditions must remain essentially stable; variations that do occur must be within the narrow range that can be tolerated by the emerging organisms.

FIGURE 19–1 *When double stars are widely separated, planets with stable conditions may be in orbit around each of them while the stars move around each other. The "cloud" around the stars is the so-called "life zone," the region within which conditions essential to life would prevail.*

Close to a star, temperatures would be too high. At great distances, temperatures would be too low. If a planet were associated with a system of two or more stars, the orbit of the planet would be greatly perturbed. Such a planet would very likely move into, and out of, regions favorable to the emergence and evolution of an organism, never remaining long enough for development to occur.

This argument was presented in the early 1960s by the Chinese-born American astronomer Su-Shu Huang, who pointed out that this negative condition would prevail only if the stars in the system were between 0.5 and 2 astronomical units apart. If the stars are close together (within 0.05 a.u.) or far apart (separated by more than 10 a.u.), a hypothetical planet, or planets, associated with the system could remain inside the "life zone" continually.

In Figure 19–1 we show a two-star system in which the stars are widely separated, and planets move in essentially circular orbits about each of the stars. It is quite apparent that the planets are well within the life belt at all times.

In Figure 19–2 a close two-star system is represented, as are the orbits of two hypothetical planets. The one in a nearly circular orbit stays within the life belt, while the planet that follows an elliptical orbit is carried out of the belt during part of its journey.

FIGURE 19–2   When double stars are close together, associated with them may be planets with stable conditions as well as planets with radically changing conditions.

When all factors are considered, probably only 1 or 2 percent of double or multiple stars can support planets upon which life may have evolved. But even this small percentage means there are a million possibilities within our galaxy. We do not know how many multiple-star systems there are, but it is reasonable to suppose that they comprise 25 percent of the stars in our galaxy, or some 25 to 35 billion systems.

It would appear, then, that solitary stars, or those that appear to be solitary, could support viable planets, since forces affecting motions would be in equilibrium. But it turns out that the conditions considered essential for life to evolve exist around only certain of the "single" stars—those that have masses close to that of the Sun. On the H-R diagram these would be late F-type to early K-type stars.

The more massive stars produce prodigious amounts of energy, energy that extends far into space and so creates extensive life belts. But massive stars are profligate. They remain on the main sequence where energy production is steady only 10 million years at the most, not long enough for life to develop on any planets that might be associated with them. The stars become red giants; they blow themselves apart as novae, or supernovae, and deteriorate to white dwarfs, black dwarfs, and perhaps ultimately become neutron stars. Whatever organisms may have developed on associated planets would have been destroyed. Indeed, the planets themselves would have been reduced to vapors.

Stars less massive than the Sun, the K and M stars, remain on the main sequence for billions of years—perhaps 100 billion. From the standpoint of time, therefore, intelligent life could emerge on a planet that might be associated with such a star. But these low-mass stars produce limited energy. The life belt of such a star is extremely narrow. Even if a planet were inside the "belt," the appearance of life upon it would be questionable. Very likely substantial energy in the form of electrical discharges, ultraviolet radiation, or X rays is needed to put the spark of life to inert matter. Such energy bursts are not associated with cool, red stars. Planets might revolve around these slow-burning, long-lived stars; but chances are they would be barren, desolate worlds.

From the standpoints of time and temperature (energy), it appears that only Sun-type stars could have planets populated by intelligent creatures. About 10 percent of the stars in our galaxy, or about 10 billion, are classified between late F and early K, which means there are some 10 billion possible locations. According to the accretion theory of star formation, it is reasonable to suppose that the so-called solitary stars are not lone sentinels. Very likely they have companions that so far defy our efforts to detect them.

OTHER STARS AND OTHER PLANETS

As discussed earlier, the accretion theory for the formation of stars may be modified slightly to suppose that it would be most rare for all the materials to assemble into a single, solitary formation. More likely two or more stars would evolve. Or stars would utilize only the bulk of the basic material. Leftover material, the debris of stellar creation, would become consolidated into planets—satellites of the stars.

The argument is logical, so it was reasonable to expect astronomers to search the stars for the presence of planets. However should they exist, planets associated with other stars could not be seen. The planets are too far from Earth for their reflected light to reach us, and also optical equipment could not separate a star and its planet to make them appear as discrete points. However, the presence of masses associated with stars can sometimes be detected using other techniques.

Stars move radially from Earth; that is, they move farther from us or closer to us along lines of sight. Also, stars move left or right, up or down; they exhibit what is called "proper motion." If a star is unaffected by nearby masses, its proper motion will be along a smooth path. Should there be another mass associated with the star the path of the star will be wavy. There will be a side to side motion. For several decades the study of the proper motions of stars has been a specialty of Sproul Observatory at Swarthmore College.

The observed amount of side-to-side motion of a given star is extremely small, not more than 0.001 millimeter. The measurement of these small variations requires delicate equipment and painstaking application. Nevertheless, the observers at Sproul, especially Kaj Strand and Sara Lee Lippincott, measured the amount of motion (the perturbation) of several stars, particularly 61 Cygni and Lalande 21185. From their measurements they could compute the mass of the bodies that were causing the variations. The masses always turned out to be greater than 10 times that of Jupiter, so great that the unseen bodies were classified as stars.

Another interesting star had been first observed in 1916 by Edwin E. Barnard, a skilled observer working at the Yerkes Observatory. It had a greater proper motion than any other star, 10.3 seconds of arc per year. Thus it would take 170 years to move an angle equal to the Moon's diameter. The star is in Ophiuchus and is called Barnard's star. It is nearby, only 5.9 light-years away, the second nearest star to Earth.

Photographs of Barnard's star revealed that it was not moving along a smooth path, but rather one that was very slightly wavy. It seemed to be pulled from side to side in a period of 25 years, indicating that the star was not alone in space. A companion was associated with it, probably another star. Interest in the star and its motion diminished, until 1937 when astronomers at Sproul began a study to learn more about that companion. For 30 years additional photographs of the star were made—some 10 000 altogether. Movements of the star were measured. The movement on the photographs was slight, 0.03 seconds of arc. However, the task was done, and when the data were assembled, it became possible to compute the mass as well as other characteristics of the companion.

It was thought that the perturbations might be best explained by supposing there were two masses in motion around Barnard's star with periods of 12 and 26 years, and at distances from the central star of 2.8 and 4.7 astronomical units. Their combined masses work out to be closer to that of Jupiter than when the perturbations are explained by a single mass.

Unfortunately, Barnard's star may turn out to be a false alarm. The slight waviness in its path may be due to slight distortions in the telescope which made the photographs. Still astronomers look to other nearby stars for evidence of planets. To date, none has been seen.

In our local area of space, out to 17 light-years, there are only 40 stars. Of those 40 stars only one, Tau Ceti, some 10.8 light-years away, appears to satisfy all those conditions believed necessary for a star to have a life-supporting planet moving around it. No evidence has as yet been obtained of the presence of such a planet. As of now the only planets we have observed are those in our own solar system.

SEARCH FOR LIFE

There are certainly hundreds of thousands, if not millions, of possibilities for the existence of planets upon which organisms could emerge and evolve into advanced forms. The search for such a planet, and the possible discovery of life existing at some other location in the universe, fascinates scientists and laymen alike.

It has always been so. In the fourth century B.C., the Epicurean philosopher Metrodorus said, "To consider Earth as the only populated world in infinite space is as absurd as to assert that in an entire field sown with millet only one grain will grow."

This was a prophetic statement, one long before its time. Man's senses told him that the universe was small: it was certainly Earth-centered, made of his immediate surroundings plus some provision in the sky for certain gods. For millenniums man did not move far from this viewpoint. Although there were isolated voices that said otherwise, man persisted until the sixteenth century in this narrow, Earth-centered view of his world.

In 1543, Nicholaus Copernicus published *De revolutionibus orbium coelestium (On the Revolutions of the Heavenly Spheres)*. In it Copernicus clearly stated that all observations of sunrise and sunset as well as direct and retrograde motions of the planets could be explained by supposing that Earth, and all other planets, moved in orbits around the Sun. The idea was bitterly contested, but was finally set on a firm foundation by Galileo, Kepler, and Newton.

Man's horizons had been expanded tremendously. Disciples of Copernicus advocated the heliocentric concept throughout the academic world in spite of the growing opposition of the Church. At the close of the century, the Dominican monk Giordano Bruno proclaimed an equally startling idea. He said, "Innumerable Suns exist. Innumerable Earths revolve about these Suns in a manner similar to the way the planets revolve around the Sun. Living beings inhabit those worlds."

This was blasphemy. In 1600 Bruno was burned at the stake as a heretic. But the belief that other worlds moved around other "Suns" persisted through the centuries. A few decades later Cyrano de Bergerac described fabulous ways to journey to the Moon, including cups of dew. The traveler fastened a belt of cups around his waist. The cups were filled with dew, which arose into space with the morning Sun, providing the propulsion that carried him to the Moon.

In earlier chapters we read about the "Moon hoax" perpetrated by Richard Locke in 1835, and the beliefs in Martians held by Schiaparelli and Percival Lowell. Even today people persist in believing that they are not alone in the solar system. They grasp even the slightest indication that might corroborate the belief.

Even with the photographs of Mars now available, there still are doubters who maintain that if men were to land on Mars and explore firsthand they would discover fossils of primitive Martian life. Man's desire to find that he is not alone is not easily destroyed.

The solar system has turned out to be a barren hunting ground. It is generally conceded that the possibility of finding life forms of any kind within its boundaries is slim. But the possibility of life existing outside our system of planets is as broad as the universe itself. In 1972 the Astronomy Survey Committee of the National

Academy of Sciences stated in its report: "At this instant, through this very document, are perhaps passing radio waves bearing the conversations of distant creatures, conversations that we could record if we but pointed a radio telescope in the right direction and tuned to the proper frequency."

Organic Molecules

In 1952, Stanley Miller, then working with Harold Urey at the University of Chicago, set up a rather simple apparatus to determine whether organic molecules could be created from inorganic substances. Methane, ammonia, hydrogen, and water (substances probably present in the early stages of Earth's evolution) were introduced into the apparatus. They were subjected to an electric-spark discharge of 60 000 volts. After several days, the gases were condensed out. Analysis revealed that several amino acids and urea were present, substances prevalent in proteins and in living organisms. It appeared that life could have arisen spontaneously here on our own planet at a time when our atmosphere contained methane and ammonia, and at a time when the planet was subjected to severe electrical storms. Miller's experiment has been repeated several times, using both electric discharges and ultraviolet radiation (Figure 19–3).

We know that these gases exist in other planets. And in recent years various combinations of carbon, nitrogen, hydrogen, and oxygen have been identified in outer space. It appears that the raw materials for organic molecules are common in the universe.

This being so, it is logical to assume that somewhere, sometime, the substances would have combined to make ammonia (NH_3) and methane (CH_4), and that they would have been exposed to energetic discharges. Belief in the possibility is supported by scientists, as witness the earlier statement by the National Academy.

The search for interstellar molecules continues, especially in this country and the Soviet Union. Methane is strongly suspected, since one would expect it to be present where the compound formaldehyde (H_2CO) is found. In some terrestrial systems, formaldehyde appears as a breakdown product from amino acids. This leads some investigators to suspect, and perhaps hope, they will eventually find amino acids in the gas clouds that exist in interstellar space. Such a discovery might help solve a perplexing puzzle, one that fascinates biochemists: the right and left symmetry displayed by organic molecules. When polarized light is passed through a solution, the plane of the light is shifted. In the case of laboratory-produced molecules, the light may be turned either left or right. There is no pattern. But when the molecules originate in a living organism, the polarization is always "left-handed."

If amino acids can be detected in space, it should be possible to determine the distribution of right- and left-handed molecules. Biochemists will then have some answers, but also more questions. If they should be found only on Earth, do left-handed molecules result from some life process, or did life emerge from the only kind of molecules available at the time—the left-handed variety? If left-handed molecules occur in space, could this mean processes akin to those in organic systems are occurring there?

FIGURE 19–3 *Under proper conditions, inorganic substances will produce amino acids, the building blocks of proteins. Steam introduced at lower left was combined with methane (CH₄), and ammonia (NH₃), and hydrogen. After exposure to electrical discharges, amino acids were produced.*

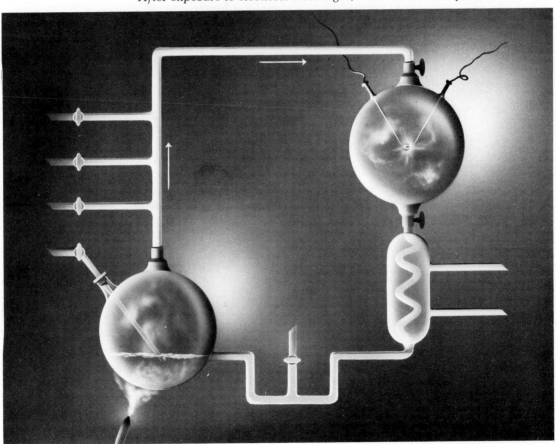

EXTRATERRESTRIAL COMMUNICATION

Two centuries ago Immanuel Kant, the German philosopher, believed that the planets were inhabited. Furthermore, those creatures that lived closer to the Sun than we, inhabitants of Mercury and Venus, were of low moral character. Apparently the intense heat had harmful effects on the mind. Inhabitants of Jupiter were in a much better situation. Being farther from the dangerous effects of the Sun, they lived pleasant lives, free of turmoil and disruption.

Kant proposed no way of conversing with residents of other planets, but in the last century Karl Friedrich Gauss, among the greatest of all mathematicians, made several suggestions. He was convinced there were Martians, and we should provide them with clues to our presence. A large right triangle, he proposed, would be discernible. Wide lines of trees should be planted in the steppes of Siberia to outline the triangle. Inside, the area should be covered with wheat to make a strong contrast.

The triangle idea could be expanded to illustrate the Pythagorean theorem. Squares of wheat should be planted using each side of the triangle as a base line. A civilization of even limited advancement would recognize that the square of the hypotenuse equals the sum of the squares of the other two sides.

Another suggestion was to dig channels in the form of a triangle in the Sahara desert. At night the channels would be flooded with kerosene. When ignited, the figure would be visible from Mars. Its geometric shape would be a positive clue to our presence.

Before the discovery of radio, any plan for communication, feasible or not, had to be visual. Radio opened a new horizon. In the early days, Marconi was quite certain that "strange signals" he picked up on his receiver originated with the Martians.

Nikola Tesla, the genius who was to contribute so much to our knowledge of electricity, was certain he had picked up communication attempts from Mars while he was experimenting with a high transmission tower at Colorado Springs.

In 1924, during a close approach of Mars, the Army and Navy were directed to keep radio silence so that, should the Martians be sending out signals, we would be more able to detect them.

For decades the question of communication with other planets rested. Indeed, the presence of intelligent life on any of them was not seriously suspected. However, in the 1950s the existence of other planets going around other stars was considered extremely reasonable. Should there be such planets, and should there be intelligent creatures upon them, communication could be made via radio, using the new radio telescopes—powerful transmitters and receivers.

About this time the 21-centimeter radiation of the hydrogen

atom was detected. It was found to be a "universal" wavelength; thus it should be the one employed for pulsing signals into space, and the one to which receivers were tuned. If attempts were made to contact us, it was reasonable to suppose modulations would be put on the 21-cm band.

From May to July in 1960, Frank Drake spent 150 hours scanning the region close to those nearby stars that had characteristics believed necessary for a star to support planets. Prominent were Tau Ceti and Epsilon Eridani. He noted no signals of any kind.

More recently, the Russians have been scanning some 50 stars with much more sophisticated equipment and are believed to be continuing the investigations. Presently it is hoped the United States will undertake the construction of a very large radio telescope to increase sensitivity and make it possible to receive extremely faint signals. There would be an array of 27 telescopes arranged in a Y-shaped track having arms 20 kilometers long. The combination would be as effective as a single disk 30 kilometers in diameter.

If we were ever to receive a coherent signal from beyond the Earth, what could be done about it? Could we communicate with its source?

There are a good many stars within, say, 30 light-years that could support planets. Were we to send a signal to any one of them, the pulse would take 30 years to get there, and the reply another 30 years to return. It is conceivable that a two-way communication could be made within a lifetime.

Even a meager contact such as this would be sensational, enough to affect strongly man's philosophy and his interactions. Actual discourse probably could not be accomplished at our present level of technological achievement. But our technology moves rapidly. In less than a century air travel has progressed from the Wright brothers to the 747; and in many other areas the achievements of the last 50 years are greater than all those made in the history of man on Earth. Who knows what will be possible before the end of this century?

Perhaps hardware will be used for communications. One such agent of interplanetary communication is already on its way. In the early 1970s, Pioneer 10 was launched from Earth. Its orbit goes outside the solar system. Aboard Pioneer is a plaque containing data about our planet, its place in the Galaxy, and indications of the size and shape of human beings (Figure 19–5).

A far-off civilization may someday retrieve that plaque and use it as a rosetta stone for understanding something about our world. It is also conceivable that civilizations out there may have sent automated probes into deep space, and that one of these explorers may be destined for Earth.

FIGURE 19–4
(opposite page)
Radio telescopes have revealed the presence of some two dozen kinds of molecules in interstellar space, and they are presently scanning nearby stars for the presence of planets.

No one can say how it will come about, but sooner or later we will have positive evidence of life beyond Earth—of the existence of intelligent beings living on planets going around other stars.

Presently, there are few hard facts to support such a statement. But philosophically it is sound, and man's acquisition of knowledge in any field must begin with a positive philosophy, one that urges him to further understanding. Professor Philip Morrison summed it up well in a published symposium, *Life Beyond the Earth and the Mind of Man.*

> I think, therefore, that we will get a message, but it will not be simple. . . . I think the most important thing the message will bring us . . . will be a description, if one exists at all, of how these beings were able to fashion a world in which they could live, persevere, and maintain something of worth and beauty for a long period of time. . . . So I am neither fearful nor terribly expectant. I am anxious for that first acquisition, to make sure we are not alone. But once that is gained—it might be gained in my lifetime—then I think we can rest with some patience to see what complexities have turned up on other planets. And if after considerable search we do not find that our counterparts exist somewhere else, I cannot think that would be wrong either, because that would give us even a heavier responsibility to represent intelligence in this extraordinarily large and diverse universe.

FIGURE 19–5
Identification plaque aboard Pioneer 10. At left are shown positions of 14 pulsars arranged to indicate the home star of the launching civilization. Symbols at the ends of the lines represent frequencies relative to that of hydrogen shown at upper left. The Pioneer vehicle is shown behind the figures. The solar system and the trajectory of Pioneer are shown at the bottom.

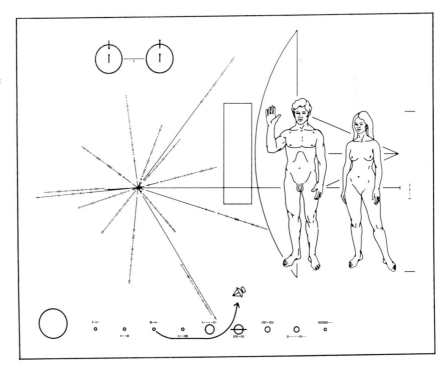

REVIEW

1 Historically the skies were considered the abode of the gods, and today the idea of beings living outside Earth persists.

2 Statistically there must be billions of planets in the universe; indeed, as some astronomers believe, it may be unusual to find a star that does not have another star (or stars) associated with it, or planets and planetary systems.

3 Billions of stars are similar to the Sun, having essentially the same temperature and remaining stable for a billion years or more—long enough for life to emerge.

4 Since explorations of the Moon, Mars, Venus, and Mercury have revealed no signs of life, one must be extremely optimistic to expect that life exists anywhere in the solar system except on our planet.

5 Studies of the proper motions of stars reveal the presence of masses that cause perturbations, some of which may be produced by masses small enough to be classified as planets.

6 Radio telescopes scan the region near stars searching for a modulated signal.

7 Complex molecules that may be the raw materials for organic molecules have been identified in interstellar clouds, strengthening the belief that life has arisen elsewhere in the universe.

8 Under the proper conditions, methane, ammonia, hydrogen, and water combine to form amino acids, the basic substance of proteins.

QUESTIONS

1 What are the obstacles to interstellar communication, and how might such communication become established?

2 Consider each of the planets of the solar system with respect to its ability to harbor both (a) life of some kind and (b) intelligent life.

3 What are the requirements for life to flourish on Earth?

4 How do you expect life on Earth would be altered if an extraterrestrial civilization were discovered?

5 Discuss the statisticians' arguments for the existence of Earthlike creatures versus those of the evolutionists.

6 According to the accretion theory, what are the arguments in favor of the existence of numerous planets?

7 Briefly trace the life history of stars at various positions in the H-R diagram with reference to the possibility that life-supporting planets might be associated with them.

8 What factual evidence is there to support the argument that there are other planets going around other stars?

9 The biographical history of man's belief in the existence of extra-terrestrial life is fascinating. You might begin your research with Metrodorus, Copernicus, Bruno, de Bergerac, Verne, Marconi, Gauss, and Tesla.

10 Look up the books by Percival Lowell that are listed in the bibliography. They are fascinating examples of how a few questionable observations can be built into "firm" statements and beliefs.

READINGS

Berry, Adrian. *The Next Ten Thousand Years.* New York: Saturday Review Press, 1974.

Dole, Stephen. *Habitable Planets for Man.* New York: American Elsevier Publishing Co., Inc., 1970.

Lowell, Percival. *Mars as the Abode of Life.* New York: The Macmillan Company, 1908.

Lowell, Percival. *Mars and its Canals.* New York: The Macmillan Company, 1906.

Ponnamperuma, Cyril, and A. G. W. Cameron. *Interstellar Communication.* New York: Houghton Mifflin Company, 1974.

Sagan, Carl, ed. *Communication with Extraterrestrial Intelligence.* Cambridge, Mass.: The M.I.T. Press, 1973.

APPENDICES

A. EXPONENTS AND LOGARITHMS

Exponents

In science, and particularly in astronomy, it is often necessary to work with both very large and very small numbers. Many zeros and decimal places are needed to express the numbers, which is inconvenient and time consuming. Therefore, numbers are expressed as powers of ten. (Ten is used since it is the base of our usual number system.)

The number 100 is written as 10^2; this notation reads as "ten to the two" or "ten to the second power." It means that 10 is multiplied by 10. The 2 is called the *exponent*. This power of ten notation is convenient to use, for the exponent is the number of zeros in the number. Other examples are the following:

$$10^3 = 1000 \qquad 10^4 = 10\ 000 \qquad 10^5 = 100\ 000$$

When the number is less than 1, negative exponents are used, indicating that the number is a fraction. Thus,

$$\frac{1}{10} = 10^{-1} = \frac{1}{10}^1$$
$$\frac{1}{1000} = 10^{-3} = \frac{1}{10}^3$$

In all cases the significant figures of the number are multiplied by 10 to some power, or by $\frac{1}{10}$, $\frac{1}{100}$, and so on, to express the quantity. For example,

1 day $= 8.64 \times 10^4$ seconds $= 84\ 400$ seconds

1 year $= 3.16 \times 10^7$ seconds $= 31\ 600\ 000$ seconds

(The exponent indicates the number of places the decimal point is moved to the right.)

1 millimeter $= 10^{-3}$ meters $= 0.001$ meters

Mass of proton $= 1.6724 \times 10^{-24}$ gram $=$
$$0.0000000000000000000000016724 \text{ gram}$$

(The exponent indicates the number of places the decimal point is moved to the left.)

To *add* or *subtract* numbers that are expressed in powers of ten, the numbers must be made to have the same power, as below:

$$1.02 \times 10^5$$
$$+3.0 \ \times 10^4 \qquad \text{Since } 3.0 \times 10^4 = 0.30 \times 10^5, \text{ then}$$

$$\begin{aligned} 1.02 \times 10^5 \\ 0.30 \times 10^5 \\ \hline 1.32 \times 10^5 \end{aligned}$$

To *multiply* two numbers expressed in this notation, multiply the significant figures, and add the exponents of the tens. Thus,

$$400 \times 3400 = (4 \times 10^2) \times (3.4 \times 10^3)$$
$$= 13.6 \times 10^{2+3} = 13.6 \times 10^5 = 1.36 \times 10^6$$

To *divide* one number by another, divide the significant figures, and subtract the exponent of the divisor from the exponent of the dividend. For example,

$$\frac{6 \times 10^7}{3 \times 10^4} = \frac{6}{3} \times 10^{7-4} = 2 \times 10^3$$

In this way the writing of large and small numbers and their manipulation is facilitated.

Logarithms

We know that $10^1 = 10$, and $10^2 = 100$. The number 50 lies between 10 and 100, and thus there must be some number between 1 and 2 that we can use as an exponent of the 10 to get 50. This exponent is the *logarithm* of 50. In the same way we could say that 1 is the logarithm of 10, and 2 is the logarithm of 100. Both of these are to the *base* 10.

It turns out that the logarithm (or log) of 50 is

$$\log 50 = 1.6990$$

The number 500 is between 100 and 1000, so there must also be a number between 2 and 3 (since $100 = 10^2$ and $1000 = 10^3$) that is the log of 500. It can be computed that

$$\log 500 = 2.6990$$

Notice that the part of the number after the decimal point is the same for log 50 and log 500. This part of the logarithm is called the *mantissa*, and is the same for any numbers which have the same significant figures. For example,

$\log 3.7 = 0.5682$	$\log 370 = 2.5682$
$\log 37 = 1.5682$	$\log 3\ 700\ 000 = 6.5682$

For numbers less than one, the logarithm is negative. Thus,

$\log 0.37 = -0.4318 = 0.5682 - 1$

(sometimes expressed $9.5682 - 10$)

$\log 0.037 = -1.4318 = 0.5682 - 2 = 8.5682 - 10$

$\log 0.000037 = -4.4318 = 0.5682 - 5 = 5.5682 - 10$

The number to the left of the decimal point is called the *characteristic* of the logarithm. For numbers greater than 1, the characteristic is 1 less than the number of significant figures to the left of the decimal point. For numbers less than 1, it is the number of zeros before the first significant figure. Remember that if the log is in the form, say, $6.5682 - 10$, you must first convert to the form of the log without the 10 in it. Negative numbers do not have logarithms.

Logarithms enable complicated multiplication and division calculations to be performed without tedious labor. The multiplication of two numbers is accomplished by adding together their logarithms. Division becomes a process of subtracting the log of the divisor from the log of the dividend. Before the invention of high-speed computing machines, logarithms were the method used for long calculations. Tables of logarithms of numbers, and explanations of how to use the tables, may be found in many references. With the advent of small calculators, the use of logarithms is waning, but knowledge of the *concept* behind them is necessary for everyone studying science.

B. THE METRIC SYSTEM

Throughout this book the metric system of measurement is used. It is the worldwide system of science, and is in general use in all the world, except the United States and a few small nations. The advantages of its use are that it is consistent and is based on the number 10 as is our number system. In the book occasional mention is made of the customary system, which used to be called the English system. It is used for comparison, and because the reader is more likely to feel at home using customary units.

The metric system is more properly called the *Système Internationale*, or SI. It makes use of a few basic units and multiples of these units. The multiples are produced by using a prefix and the name of the unit. In SI, the preferred prefixes are in multiples of 1000. That is, one uses units of, say, a meter and a kilometer, but not a decameter or decimeter. The units are the following:

Length. meter (m). 1 m = 1 650 763.73 wavelengths of the orange-red spectral line of krypton-86.

Mass. kilogram (kg). 1 kg = the mass of the standard kilogram at the International Bureau of Weights and Measures in Paris.

Time. second (sec). 1 sec = 9 192 631 770 cycles of the radiation from a cesium-133 atom.

Temperature. Kelvin (K). The Kelvin, or degree Kelvin (°K), is 1/273.16 of the temperature of the triple point of water (the temperature at which water can coexist as a solid, a liquid, and a gas simultaneously).

Force. Newton (N). 1 N is the force that will accelerate a 1-kg mass by 1 m/sec each second.

Work. Joule (J). 1 J is the work done when a force of 1 N moves an object 1 m.

Power. Watt (W). 1 W = 1 J/sec.

Prefixes

The following prefixes are used with the names of the basic units to give larger multiples:

Tera (T)	10^{12}
Giga (G)	10^{9}
Mega (M)	10^{6}
Kilo (k)	10^{3}

These prefixes are for fractions of the basic unit:

milli (m)	10^{-3}
micro (μ)	10^{-6}
nano (n)	10^{-9}
pico (p)	10^{-12}

A few other prefixes are used, although they are not properly part of the SI:

hecto (h)	10^{2}
centi (c)	10^{-2}
deca (da)	10^{1}
deci (d)	10^{-1}

Combining the names of the units and the prefixes gives us units such as kilometer, nanosecond, Megawatt, and so forth.

Because of the vast distances involved in astronomy, two other distance units are used in SI:

astronomical unit (a.u.) $= 1.496 \times 10^8$ km

parsec $= 3.084 \times 10^{13}$ km

Also common, but not a part of SI, is the Ångstrom (Å or A). $1 A = 10^{-10}$ m.

Conversion factors

The following conversion factors are useful:

1 inch $= 0.0254$ m

1 yard $= 0.9144$ m

1 mile $= 1.61$ km

1 light-year $= 9.461 \times 10^{12}$ km

1 pound $= 0.4536$ kg

1 short ton $= 0.91$ Mg $= 0.91$ metric tons

To convert temperature scales of Fahrenheit (°F), Celsius (°C), and Kelvin (°K), use the following:

$°C = °K - 273$

$°C = \frac{5}{9} (°F - 32)$

$°F = (\frac{9}{5})°C + 32$

Other conversion factors may be found in mathematical handbooks and books on the metric system.

C. ASTRONOMICAL CONSTANTS AND OTHER UNITS

Velocity of light (c)	3.00×10^5 km/sec
	3.00×10^{10} cm/sec
	186 000 miles/sec
Ångstrom unit (Å or A)	10^{-8} cm
Mean radius of Earth	6371 km
Equatorial radius of Earth	6378 km

Polar radius of Earth	6357 km
Radius of Sun	696 000 km
Astronomical unit (a.u.)	1.50×10^8 km 0.930×10^8 miles
Light-year (l.y.)	0.946×10^{18} cm 5.88×10^{12} mi
Parsec (pc)	3.08×10^{18} cm 206 265 a.u. 3.26 l.y.
Period of Earth's rotation	86 164 sec
Mass of Earth	5.98×10^{27} g 5.98×10^{21} metric tons
Mass of Sun	1.99×10^{33} g (333 000 Earth masses)
Luminosity of Sun	3.86×10^{33} erg/sec
Absolute zero	$0°K = -273.15°C = -459.67°F$
Mass of Moon	7.35×10^9 metric tons
Radius of Moon	1 740 km

D. MEASURING ANGLES; PARALLAX

The angle between two objects or positions is a measurement often used in astronomy. For this reason, it is important that we understand how angles are measured and used.

A circle is divided evenly into 360 degrees (360°). Each degree is divided into 60 parts, called minutes, symbolized by ('). Each minute, in turn, is divided into 60 seconds of arc, symbolized by ("). Thus,

$$1° = 60' = 3600''$$

$$360° = 21\,600' = 1\,296\,000''$$

(In dealing with a position on the celestial sphere, we often use the coordinate called *right ascension* (R.A.) which is measured in hours, minutes, and seconds of *time*. It is important not to confuse minutes of time with minutes of arc, and seconds of time with seconds of arc. The relationship between them is

$$1^m = 15' \qquad 1^s = 15''$$

One second of arc is a very small angle. It is the angular size of a quarter (which is 24 millimeters in diameter) at a distance of 4½ km (3 miles). One minute of arc is the angular size of a quarter at a distance of 100 meters. Note that there is a relationship between actual size (the diameter of the coin), angular size, and distance. This enables us to measure distances in astronomy if both the actual size and the angular size of an object are known.

The Distance to the Moon

Suppose that an observer in Greenwich, England, and an observer at the Cape of Good Hope both observe the Moon at the same time. These two observatories are 8640 kilometers apart. Because the observers are in different locations, they will see the Moon in a slightly different location on the celestial sphere. This shift is called *parallax*. The observer at the Cape of Good Hope will see the Moon at a location 1°20′ north of where the observer at Greenwich will see it.

FIGURE D–1
Obtaining the distance to the moon by parallax. The distance between the observatories is the chordal distance (a straight line through Earth, not the distance along the surface).

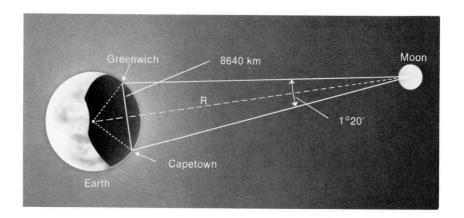

Now refer to Figure D-1. That 1°20′ is the angle between the two observatories as seen by someone on the Moon. So the angular separation of the two places is 1°20′, and the actual separation is 8640 kilometers. Now let the distance to the Moon be R. Because the distance to the Moon is so large, the distance between the two observatories is very close to being part of a circle centered on the Moon. The total circumference of the circle is $2\pi R$, so we can set up an equation utilizing the following ratio:

$$\frac{8640 \text{ km}}{2\pi R} = \frac{1°20′}{360°}, \text{ or } \frac{80′}{21\,600′}$$

$$\frac{8640 \text{ km}}{2\pi R} = \frac{80}{21\,600} \quad \text{(the minutes units cancel out)}$$

$$8640 \text{ km} \times 21\,600 = 2\pi R \times 80$$

$$\frac{8640 \text{ km} \times 21\,600}{2\pi \times 80} = R$$

Substituting 3.1416 for π, we find that $R = 371\,000$ kilometers.

E. SYMBOLS, ABBREVIATIONS, PRONUNCIATIONS*

Sun, Moon and Planets

⊙ The Sun	☾ The Moon generally	♃ Jupiter
● New Moon	☿ Mercury	♄ Saturn
◉ Full Moon	♀ Venus	♅ Uranus
☽ First quarter	⊕ Earth	♆ Neptune
☾ Last quarter	♂ Mars	♇ Pluto

Signs of the Zodiac

♈ Aries .... 0°	♌ Leo ....120°	♐ Sagittarius ..240°
♉ Taurus ......30°	♍ Virgo ......150°	♑ Capricornus .270°
♊ Gemini ......60°	♎ Libra ....180°	♒ Aquarius ...300°
♋ Cancer ......90°	♏ Scorpius ....210°	♓ Pisces ......330°

The Greek Alphabet

A, α	Alpha	I, ι	Iota	P, ρ	Rho
B, β	Beta	K, κ	Kappa	Σ, σ	Sigma
Γ, γ	Gamma	Λ, λ	Lambda	T, τ	Tau
Δ, δ	Delta	M, μ	Mu	Y, υ	Upsilon
E, ε	Epsilon	N, ν	Nu	Φ, ϕ	Phi
Z, ζ	Zeta	Ξ, ξ	Xi	X, χ	Chi
H, η	Eta	O, o	Omicron	Ψ, ψ	Psi
$\Theta, \theta, \vartheta$	Theta	Π, π	Pi	Ω, ω	Omega

The Constellations

Latin Names with Pronunciations and Abbreviations

Andromeda, ăn-drŏm′ĕ-dȧ ....And	Andr	
Antlia, ănt′lĭ-ȧ ....Ant	Antl	
Apus, ā′pŭs ....Aps	Apus	
Aquarius, ȧ-kwâr′ĭ-ŭs ....Aqr	Aqar	
Aquila, ăk′wĭ-lȧ ....Aql	Aqil	
Ara, ā′rȧ ....Ara	Arae	

Aries, ā′rĭ-ēz ....Ari	Arie
Auriga, ô-rī′gȧ ....Aur	Auri
Boötes, bō-ō′tēz ....Boo	Boot
Caelum, sē′lŭm ....Cae	Cael
Camelopardalis, kȧ-mĕl′ō-pär′dȧ-lĭs .....Cam	Caml
Cancer, kăn′sẽr ....Cnc	Canc

* Listings on pages 541-544 adapted from *The Observer's Handbook*, published annually by the Royal Astronomical Society of Canada.

Canes Venatici,
  kā'nēz vĕ-năt'ĭ-sī ......CVn CVen
Canis Major,
  kā'nĭs mā'jẽr ....CMa CMaj
Canis Minor,
  kā' nĭs' mĭ'nẽr ....CMi CMin
Capricornus,
  kăp'rĭ-kôr'nŭs ....Cap Capr
Carina, kȧ-rī'nȧ ....Car Cari
Cassiopeia, kăs'ĭ-ō-pē'yȧ' ..Cas Cas
Centaurus, sĕn-tô'rŭs .....Cen Cent
Cepheus, sē'fūs ....Cep Ceph
Cetus, sē' tŭs ....Cet Ceti
Chamaeleon, kȧ-mē'lē-ŭn..Cha Cham
Circinus, sûr'sĭ-nŭs ....Cir Circ
Columba, kȯ-lŭm'bȧ .....Col Colm
Coma Berenices,
  kō'mȧ bĕr'ē-nī'sēz ......Com Coma
Corona Australis,
  kȯ-rō'nȧ ôs-trā'lis ......CrA CorA
Corona Borealis,
  kȧ-rō'nȧ bō'rē-ā'lĭs .....CrB CorB
Corvus, kôr'vŭs ....Crv Corv
Crater, krā'tẽr ....Crt Crat
Crux, krŭks ....Cru Cruc
Cygnus, sĭg'nŭs ....Cyg Cygn
Delphinus, dĕl-fī'nŭs ....Del Dlph
Dorado, dȯ-rä'dō ....Dor Dora
Draco, drä'kō ....Dra Drac
Equuleus, ē-kwōō'lē-ŭs ...Equ Equl
Eridanus, ē-rĭd'ȧ-nŭs ....Eri Erid
Fornax, fôr'năks ....For Forn
Gemini, jĕm'ĭ-nī ....Gem Gemi
Grus, grŭs ....Grų Grus
Hercules, hûr'kŭ'lēz .....Her Herc
Horologium,
  hŏr'ȯ-lō'jĭ-ŭm ....Hor Horo
Hydra, hī'drȧ ....Hya Hyda
Hydrus, hī'drŭs ....Hyi Hydi
Indus, ĭn'dŭs ....Ind Indi
Lacerta, lȧ-sûr'tȧ ....Lac Lacr
Leo, lē'ō ....Leo Leon
Leo Minor, lē'ō mĭ'nẽr ....LMi LMin
Lepus, lē'pŭs ....Lep Leps
Libra, lī'brȧ ....Lib Libr
Lupus, lū'pŭs ....Lup Lupi

Lynx, lĭngks ....Lyn Lync
Lyra, lī'rȧ ....Lyr Lyra
Mensa, mĕn'sȧ ....Men Mens
Microscopium,
  mī'krȯ-skō'pĭ-ŭm ......Mic Micr
Monoceros, m-ŏnŏs'ēr-ŏs ..Mon Mono
Musca, mŭs'kȧ ....Mus Musc
Norma, nôr'mȧ ....Nor Norm
Octans, ŏk'tănz ....Oct Octn
Ophiuchus, ŏf'ĭ-ūkŭs .....Oph Ophi
Orion, ȯ-rī'ŏn ....Ori Orio
Pavo, Pä'vō ....Pav Pavo
Pegasus, pĕg'ȧ-sŭs ....Peg Pegs
Perseus, pûr'sūs ....Per Pers
Phoenix, fē'nĭks ....Phe Phoe
Pictor, pĭk'tẽr ....Pic Pict
Pisces, pĭs'ēz ....Psc Pisc
Piscis Austrinus,
  pĭs'ĭs ôs-trī'nŭs ....PsA PscA
Puppis, pŭp'ĭs ....Pup Pupp
Pyxis, pĭk'sĭs ....Pyx Pyxi
Reticulum,
  rē-tĭk'ū-lŭm ....Ret Reti
Sagitta, sȧ-jĭt'ȧ ....Sge Sgte
Sagittarius, săj'ĭ-tā'rĭ-ŭs ...Sgr Sgtr
Scorpius, skôr'pĭ-ŭs ......Sco Scor
Sculptor, skŭlp'tẽr ....Scl Scul
Scutum, skū'tŭm ....Sct Scut
Serpens, sûr'pĕnz ....Ser Serp
Sextans, sĕks'tănz ....Sex Sext
Taurus, tô'rŭs ....Tau Taur
Telescopium,
  tĕl'ē-skō'pĭ-ŭm ....Tel Tele
Triangulum,
  trī-ăng'gŭ-lŭm ....Tri Tria
Triangulum Australe,
  trī-ăng'gŭ-lŭm ôs-trā'lē ..Tra TrAu
Tucana, tū-kā'nȧ ....Tuc Tucn
Ursa Major,
  ûr'sȧ mā'jẽr ....UMa UMaj
Ursa Minor,
  ûr'sȧ mĭ'nẽr ....UMi UMin
Vela, vĕ'la ....Vel Velr
Virgo, vûr'gō ....Vir Virg
Volans, vō'lănz ....Vol Voln
Vulpecula, vŭl-pĕk'ū-lȧ ...Vul Vulp

Pronunciation key:
ā fāte; ȧ chȧotic; ă tăp; ă finȧl; ȧ ȧsk; â câre; ä älms; au aught; ē bē; ē crēatē;
ĕ ĕnd; ĕ angĕl; ē makēr; ī tīme; ĭ bĭt; ĭ anĭmal; ō nōte; ō anatȯmy; ŏ hŏt; ŏ ŏccur;
ô ôrb; ōō mōōn; ŏŏ bŏŏk; ou out; ū tūbe; ů ůnite; ŭ sŭn; ų sųbmit; û hûrl.

The Constellations: Latin Names and English Meanings

Andromeda	Andromeda (chained lady)*	Leo	Lion
		Leo Minor	Little Lion
Antlia	Pump	Lepus	Hare
Apus	Bird of Paradise	Libra	Balance
Aquarius	Water Bearer	Lupus	Wolf
Aquila	Eagle	Lynx	Lynx
Ara	Altar	Lyra	Harp
Aries	Ram	Mensa	Table Mountain
Auriga	Charioteer	Microscopium	Microscope
Bootes	Herdsman	Monoceros	Unicorn
Caelum	Chisel	Musca	Fly
Camelopardalis	Giraffe	Norma	Level
Cancer	Crab	Octans	Octant
Canes Venatici	Hunting Dogs	Ophiuchus	Snake Bearer
Canes Major	Big Dog	Orion	Orion (the hunter)*
Canes Minor	Little Dog		
Capricornus	Horned Goat	Pavo	Peacock
Carina	Ship's Keel	Pegasus	Pegasus (winged horse)*
Cassiopeia	Cassiopeia* (lady in chair)		
		Perseus	Perseus*
Centaurus	Centaur*	Phoenix	Phoenix (mythical bird)*
Cepheus	Cepheus (the King)*		
		Pictor	Easel
Cetus	Whale	Pisces	Fish
Chamaeleon	Chameleon	Piscis Austrinus	Southern Fish
Circinus	Pair of Compasses	Puppis	Ship's Stern
Columba	Dove	Pyxis	Ship's Compass
Coma Berenices	Berenice's Hair*	Reticulum	Net
Corona Australis	Southern Crown	Sagitta	Arrow
Corona Borealis	Northern Crown	Sagittarius	Archer
Corvus	Crow	Scorpius	Scorpion
Crater	Cup	Sculptor	Sculptor's Workshop
Crux	(Southern) Cross		
Cygnus	Swan	Scutum	Shield
Delphinus	Dolphin	Serpens	Snake
Dorado	Dorado (a fish)	Sextans	Sextant
Draco	Dragon	Taurus	Bull
Equuleus	Little Horse	Telescopium	Telescope
Eridanus	River Eridanus	Triangulum	Triangle
Fornax	Furnace	Triangulum Australe	Southern Triangle
Gemini	Twins		
Grus	Crane	Tucana	Toucan (a bird)*
Hercules	Hercules*	Ursa Major	Great Bear
Horologium	Clock	Ursa Minor	Little Bear
Hydra	Hydra*	Vela	Ship's Sails
Hydrus	Water Snake	Virgo	Virgin
Indus	Indian	Volans	Flying Fish
Lacerta	Lizard	Vulpecula	Little Fox

* Creatures or personalities from mythology.

List of Named Stars

Name	Con.	R.A.	Name	Con.	R.A.
Acamar, ā′kȧ-mär	θ Eri	02	Gienah, jē′nȧ	γ Crv	12
Achernar, ā′kĕr-när	α Eri	01	Hadar, hăd′är	β Cen	14
Acrux, ā′krŭks	α Cru	12	Hamal, hăm′ăl	α Ari	02
Adhara, ȧ-dā′rȧ	ε CMa	06	Kaus Australis,		
Al Na'ir, ăl-nâr′	α Gru	22	kôs ôs-trā′lĭs	ε Sgr	18
Albireo, ăl-bĭr′ē-ō	β Cyg	19	Kochab, kō′kăb	β UMi	14
Alcyone, ăl-sī′ō-nē	η Tau	03	Markab, mär′kăb	α Peg	23
Aldebaran, ăl-dĕb′ȧ-răn	α Tau	04	Megrez, mē′grĕz	δ UMa	12
Alderamin, ăl-dĕr′ȧ-mĭn	α Cep	21	Menkar, mĕn′kär	α Cet	03
Algenib, ăl-jē′nĭb	γ Peg	00	Menkent, mĕn′kĕnt	θ Cen	14
Algol, ăl′gŏl	β Per	03	Merak, mē′răk	β UMa	10
Alioth, ăl′ĭ-ŏth	ε UMa	12	Miaplacidus,		
Alkaid, ăl-kād′	η UMa	13	mī′ȧ-plăs′ĭ-dŭs	β Car	09
Almach, ăl′măk	γ And	02	Mira, mī′rȧ	o Cet	02
Alnilam, ăl-nī′lăm	ε Ori	05	Mirach, mī′răk	β And	01
Alphard, ăl′färd	α Hya	09	Mirfak, mĭr′făk	α Per	03
Alphecca, ăl-fĕk′ȧ	α CrB	15	Mizar, mī′zär	ζ UMa	13
Alpheratz, ăl-fē′răts	α And	00	Nunki, nŭn′kē	σ Sgr	18
Altair, ăl-târ′	α Aql	19	Peacock	α Pav	20
Ankaa	α Phe	00	Phecda, fĕk′dȧ	γ UMa	11
Antares, ăn-tā′rēs	α Sco	16	Polaris	α UMi	01
Arcturus, ärk-tū′rŭs	α Boo	14	Pollux, pŏl′ŭks	β Gem	07
Atria, ā′trĭ-ȧ	α TrA	16	Procyon, prō′sĭ-ŏn	α CMi	07
Avior, ă-vĭ-ôr′	ε Car	08	Ras-Algethi, ràs′ăl-jē′the	α Her	17
Bellatrix, bĕ-lā′trĭks	γ Ori	05	Rasalhague, ràs′ăl-hā′gwē	α Oph	17
Betelgeuse, bĕt′ĕl-jŭz	α Ori	05	Regulus, rĕg′ū-lŭs	α Leo	10
Canopus, kȧ-nō′pŭs	α Car	06	Rigel, rī′jĕl	β Ori	05
Capella, kȧ-pĕl′ȧ	α Aur	05	Rigil Kentaurus		
Caph, kăf	β Cas	00	rī′jĭl kĕn-tô′rŭs	α Cen	14
Castor, kàs′tĕr	α Gem	07	Sabik, sā′bĭk	η Oph	17
Deneb, dĕn′ĕb	α Cyg	20	Scheat, shē′ăt	β Peg	23
Denebola, dĕ-nĕb′ō̇-lȧ	β Leo	11	Schedar, shĕd′ȧr	α Cas	00
Diphda, dĭf′dȧ	β Cet	00	Shaula, shô′lȧ	λ Sco	17
Dubhe, dŭb′ē	α UMa	11	Sirius, sĭr′ĭ-ŭs	α CMa	06
Elnath, ĕl′năth	β Tau	05	Spica, spī′kȧ	α Vir	13
Eltanin, ĕl-tā′nĭn	γ Dra	17	Suhail, sŭ-hāl′	λ Vel	09
Enif, ĕn′ĭf	ε Peg	21	Vega, vē′gȧ	α Lyr	18
Fomalhaut, fō′măl-ôt	α PsA	22	Zubenelgenubi,		
Gacrux, gă′krŭks	γ Cru	12	zōō-bĕn′ĕl-jĕ-nū′bē	α Lib	14

Pronunciations are generally as given by G. A. Davis, *Popular Astronomy*, 52, 8 (1944). See page 542 for pronunciation key.

GLOSSARY

Abberration. Refers to starlight—apparent displacement due to Earth's revolution; in optics refers to defects in the image.

Absolute magnitude. A measure of the luminosity of a star. By definition, the absolute magnitude of a star is equal to what its apparent magnitude would be if the star were 10 parsecs away.

Absorption spectrum. A continuous spectrum from which certain wavelengths have been removed as a result of absorption by a cool gas.

Aerolites. Stony meteorites mostly composed of silicates; rare because they weather rapidly and are not readily discernible from terrestrial stones.

Albedo. The fraction of sunlight that a planet, or other object, reflects (albedo of Uranus, 0.93; of Moon, 0.067).

Algol star. See *eclipsing binary.*

Angular momentum. That property of matter which tends to keep a rotating body rotating.

Annular eclipse. An event that occurs when a ring (annulus) of the Sun surrounds the Moon, which obscures the rest of the Sun.

Aphelion. The point in an elliptical orbit around the Sun which is farthest from the Sun.

Apparent magnitude. The magnitude an object appears to be; usually denoted by *m*. See *absolute magnitude.*

Apparent solar time. Time based upon the position of the Sun; the hour angle of the solar center plus 12 hours, or sundial time.

Apsides. The nearest approach and farthest approach in an orbit: in the case of Earth, perigee and apogee; in the case of the Sun, perihelion and aphelion.

Apsides, line of. Line connecting nearest and farthest approach in an orbit; the major axis of an orbit.

Association. A very loose cluster of luminous O and B stars, or of T Tauri stars. These are stars which have been formed recently from a cloud of interstellar gas and dust and are now dispersing.

Astrology. A belief which holds that events on Earth are influenced by events in the sky.

Astrometric binary. A type of binary star whose binary nature is deduced not from observation of the second star but from the fact that the path of the observed star in the sky is not a straight line.

Astronomical unit (a.v.). The mean distance of Earth from the Sun. The approximate value of the astronomical unit is 1.50×10^8 km., or $.930 \times 10^8$ miles.

Astronomy. The science dealing with the universe and its parts.

Atom. The smallest unit of a chemical element. An atom is composed of a nucleus and is surrounded by a cloud of electrons. The nucleus consists of protons and neutrons. In a neutral atom, the number of electrons is equal to the number of protons. See ion.

Aurora australis. The "southern lights" produced when subatomic particles from the Sun bombard ions in Earth's upper atmosphere.

Aurora borealis. The "northern lights" which occur in arctic regions when charged solar particles bombard ions in the atmosphere.

Barycenter. Center of mass of two bodies; the "point" around which they turn. The barycenter of Earth-Moon is about 1600 km below Earth's surface.

Big Bang Theory. A cosmological theory which holds that the universe was created from a cosmic fireball some 10 to 20 billion years ago.

Binary star. A system of two stars gravitationally bound to one another.

Black body. A hypothetical object which absorbs all wavelengths falling upon it, and radiates all wavelengths.

Black hole. A region in space surrounding a superdense object from which light cannot escape. A region in which the escape velocity is greater than the speed of light.

Blink comparator. An instrument for comparing two photographs of a star field to reveal the displaced position of an object (perhaps an asteroid or comet) in that field.

Bright-line spectrum. See *emission spectrum*.

Carbon cycle (carbon-nitrogen-oxygen cycle). Chain of nuclear reactions which produces helium and energy from hydrogen in stars with central temperatures hotter than about 15 million degrees Kelvin.

Cepheid variable. A type of variable star having periods ranging from 1 day to about 50 days, and having a relationship between average absolute magnitude and period. There are two types of Cepheids.

Chemosphere. That part of Earth's atmosphere (50 to 80 km above the surface) in which molecules separate and recombine.

Chromosphere. The bright red "color" layer of the atmosphere of the Sun (or a star) that lies just above the photosphere.

Circle. An ellipse with an eccentricity of 0.

Co-latitude. The difference between latitude reading and 90°; the co-latitude of 40° is 50°.

Color-magnitude diagram. A graph, similar to an H-R diagram, in which color index is plotted instead of spectral type. If a single cluster is being considered, apparent magnitude may be used instead of absolute magnitude, since all stars are at the same distance.

Color index. The difference in the magnitude of an object measured in two different parts of the spectrum. The color index is an approximate way of measuring the temperature of the object.

Constellation. One of some 88 configurations of stars (and regions of the sky) into which the celestial sphere is divided.

Contact binary. A type of binary star in which the two stars are so close together that one or both are distorted and extended toward each other, to the point where matter can be exchanged between them.

Continuous spectrum. A spectrum that contains all wavelengths and is given off by hot, dense objects.

Continuum. See *continuous spectrum*.

Convection. The bulk motion of a gas or liquid caused by regions of different density. This is an important method of energy flow in stars.

Coordinate system. A pattern of lines in the sky that enables us to refer to the position of an object with respect to certain fixed standard points and planes.

Corona. The outermost atmosphere of the Sun, perhaps extending beyond Earth. The inner portion is seen as a pearly glow during a total eclipse. Also see *galactic halo*.

Cosmic ray. Any of the high-energy elementary particles that move through space, or the secondary particles that reach Earth and are produced in collisions between primary cosmic rays and atoms in the atmosphere.

Cosmology. The study of the properties and evolution of the universe as a whole.

Cusps. The points of a crescent, usually used with reference to the Moon or a phasing planet, such as Mercury or Venus.

Dark-line spectrum. See *absorption spectrum*.

Declination. Angular distance of an object north or south of the equator.

Deferent. A circle upon which another circle is centered; used in connection with geocentric universe.

Degenerate gas. A gas in which the molecules are so close together that they do not behave independently of one another, but in concert.

De revolutionibus. Volume published in 1543 by Copernicus, properly called *De revolutionibus orbium coelestium*.

Diatomic. A molecule made of two atoms of a substance, for example, O_2, H_2.

Differential rotation. The state of motion of an object, such as the Galaxy, in which objects at different distances from the center move about the center in different periods.

Disk. See *galactic disk*.

Distance modulus. The difference between the apparent and absolute magnitudes of an object, $m - M$. It is an indicator of the distance, d, to the object, since $m - M = 5 \log d - 5$.

Doppler shift (or effect). The effect which causes a shift in the received wavelength or frequency of a wave emitted by an object moving along the line of sight. An object moving away will be seen to have its spectral lines shifted to the red.

Double-lined spectroscopic binary. A type of binary in which, although only one star is seen, spectral features of two stars are noted to vary, indicating the motion of one about the other.

Double star. Two stars close together in the sky. There are optical doubles, which are merely coincidences in direction to two separate stars, and physical doubles, which are stars gravitationally bound to one another.

Dwarf star. A star on the lower main sequence of the H-R diagram. Not to be confused with *white dwarf*.

Eastern elongation. Angle made by intersection of Earth-planet line and Earth-Sun line when planet is east of Sun (seen after sunset). See *western elongation*.

Eccentric. Point removed from the center of a circle and around which an object moves.

Eccentricity. In an ellipse, the ratio of distance between the foci to the major axis. ($e = 0$, circle; $e = 1$, parabola; $e > 1$, hyperbola; $e < 1$, ellipse)

Eclipsing binary. A type of binary star whose observed light varies as a result of the periodic eclipse of one star by the other.

Ecliptic. The apparent path of the Sun on the celestial sphere; the plane of Earth's orbit. Eclipses can occur only when Sun and Moon are on or close to the ecliptic.

Electromagnetic radiation. A type of radiation, including light, X-ray, infrared, ultraviolet, and radio waves, which consists of vibrations of electrical and magnetic fields in space.

Electron. One of the elementary atomic particles. It has a negative charge and a mass of 9.1×10^{-28} g. The electron structure of an atom is responsible for the chemical and spectral properties of the atom.

Element. One of 106 basic substances, each characterized by an atom of a particular structure.

Ellipse. A geometric figure in which the sum of the distances from two fixed points (the foci) to any point on the curve is constant.

Ellipsoid. Solid that is produced when an ellipse is rotated around one of its axes.

Ellipsoidal galaxy. See *elliptical galaxy*.

Elliptical galaxy. Large collections of population II stars having an ellipsoidal shape and no arms; also called *ellipsoidal galaxies*.

Emission nebula. A nebula in which the nearby stars, which must be earlier than spectral type B0, excite the gas component of the cloud to fluoresce. Occasionally fluorescence is produced by interaction of the cloud with other interstellar matter.

Emission spectrum. The spectrum given off by a hot gas, consisting of a series of bright lines at specific wavelengths characteristic of the composition of the gas.

Energy. That property of an object which enables it to do work.

Epicycle. A circle moving in a deferent, and used to explain retrograde motion in early conceptions of the universe.

Equant. In early ideas of the solar system, a point removed from Earth and opposite the eccentric, around which planets, Sun, and Moon moved with steady angular velocity.

Equation of time. The difference between apparent solar time and mean solar time.

Eruptive variable star. A type of variable star in which the change in brightness is caused by an explosion (either mild or catastrophic) of the star.

Facula. A bright region near the limb (edge) of the Sun.

False dawn. Appearance of the zodiacal light in the eastern sky preceding sunrise.

Focal light. The distance from a lens or mirror to the point at which parallel incoming light rays are brought to a focus.

Forbidden lines. Spectral lines produced by atoms in energy states which cannot be achieved in a laboratory, but which are common in interstellar space.

Force. An effect on an object which causes the object to change its state of motion. Force is defined by Newton's second law, $F = ma$, in which F is the force, m the mass of an object, and a the acceleration it undergoes because of the action of the force.

Fraunhofer lines. The dark lines in the spectrum of the Sun, caused by absorption in the Sun's atmosphere.

Frequency. The number of vibrations, or crests, of a wave per second.

Galactic cluster. See *open cluster*.

Galactic disk. The flat distribution of material in the Galaxy. It is composed of population I objects.

Galactic equator. The plane of the Galaxy, and its intersection with the celestial sphere. It runs along the band of the Milky Way.

Galactic halo. The portion of the Galaxy surrounding the nucleus and disk. It is composed of population II objects. Also called the *galactic corona*.

Galaxias Kyklos. Greek for "milky circle"; the name given by the Greeks to the Milky Way, the star-filled belt of the sky.

Galaxy. A large assemblage of stars, gas, and dust gravitationally associated with each other. Galaxies occur in several forms.

Gamma rays. A range of wavelengths of electromagnetic radiation shorter than X rays.

Gegenschein. Faint glow opposite the Sun and believed to be produced when sunlight reflects from interplanetary dust particles.

Geocentric. Centered upon Earth.

Giant star. A star tens to hundreds of times larger than the Sun. These stars have completed their main-sequence lives and are seen in the upper right portion of the H-R diagram.

Globular cluster. A star cluster having several tens or hundreds of thousands of stars in a roughly spherical shape. These clusters occur in the halo of the Galaxy.

Gravitational constant. The quantity, usually symbolized by G, relating the force between two objects. Its value is 6.67×10^{-8} when masses are expressed in grams and separation in centimeters.

Greenhouse effect. Heating that results when shorter wavelength radiation passes through a surface but longer wavelength radiation does not. Responsible for heating of atmosphere-covered bodies, especially Venus and Earth.

Heliocentric. Centered upon the Sun.

H-R diagram. See *Hertzsprung-Russell diagram*.

HII region. A nebula in which the hydrogen is ionized, usually by ultraviolet radiation from nearby stars.

Halo. See *galactic halo*.

Hertzsprung gap. The region of the H-R diagram lying between the main sequence and the red giant branch. Very few stars are plotted here because they pass through this stage of their lives very rapidly.

Hertzsprung-Russell diagram. A graph of absolute magnitude versus spectral type of stars. See *main sequence, giant, dwarf, supergiant, white dwarf*.

Horizontal branch. A portion of the color-magnitude diagram of globular clusters lying to the left of the red giant branch and above the main sequence.

Hubble constant. The constant of proportionality between the velocity of recession of a galaxy and its distance. Usually symbolized by H, its present value is about 50 km/sec/Mpc (megaparsec).

Hubble law. The relationship between the speed of recession and the distance of a galaxy, $V = H \times d$, where H is the Hubble constant, V is in km/sec, and d is in Mpc.

Hydrostatic equilibrium. The balance between the forces of pressure and weight on a typical cube of material inside a star.

Hyperbola. Conic section with eccentricity greater than 1 ($e > 1$); curve of intersection produced by a plane cutting through a circular cone, such that the plane is not parallel to the face of the cone.

Ideal gas. A gas which obeys the ideal gas law $p = dkT$, where p and T are the pressure and temperature of the gas, d is its density, and k is a constant. The necessary condition which must exist is that the gas molecules be far apart.

Inferior conjunction. The condition that exists when an inferior planet is in line between Earth and the Sun.

Infrared waves. A range of wavelengths of electromagnetic radiation longer than red waves but shorter than radio waves.

Intensity. The strength of a beam of light, measured in units of energy per second.

Intercalary. Inserted into the calendar, as an extra day, week, or month. Intercalary adjustments are often used to make the calendar keep in step with the seasons.

Interstellar matter. The gas and dust lying between the stars. It exists in discrete nebulae and in a thin, all-pervading "haze" in space.

Inverse square law. A relationship between two quantities such that increasing one by a certain factor results in decreasing the other quantity by the square of that factor. For example, if one is doubled, the other is quartered: $(1/2)^2 = 1/4$.

Ion. An atom whose electrons do not equal the number of protons in its nucleus. An ion always has a negative or positive charge.

Ionosphere. That part of the atmosphere (80 to 380 km) where solar ultraviolet radiation strips atoms of electrons.

Irregular variable star. A type of variable star in which the light variation has no pattern or regularity.

Julian calendar. Calendar introduced by Julius Caesar and designed after that of the Egyptians; a great improvement over calendar then used.

Julian day. Number of days since Jan. 1, 4713 B.C.; originated in 1582 by Joseph Scaliger and named by him after his father.

Kepler's laws. Three statements which relate to the behavior of an object in orbit about another. See Chapters 7 and 16.

Law of gravity. The relationship giving the force between the masses and separation of two bodies: $F = Gm_1m_2/d^2$. F is the force, G the gravitational constant, m_1 and m_2 the masses, and d their separation.

Lenticular galaxy. A galaxy shaped like a lens, intermediate between the spiral galaxies and the elliptical galaxies.

Libration. Phenomena that enable an observer to see more than one-half of the Moon:

libration in latitude, in longitude, and daily libration.

Light. Electromagnetic radiation in one of its many forms. Strictly speaking, light is radiation visible to the human eye.

Light curve. A graph of the light variations of a variable star with time.

Light-year. The distance light can travel in one year, equal to about 9½ trillion kilometers.

Limb. The apparent edge of an object seen against the dark sky background.

Local group. A cluster of about 20 galaxies which includes the Milky Way, the Andromeda galaxy, and others.

Logarithm. The power to which a number, called the base number, must be raised to equal another number. Thus, since $100 = 10^2$, 2 is the logarithm of 100 to the base 10.

Long-period variable star. A type of variable star with a period of variation of several hundred days. These are spectral class M supergiants.

Luminosity. The total energy output of a star; for example, the luminosity of the Sun is 4×10^{33} ergs/sec.

Magnetic declination. The difference between magnetic and geographic north, expressed in degrees east or west.

Magnetosphere. That region of space (out some 90 000 km) into which Earth's magnetic field extends.

Magnification. The amount by which a lens or mirror enlarges the apparent size of an object.

Magnitude. A logarithmic system of measuring brightnesses of stars. A difference of one magnitude corresponds to 2.512 on the logarithmic scale. Fainter objects have a numerically higher magnitude.

Main sequence. The locus of most stars on the H-R diagram. It is the locus of stars producing energy by conversion of hydrogen to helium.

Major axis. Maximum dimension of an ellipse; the line connecting the apsides (the closest and farthest positions from a focus of the ellipse).

Mean solar time. Hour angle of the mean Sun plus 12 hours. (The mean Sun is a fictitious Sun that moves an equal number of degrees every 24 hours.)

Messier object. One of the list of nonstellar objects compiled by Charles Messier in the eighteenth century during his searches for comets. Objects are referred to by an "M" and a number, for example, M31.

Metals. In astronomy, the elements heavier than hydrogen and helium.

Meteor. Streak of light produced when meteoroids pass through Earth's atmosphere.

Meteorite. A meteoroid that has fallen upon Earth or upon another body in the solar system.

Meteoroid. Solid particles in space which produce meteors when they pass through Earth's atmosphere.

Milky Way. The name of our galaxy, or of the band of hazy light running across the sky due to the myriad stars of the Galaxy.

Model. A physical and mathematical description of a system or process which allows calculation and prediction of properties and behavior.

Molecule. A chemical compound consisting of more than one atom bound together.

Monatomic. A single atom of a substance, for example, O,H.

Moving cluster distance. A method of obtaining the distance to a nearby cluster in which the individual proper motions may be observed to converge to a point.

Nadir. The point opposite (or 180° removed from) the zenith. When a person stands, his head is toward the zenith, his feet toward the nadir.

Neap tides. The lowest tides in a month, occurring when the Moon is at or near quadrature (first and third quarter).

Nebula. A cloud of dust and gas in space.

Neutrino. One of the elementary atomic particles. It has no mass or charge but carries energy. It reacts very little with other matter.

Neutron. One of the elementary atomic particles. It has a neutral charge and a mass of 1.7×10^{-24} g.

Neutron star. A very small, dense star, in which the electrons and protons have been forced together to produce neutrons. Some rotating neutron stars may be pulsars.

Nodes. Points in an orbit where the plane of the orbit intersects some other orbital plane: Earth and Moon, for example.

Nova. A star which explodes, increasing its light output by a factor of hundreds of thousands of times and throwing a small percentage of its mass into space.

Nuclear fusion. The process of building up heavier elements from lighter ones by combining their nuclei.

Nucleus. The core of an atom, containing protons and neutrons.

Nutation. The "nodding" of Earth's axis; a slight movement which is added to the precession motion.

Objective. The general term for the main light-gathering element of a telescope, usually a lens or a mirror.

Objective prism. A thin wedge of glass placed over the objective of a telescope to spread the light of each star in the field of view into a spectrum.

Open cluster. A star cluster, also called a galactic cluster, in which there are from several dozen to several hundred stars loosely bound to one another. They occur close to the plane of the Galaxy.

Orbit. The path of one object about another.

Oscillating universe theory. A modification of the Big Bang theory which holds that the expansion of the universe will eventually stop and reverse, producing a contracting universe. It will contract to another fireball and explode again.

Ozone. An oxygen gas in which the molecules contain three atoms.

Parabola. Conic section with eccentricity of 1; curve of intersection of a plane that is parallel to the surface of the cone.

Parallax. The amount of shift in the position of a star as a result of Earth's orbital motion. Also a measure of distance of an object, whether or not its shift can actually be observed.

Parsec. A distance of 3.2 light-years, or 3.084×10^{12} km. It is the distance at which a star would have a *parallax* of one *second* of arc.

Peculiar galaxy. A galaxy which shows characteristics unlike those common to its particular type.

Penumbra. The partial shadow cast by an opaque object.

Perihelion. In an elliptical orbit around the Sun, that point which is closest to the Sun.

Periodic comet. A comet that reappears at a regular interval. For example, Halley's comet is a periodic having a period of about 75 years.

Photometer. An instrument for measuring the intensity of a beam of light. It may be equipped with filters and other devices for limiting the portion of the light beam measured.

Photon. A "particle" of light. Each photon is characteristic of a particular wavelength of light and carries a certain definite amount of energy. Also called a *quantum*.

Photosphere. The light surface of the Sun; the part that we see.

Plage. A bright region of the solar surface usually observed at some particular wavelength (monochromatic light).

Planck's constant. The unit of action, usually denoted by h, equal to 6.67×10^{-27} erg-sec.

Planet. A cool, relatively solid body not massive enough to produce energy by fusion.

Planetary nebula. A nebula which is roughly spherical in shape and thus appears similar to a planetary disk in a telescope. Such nebulae are clouds of gas thrown off by exploding stars.

Polarization. That property of light which is characteristic of its vibration in a plane. Unpolarized light has waves in planes of all orientations, whereas polarized light has more waves in some planes than in others.

Population. A classification of stars and other objects into two groups: population I objects are young and associated with interstellar matter; population II objects are old and found mostly in the galactic halo and nucleus.

Position angle. The angle indicating the direction one object is from another, such as a double star primary and its companion. The angle is measured in the plane of the sky from north toward east.

Power-of-ten-notation. A way of expressing large or small numbers by counting the zeros and using this number as an exponent, for example, 1 000 000 = 10^6.

Precession. Conical motion of Earth's axis caused by gravitation of the Sun and Moon (primarily the Moon); gyroscopic effect produced when Moon pulls on bulge of Earth and Earth responds to the attraction. This produces a westward motion (about 50″ a year) of the equinoxes along the ecliptic.

Principia. Book whose full title is *Philosphiae naturalis principia mathematica,* written by Isaac Newton in 1687 to set forth his findings.

Prominence. Mass of bright gases suspended in the solar corona; often extends thousands of kilometers beyond the solar surface.

Proper motion. The angular motion of an object in the plane of the sky, expressed in seconds of arc per year. See *transverse motion* and *space motion.*

Proton. One of the elementary atomic particles. It has a positive charge and a mass of 1.7×10^{-24} g. The number of protons in an atomic nucleus is the atomic number of the element.

Proton-proton cycle. The chain of nuclear reactions which produces helium and energy from hydrogen at temperatures between 5 and 15 million degrees K.

Pulsar. Believed to be a rotating neutron star which sends out bursts of radio (and, in one case, light) waves from localized sources on or near its surface.

Pulsating variable star. A type of variable star which changes its brightness as a result of radial pulsations with consequent changes in surface temperature of the star.

Quadrature. A condition that exists when two bodies (Sun and Moon, for example) are 90° apart as seen from Earth.

Quantum theory. A physical theory which states that, on an atomic scale, objects can have only certain quantized amounts of energy.

Quasar (quasi-stellar radio source). Believed to be a young galaxy which emits much radio radiation; an object at cosmological distance which is a source of radio waves and which appears almost stellar in size. Also called *QSRS.*

Radial velocity. The component of an object's space motion which is along the line of sight to the object. Measured in km/sec, it is called positive if the motion is away from us, negative if toward us.

Radar. An acronym for radio direction and ranging. A special antenna sends out a signal in a specific direction and times the return of the echo. The time of the flight gives the range since the speed of the wave is known.

Red shift. The shift of the spectral features to the red end of the spectrum due to radial velocity away from the observer. It particularly refers to the recession of the galaxies as a result of the expansion of the universe.

Reflection nebula. A type of nebula in which the nearby stars are not sufficiently hot to cause the gas to fluoresce so that the light observed from the nebula is starlight reflected by the dust component of the cloud.

Reflector. A type of telescope having a mirror as its primary light-gathering element.

Refraction. The bending of a ray of light which occurs when it passes from one substance to another.

Refractor. A type of telescope using a lens as its primary light-gathering element.

Resolving power. The ability of an optical instrument to distinguish two nearby objects as separate; the ability to see detail.

Retrograde motion. Apparent westward (backward) motion of the planets on the celestial sphere; an observation the ancients accounted for by many different theories.

Revolution. Movement of an object around another, for example, the Moon around Earth, planets around the Sun.

Right ascension. Angular position of an object east or west of the vernal equinox, measured eastward.

Roche's limit. Distance (2.44 times radius of the larger) within which a body of the same density as a larger body would be pulled apart by gravitation. Named after Edouard Roche, French astronomer.

Rotation. Motion of an object around its axis.

RR Lyrae variables. A type of variable star with a period of less than a day and with an average absolute magnitude +0.5. These variables may be used as distance indicators because of these characteristics.

Schmidt-type telescope. A type of telescope-camera which uses both a mirror and a lens to obtain photographs of large areas of the sky at one time.

Semimajor axis. Distance from the center of an ellipse to an end (an apsis) of the ellipse.

Seyfert galaxy. A type of spiral galaxy having a very bright nucleus.

Sidereal. Of the stars; also refers to measurement by the apparent motion of fixed stars. *See sidereal month, sidereal period,* and *sidereal time.*

Sideral month. Time required for the Moon to go around Earth and realign with a star—27⅓ days.

Sidereal period. Time required for a body to revolve around another body, and to line up with a given star.

Sidereal time. The local hour angle of the vernal equinox.

Siderites. Meteorites that contain about 95 percent iron and 5 percent nickel.

Siderolites. Meteorites containing a mixture of stone and iron. Siderolites are rare, probably because the stony portion weathers rapidly.

Single-lined spectroscopic binary. A type of binary in which the spectral features of only one star are seen; these features are alternately shifted to the red and blue, indicating orbital motion about another star too faint to be seen.

Solar constant. The amount of energy received from the Sun by Earth at the top of the atmosphere. It is equal to 1.93 g-cal/cm^2/min.

Solar flare. A sudden, short-lived brightening of the solar surface; it grows rapidly, then subsides.

Solar surface. Usually given as the depth to which vision penetrates. Because the Sun is entirely gaseous, there is no surface as such. All above the solar "surface" is the solar atmosphere.

Solar wind. The stream of high-energy atomic particles coming from the Sun.

Space motion. The true motion of an object through space, usually measured in km/sec. See *radial velocity* and *transverse motion.*

Spectral type. One of the classifications of a star according to the pattern of absorption lines in its spectrum. The spectral classes or types are denoted by the letters O, B, A, F, G, K, and M, in order of decreasing temperature.

Spectroscopic binary. A type of binary star in which the double nature is deduced from shifts of the spectral lines, indicating orbital motion of one star about the other.

Spectrograph. A device which separates a beam of light into a spectrum and makes a photograph of it. If the instrument is used only for visual inspection of the spectrum, it is called a *spectroscope.*

Spectrum. The distribution of a source of radiation into different frequencies or wavelengths.

Spiral arms. The pattern of bright stars in our galaxy and other spiral galaxies produced by a slightly higher density of stars and gases in certain regions.

Spiral nebulae. The old, incorrect term for spiral galaxies.

Spring tide. The highest tides during a given month, occurring when the Moon is full and new.

Star. A gravitationally bound, roughly spherical form of gas which produces energy by nuclear fusion.

Star gauging. A method of determining the distribution of stars in space by counting the number of stars to different limiting magnitudes, and correcting for interstellar absorption.

Steady-state theory. A theory which holds that the universe had no beginning or ending. The average properties of space are the same for all time as seen by any observer.

Stellar evolution. The aging process of a star, from formation out of an interstellar cloud to main sequence, to red giant stage, to extinction.

Stratosphere. Layer of the atmosphere (12 to 30 km) which is distinguished as a region of the jet stream. The temperature in the region remains constant throughout (about $-55°C$).

Subgiant star. A star slightly larger than main-sequence stars but smaller than giants.

Sunspot. A dark, cool region in the photosphere. It may last several days and expand to considerable size.

Supercluster. A cluster of clusters of galaxies.

Supergiant. A star hundreds to thousands of times the size of the Sun. These stars are dozens of times more massive than the Sun and are near the end of their lives. They are found in the uppermost parts of the H-R diagram.

Superior conjunction. Position in an orbit when Earth, Sun, and an inferior planet are in line and in that order.

Supernova. An eruptive variable star which increases its light output by a factor of 10^6 in a day, almost destroying the star. It fades over a period of months.

Synchrotron radiation. A type of electromagnetic radiation produced by a charged particle moving at near the speed of light in a magnetic field.

Synodic month. Period required for the Moon to complete a phase cycle from new moon to new moon ($29\frac{1}{2}$ days), or for a planet to complete the cycle from opposition to opposition.

Syzygy. Condition that exists when the Earth, Moon, and Sun are aligned, as at new and full Moon.

Temperature. A number describing the physical state of a system in terms of its energy of motion, ionization, brightness, or some other property having to do with energy. The most commonly used temperature measures the kinetic motion of atoms or molecules.

Terminator. The line that separates the lighted portion of the Moon (or some other object) from the unlighted portion; the line of sunrise or sunset.

Theory of relativity. A physical model which postulates that all motion is relative to some coordinate system (that is, that there exists no basic coordinate system with respect to which everything can be measured). Another postulate is that the speed of light is a constant.

Thermal radiation. The characteristic radiation of a black body, for which the spectrum is described by a Planck curve.

Three-degree background radiation. Radio radiation coming equally from all directions in space and characteristic of a black body at a temperature of 3°K. It is thought to be a relic of the Big Bang.

Transit. Passage of a body across the disk of the Sun.

Transverse motion. The component of an object's motion which is perpendicular to the line of sight to the object. Measured in km/sec, it produces the proper motion observed. (See *proper motion.*)

Trigonometric parallax. A method of determining the distance to a nearby star by noting its shifting position in the sky as a result of Earth's orbital motion.

Troposphere. The layer of the atmosphere (to about 12 km) in which weather phenomena occur.

T Tauri star. A type of star showing irregular flashes of variation. These stars are believed to be still in the process of contracting to the main sequence.

Ultraviolet waves. A range of wavelengths of electromagnetic radiation lying between the blue-violet end of the visible spectrum and the X rays.

Umbra. The dark part of a shadow, or the region where penumbras overlap. In a sunspot, the umbra is the dark, central area.

Variable star. A star whose light output varies.

Wavelength. The distance between two peaks, or crests, of a wave.

Western elogation. Angle made by intersection of Earth-planet line and Earth-Sun line when inferior planet is seen west of Sun (appears before sunrise). See *eastern elongation.*

White dwarf. A type of star about the size of Earth which has exhausted all its nuclear energy sources and is living on stored heat. White dwarfs are found in the lower left portion of the H-R diagram.

Widmanstaetten figures. Crystalline structure seen when metorites are cut and polished.

X rays. A range of electromagnetic radiation lying between ultraviolet and gamma rays.

Zenith. The point directly overhead, or opposite to the direction of gravitational force: up as opposed to down.

Zero age main sequence (ZAMS). In the H-R diagram, the locus of stars which have just begun to convert hydrogen to helium.

Zodiac. Imaginary band 16° wide on the celestial sphere centered on the ecliptic. The zodiac contains 12 constellations, the zodiacal belt.

Zodiacal light. Sky glow which appears along the zodiac and is probably produced when sunlight is reflected from interplanetary dust particles.

INDEX